Scientific Method in Brief

The general principles of the scientific method, which are applicable across all of the sciences, are essential for perspective, productivity, and innovation. These principles include deductive and inductive logic, probability, parsimony, and hypothesis testing, as well as science's presuppositions, limitations, ethics, and bold claims of rationality and truth. The implicit contrast is with specialized techniques confined to a given discipline, such as DNA sequencing in biology. Neither general principles nor specialized techniques can substitute for one another, but rather the winning combination for scientists is mastery of both.

The purposes of this book are to enhance perspective on science by drawing insights from the humanities and to increase productivity by fostering a deep understanding of the general principles of scientific method. The examples and case studies span the physical, biological, and social sciences; include applications in agriculture, engineering, and medicine; and also explore science's interrelationships with disciplines in the humanities such as philosophy and law.

This book engages a great diversity of viewpoints on science, both historical and contemporary, and responds by affirming science's rationality. Informed by position papers on science from the American Association for the Advancement of Science, the National Academy of Sciences, and National Science Foundation, this book aligns with a distinctively mainstream vision of science. It is an ideal resource for anyone undertaking a systematic study of scientific method for the first time, from undergraduates to professionals in both the sciences and the humanities.

Hugh G. Gauch, Jr., is a Senior Research Specialist in the College of Agriculture and Life Sciences at Cornell University, New York. He teaches Cornell's course on scientific method, and for the last four decades his research has focused on the statistical analysis of ecological and agricultural data. He is author of *Scientific Method in Practice* (Cambridge, 2002), which is the basis for this more concise and student-focused text.

Scientific Method in Brief

Hugh G. Gauch, Jr.
Cornell University, New York

CAMBRIDGE UNIVERSITY PRESS
Cambridge, New York, Melbourne, Madrid, Cape Town,
Singapore, São Paulo, Delhi, Mexico City

Cambridge University Press
The Edinburgh Building, Cambridge CB2 8RU, UK

Published in the United States of America by Cambridge University Press, New York

www.cambridge.org
Information on this title: www.cambridge.org/9781107666726

First published 2012

Printed and Bound in the United Kingdom by the MPG Books Group

A catalogue record for this publication is available from the British Library

Library of Congress Cataloguing in Publication data
Gauch, Jr., Hugh G., 1942–
Scientific method in brief / Hugh G. Gauch, Jr.
p. cm.
ISBN 978-1-107-66672-6 (pbk.)
1. Science – Methodology. I. Title.
Q175.G3368 2012
001.4′2 – dc23 2012016520

ISBN 978-1-107-66672-6 Paperback

This book is dedicated to my esteemed Cornell colleagues,
Gary Warren Fick, Charles Christopher Fick, and
Justin David McGeary.

Contents

Foreword

Approximately halfway through her Ph.D. program in the biological sciences, my first doctoral student requested that we meet to review progress toward the degree. Knowing that both the courses and research of this student were progressing nicely, I entered the appointment confident of a glowing report both for graduate student and major professor. However, the conversation took an unexpected twist as we finished discussing the items on my agenda.

When and how, the student asked, do we get to the more philosophical part of this Doctorate of Philosophy degree in science? Was it true that this program that I and the guidance committee had designed would not include even one course in science philosophy or no structured examination of the logical underpinnings of science? The final question had an unintended sting for me as major professor – something to the effect of "Will I graduate feeling worthy of more than a technical degree?"

Stunned and befuddled, I sent the student on her way with lame excuses: there simply wasn't sufficient time in modern science training for students to become renaissance scholars and well-published researchers capable of competing successfully for grant dollars. Moreover, to venture into science philosophy required an unhealthy tolerance for time wasted in silly, perfectionistic arguments over whether or not the sun will rise tomorrow. The better path to becoming a successful scientist, I argued, was to function as an apprentice to successful researchers and get on generating data from real-world experiments. After all, I concluded, the quality of your Ph.D. program will be at least equal to my own. Had not my Ph.D. landed me a great postdoctoral experience at Cornell University and an enviable tenure-stream Assistant Professorship at Michigan State University?

To this day, my former student does not realize how that conversation awakened my conscience to the awesome responsibility that educators shoulder in passing the scientific torch across intellectual generations. Although it came too late for my first Ph.D. student, that conversation whetted my sustained appetite and commitment for seeking suitable teaching materials to put some Ph. back into the Ph.D. degree in science. For three decades now, I have taught a graduate seminar course entitled "The Nature and Practice of Science." This course seeks to leaven the minds of graduate students from across our university with sufficient science philosophy, logic, and best practices for high-impact careers as scientists and educators. Finding no suitable text for such a course, we rely on diverse readings, including John Platt, Karl Popper, Thomas Kuhn, Ronald Giere, and others. Unfortunately, these scholars use differing terminologies that can be confusing, and large gaps are left that cause students and instructors to struggle mightily when attempting to build a unified whole. Due to the limited time graduate students are typically given to pursue this type of interest, our coverage is merely introductory. We lacked an appropriate resource to which interested students could be pointed for further self-study.

It was with relief and enthusiasm that I served as a reviewer for *Scientific Method in Practice* by Hugh G. Gauch, Jr., the predecessor to *Scientific Method in Brief*. Here at last was a comprehensive and up-to-date treatise on the fundamentals of science philosophy and method between the covers of one book and written from the pragmatic perspective of a credible science practitioner with whom researchers could identify. Here would have been the book I should have handed to my first Ph.D. student at the time of her request.

Scientific Method in Brief is more precisely targeted to the needs of both developing scientists and those who have long been doing research but remain open to lifetime learning and improvement. Here is the distilled yet complete version of Gauch's lifetime search for a coherent and practical version of science philosophy for practitioners. Moreover, the current book will be a boon to science educators because it brings into focus the core issues of how and with what degree of confidence we can know scientific truths. Here is a boat to escape from that tiny and intellectually impoverished island of the "classical scientific method" that has held sway for far too long.

Scientific Method in Brief is an eminently readable, understandable, and well-grounded philosophical book in its own right. This book belongs on the shelf of scholars across the intellectual spectrum. Gauch's books represent contributions on a par with the great philosophers of science, many of whom left the field in as much disarray as they found it. As a science practitioner and educator, *Scientific Method in Brief* restores my confidence and commitment in keeping Ph. in front of the D. in Ph.D. Finally, this knowledge is now accessible to anyone who can read and reason.

Dr. James R. Miller
Distinguished Professor
Former Director of the Division of Science and Mathematics Education
Michigan State University
East Lansing, Michigan

Preface

The thesis of this book is that there exist general principles of scientific method that are applicable across all of the sciences, undergird science's rationality, and greatly influence science's productivity and perspective. These general methodological principles involve deductive and inductive logic, probability, parsimony, and hypothesis testing as well as science's presuppositions, limitations, and bold claims of rationality and truth. The implicit contrast is with specialized techniques confined to a given discipline, such as DNA sequencing in biology. Neither specialized techniques nor general principles can substitute for one another, but rather the winning combination for scientists is mastery of both.

This book has two purposes. One purpose is to increase productivity by fostering a deep understanding of the general principles of scientific method. For instance, although few scientists are aware of this tremendous opportunity, parsimonious or simple models are often more accurate than their data, and this greater accuracy can increase repeatability, improve decisions, and accelerate progress. The other purpose is to enhance perspective on science by interrelating the sciences and humanities. A humanities-rich version of science is more engaging and beneficial than a humanities-poor version. Several of the 14 chapters of this book serve both purposes, but five, Chapters 2–6, are for enhancing perspective, whereas the following five, Chapters 7–11, are for increasing productivity.

The intended audience is persons undertaking a systematic study of scientific method for the first time. This includes undergraduates to professionals in both the sciences and the humanities. Because the general principles of scientific method have been so neglected in the contemporary science curriculum, most students and professionals have not yet had a first book or course on scientific method – despite vigorous calls to prioritize method from the American Association for the Advancement of Science (AAAS), National Academy of Sciences (NAS), and National Science Foundation (NSF). No specific background knowledge or courses are required – not in statistics, philosophy of science, physics, biology, or in any other subject – but rather this book should be accessible to everyone who has at least begun their university education.

This book is suitable for classroom use or individual study. The 14 chapters, which are nearly equal in length, can be studied one chapter per week at universities having semesters of about 14 weeks (while assigning Chapters 1 and 2, or else 13 and 14, one week if need be; or discussing Chapter 11 for two weeks if need be). When professors want their students to undertake additional reading – either optional or assigned, either occasional or weekly – this book's citations can supplement the professors' own ideas about supplementary materials. For a course taught by faculty in the basic or applied sciences, occasional guest faculty from the humanities can enrich the discussion, such as a philosopher joining the discussion of science's presuppositions. Likewise, when used by faculty in the humanities, occasional guest scientists or engineers can be helpful, such as a chemist giving practical examples of parsimony gaining accuracy. Although science courses typically draw mostly science majors whereas humanities courses draw mostly humanities majors, efforts to include students with a diversity of majors promotes creative and memorable classroom conversations.

On the other hand, when this book is used for individual study, I offer one recommendation. If you are situated in the sciences, ask a friend preferably in the humanities to read and discuss this book together – and the reverse if you are in the humanities. Two minds exploring complementary insights are better than one.

Three factors have driven my choice of topics. First, philosophical consideration of scientific method identifies its basic components, which perforce are essential topics, including science's presuppositions, evidence, and logic. Second, position papers from the AAAS, NAS, and NSF specify certain topics regarding scientific method as being essential for scientific literacy. Third and finally, I have taught a course on scientific method with a colleague here at Cornell University for nearly a decade, using this book's predecessor as the textbook, *Scientific Method in Practice* (Cambridge University Press, 2002; Chinese edition, Tsinghua University Press, 2004). This course has had undergraduate and graduate students majoring in the sciences and the humanities, and most years has also included faculty sitting in on the course. This experience has increased my understanding of the needs, interests, and abilities of both students and professionals.

Understandably, readers vary in the topics they find of particular interest. For instance, one reader may be fascinated by the historical contributions of Robert Grosseteste and Albertus Magnus and be uninterested in the use of parsimonious models to gain accuracy, whereas another reader may have the opposite reaction. Therefore, readers may study or skim various sections as their interests dictate when this book is used for individual study. However, I would strongly encourage readers not to be too hasty in judging whether a given topic is interesting and relevant to their personal and professional goals.

For instance, the chapter on science's presuppositions might strike a science or engineering major as being too philosophical and impractical to merit attention, but appropriate presuppositions are indispensable for any defense of science's rationality, which is far indeed from superfluous for the scientific community. So, my principal advice to readers is: allow for surprises. It is precisely those topics that initially seem most unfamiliar and irrelevant that may surprise readers by becoming some of their favorite parts of this book because those unfamiliar topics stimulate a more complete and confident understanding of how scientific thinking works.

Satisfactory study of scientific method simply *must* be coherent and systematic because the components of scientific thinking have multiple aspects that interact deeply. For instance, science's presuppositions have multiple roles, rendering evidence admissible and thereby conclusions attainable, engendering fruitful interactions with the humanities, and largely determining the forum of worldviews in which science makes sense. Likewise, deductive logic and inductive logic are each a substantial topic, but in addition, they must interconnect and interact in scientific thinking. Consequently, the general principles of scientific method cannot possibly be learned adequately as bits and pieces gleaned from routine science courses and books. Likewise, extensive empirical evidence from science educators shows what many scientists may find surprising, that research experience is remarkably ineffective for learning the general principles of scientific method. Effective pedagogy for the general principles of scientific method offers but one choice: systematic study.

Furthermore, in historical perspective, it must be acknowledged that civilizations rose and fell around the globe for millennia before anything recognizable as scientific method emerged, somewhere between the 1200s and 1600s by various accounts. Scientific method does incorporate some crucial common-sense elements that are simple; but mastering deductive logic, inductive logic, parsimony, and so on in a coherent and working whole is anything but simple. On the other hand, from experience with our course on scientific method and from reports about similar courses elsewhere, it is feasible for students to learn the general principles of scientific method within the scope of a single course or book. This book represents my attempt to convey the essential material in as brief a book as its topic allows. Truly, a single course or book for a topic so important as scientific method seems quite reasonable, given that instruction in any particular specialty from astronomy to zoology requires four years of arduous effort for an undergraduate degree and several additional years for a graduate degree.

There are several other books on scientific method, as mentioned in Chapter 1. But this book is distinctive because it gives sustained attention to the vision of science set forth in position papers from the AAAS, NAS, and NSF. These thoughtful and careful documents distill the wisdom of literally hundreds of outstanding scientists and scholars. This wisdom is worth passing on to students.

However, I have not yet seen another book on scientific method that pays attention to these important documents. Aligning with this vision of science has two principal implications for the character of this book. First, science itself is regarded as one of the liberal arts, which fosters a rich traffic of ideas between the sciences and the humanities. Second, this book presents and defends a distinctively mainstream vision of science and its method. Of course, from antiquity, there has always been a wide spectrum of opinions on the prospects and limits of human knowledge, so this book must respond to severe critiques of science's rationality. A mainstream vision of science best prepares people for an engaging and productive experience with science.

Four reasons motivate an energetic study of scientific method. First, as documented in Chapter 1, position papers on science from leading scientific organizations in many nations express a thoughtful and compelling call for sustained and substantial study of scientific method at all levels of science education. Second, as documented in Chapter 13, science educators have conducted a huge number of empirical investigations on specific benefits obtained from explicit study of scientific method. These benefits include better comprehension, greater adaptability, greater interest, more realism, better researchers, and better teachers. Third, as contrasted with the specialized techniques employed in scientific research that are often subject to increasingly rapid development and turnover, the general principles of scientific method are broadly applicable and refreshingly enduring. Whereas new specialized techniques must be learned upon moving from one research project to another, the general principles need be mastered but once in order to be applied across all scientific disciplines and research projects. Fourth and finally, scientific discoveries routinely stimulate technological advances in agriculture, engineering, and medicine of strategic value in alleviating hunger, poverty, and disease. Consequently, much is at stake as educators train future intellectual generations. For these four reasons, understanding how scientific thinking works is a vital component of undergraduate and graduate education.

The recent worldwide increase in commitment to science education is dramatic and unprecedented, as three examples from the past year indicate. India announced plans to expand university enrollment from 12 million to 30 million students by 2025, with much of this staggering increase directed at science and technology. New York City sought proposals for a new tech campus intended to stimulate the city's economy, with the winning bid coming from a partnership of Cornell University and The Technion-Israel Institute of Technology. The Association of American Universities (AAU), which represents 59 US and two Canadian leading research universities, began a five-year initiative to improve science, technology, engineering, and mathematics (STEM) education for undergraduates, and this initiative is also supported by the Association of Public and Land-grant Universities (APLU), the President's Council of Advisors on Science and Technology (PCAST), and other influential organizations.

Besides these three examples, there are impressive commitments to strengthen world-class research universities in China, the Middle East, and elsewhere. In this context of intense interest in science education, this book's message is that scientific method is the gateway into science and technology.

Looking toward the future, my expectation is that a disproportionately large share of scientific discoveries and technological innovations will come from those researchers who have mastered specialized techniques like everyone else, but who have also mastered the general principles of scientific method. Likewise, scholars having an astute and discerning perspective on science will provide the best reflections on science's rationality, relationship with the humanities, powers and limits, and roles in culture and life.

I am grateful to several persons who provided valuable feedback on previous drafts of the entire book manuscript: Benjamin Brown-Steiner, Samuel Cartinhour, Theodore Harwood, James Pothen, Nicholas Saleh, and Justin Tyvoll. They spanned my intended audience of undergraduate to professional levels across the sciences and the humanities: three undergraduate students in biology, philosophy, and engineering; two graduate students in atmospheric sciences and the classics; and a professional in bioinformatics. Many additional persons, too numerous to list, provided feedback on individual sections or chapters of this book. I thank P. Andrew Karplus for co-authoring the section on biochemistry and pharmacology in Chapter 11. I also thank James Miller for writing the Foreword for this book and its 2002 predecessor. Cambridge University Press editors Katrina Halliday, Megan Waddington, and Hans Zauner provided extremely helpful advice. The enthusiastic support from my family and friends has been greatly appreciated. Cornell University provided a wonderfully favorable environment for writing this book, especially by virtue of its superb library system.

1

Introduction

Science and technology have immense cultural and economic significance. They transform much of what we encounter in daily life. Obvious examples are the computers, phones, and consumer electronics that change and improve noticeably on a timescale of merely a year. For instance, compared to the first flash drive for my computer that I purchased in the early 2000s, now a faster and smaller flash drive with 64 times as much memory costs only a quarter as much. This equates to the memory per dollar almost doubling annually.

Furthermore, scientific transformation is pervasive, even when not so obvious. For instance, a simple loaf of bread or bowl of rice seems like a low-tech product that is the same as it was decades ago. Not so! Wheat rust, rice blast, and other crop diseases are continually evolving new virulent strains that threaten current crop varieties. The ongoing efforts of plant breeders are necessary to protect crops from diseases, to increase yields, and to improve nutritional and other traits. The rate of change for our crops is so rapid that few varieties are still competitive after only seven or eight years. Were plant breeders to stop their work, the disease problems within a decade for wheat, rice, corn, potatoes, and other major crops would be catastrophic. So, the loaf of bread that you buy today, or the bowl of rice that you eat today, is a high-tech product that sophisticated and energetic scientific efforts have rendered quite different from its predecessors of a decade ago. My own scientific work from 1970 to the present has been developing statistical methods and software for these agricultural researchers.

Besides its obvious economic impact, science has an equally significant cultural importance. The knowledge that science has gained affects how we understand ourselves and our world. Discoveries by Galileo, Newton, Faraday, and Darwin changed science but also impacted culture. The substantial interaction between science and culture raises momentous questions about how best to integrate the sciences and the humanities in an overall approach to knowledge and life.

This book has one thesis, two purposes, and an intended audience. The thesis of this book is that scientific methodology has two components, the general principles of scientific method and the research techniques of a given specialty, and the winning combination for scientists is strength in both. This book's two purposes, set forth in its preface, are to increase productivity by understanding scientific method more deeply and to gain perspective from a distinctively humanities-rich vision of science. The intended audience is persons undertaking their first systematic study of scientific method, spanning undergraduates to professionals in both the sciences and the humanities.

The gateway into science

Given the great intrinsic, cultural, and economic significance of science, its most essential feature has tremendous importance: its gateway. Scientific method, which is the topic of this book, is the gateway into science and technology. This gateway was discovered merely a few centuries ago, between 1200 and 1600 by various accounts, long after civilizations had risen and fallen around the globe for millennia. People are not born knowing about scientific method, and many of its features are counter-intuitive and hence difficult to grasp. Consequently, scientific method requires systematic study. As the gateway into science, scientific method precedes scientific discovery, which precedes technological advances and cultural influences.

The structure of science's methodology envisioned here is depicted in Figure 1.1, which shows individual sciences, such as astronomy and chemistry, as being partly similar and partly dissimilar in methodology. What they share is a core of the general principles of scientific method. This common core includes such topics as hypothesis generation and testing, deductive and inductive logic, parsimony, and science's presuppositions, domain, and limits. Beyond methodology as such, some practical issues are shared broadly across the sciences, such as relating the scientific enterprise to the humanities, implementing effective science education, and clarifying science's ethics.

The general principles that constitute this book's topics are shown in greater detail in Figure 1.2. These principles can be described in three groups, moving from the outermost to the innermost parts of this figure.

(1) Some principles are relatively distinctive of science itself. For instance, the ideas about parsimony and Ockham's hill that are developed in Chapter 10 have a distinctively scientific character.

(2) Other principles are shared broadly among all forms of rational inquiry. For example, deductive logic is squarely in the province of scientists, as explored in Chapter 7, but deductions are also important in nearly all undertakings.

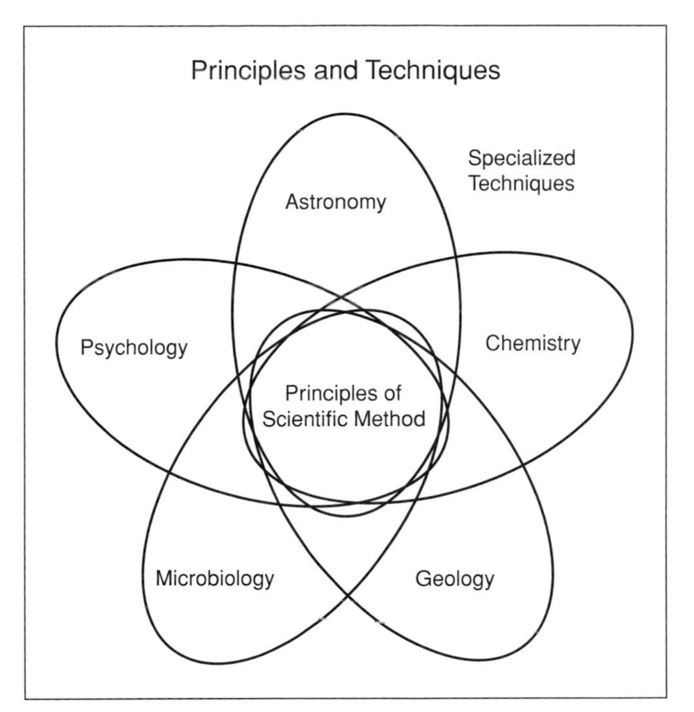

Figure 1.1 Science's methodology depicted for five representative scientific disciplines, which are partly similar and partly dissimilar. Accordingly, scientific methodology has two components. The general principles of scientific method pervade the entire scientific enterprise, whereas specialized techniques are confined to particular disciplines or subdisciplines.

(3) Still other principles are so rudimentary and foundational that their wellsprings are in common sense. This includes science's presuppositions of a real and comprehensible world, which are discussed in Chapter 5.

Naturally, the boundaries among these three groups are somewhat fuzzy, so they are shown with dashed lines. Nevertheless, the broad distinctions among these three groups are clear and useful.

There is a salient difference between specialized techniques and general principles in terms of how they are taught and learned. Precisely because specialized techniques are specialized, each scientific specialty has its own more or less distinctive set of techniques. Given hundreds of specialties and subspecialties, the overall job of communicating these techniques requires countless courses, books, and articles. But precisely because general principles are general, the entire scientific community has a single shared set of principles, and it is feasible to collect and communicate the main information about these principles within the scope of a single course or book. Whereas a scientist or technologist

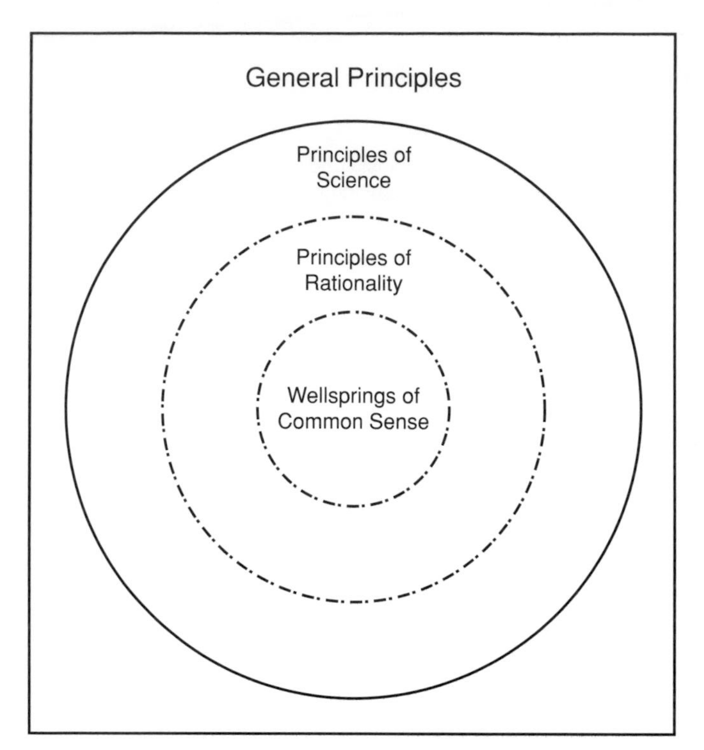

Figure 1.2 Detailed view of the general principles, which are of three kinds: principles that are relatively distinctive of science itself, broader principles found in all forms of rational inquiry, and foundational principles with their wellsprings in common sense.

needs to learn new techniques when moving from one project to another, the pervasive general principles need be mastered but once. Likewise, whereas specialized techniques and knowledge have increasingly shorter half-lives, given the unprecedented and accelerating rate of change in science and technology, the general principles are refreshingly enduring.

What a scientist or technologist needs in order to function effectively can be depicted by a resources inventory, as in Figure 1.3. All items in this inventory are needed for successful research. The first three items address the obvious physical setup that a scientist needs. The last two items are intellectual rather than physical, namely, mastery of the specialized techniques of a chosen specialty and mastery of the general principles of scientific method.

Frequently, the weakest link in a scientist's inventory is an inadequate understanding of science's principles. This weakness has just as much potential to retard progress as does, say, inappropriate laboratory equipment or inadequate training in some research technique.

Scientific Resources Inventory

- [x] Laboratory equipment to generate data
- [x] Computers and software to analyze data
- [x] Infrastructure: colleagues, libraries, Internet access
- [x] Technical training in research specialty
- [ ] General principles of scientific method

Figure 1.3 A typical resources inventory for a research group. The scientists in a given research group often have excellent laboratory equipment, computers, infrastructure, and technical training, but inadequate understanding of the general principles of scientific method is the weakest link. Ideally, a research group will be able to check off all five boxes in this inventory, and there will be no weak link.

A controversial idea

The mere idea that there exist such things as general principles of scientific method is controversial. The objections are of two kinds: philosophical and scientific. But first, a potential misunderstanding needs to be avoided. The scientific method "is often misrepresented as a fixed sequence of steps," rather than being seen for what it truly is, "a highly variable and creative process" (AAAS 2000:18). The claim of this book is that science has general principles that must be mastered to increase productivity and enhance perspective, not that these principles provide a simple and automated sequence of steps to follow.

Beginning with the philosophical objection, it is fashionable among some skeptical, relativistic, and postmodern philosophers to say that there are no principles of rationality whatsoever that reliably or impressively find truth. For instance, in an interview in *Scientific American*, the noted philosopher of science, Paul Feyerabend, insisted that there are no objective standards of rationality, so consequently there is no logic or method to science (Horgan 1993). Instead, "Anything goes" in science, and it is no more productive of truth than "ancient myth-tellers, troubadours and court jesters." From that dark and despairing philosophical perspective, the concern with scientific method would seem to have nothing to do distinctively with science itself. Rather, science would be just one more instance of the pervasive problem that rationality and truth elude us mere mortals, forever and inevitably.

Such critiques are unfamiliar to most scientists, although some may have heard a few distant shots from the so-called science wars. Scientists typically find those objections either silly or aggravating, so rather few engage such controversies. But in the humanities, those deep critiques of rationality are currently influential. By that reckoning, Figure 1.1 should show blank paper, with neither general principles nor specialized techniques that succeed in finding truth.

Moving along to the scientific objection, some scientists have claimed that there is no such thing as a scientific method. For instance, a Nobel laureate in medicine, Sir Peter Medawar, pondered this question: "What methods of enquiry apply with equal efficacy to atoms and stars and genes? What *is* 'The Scientific Method'?" He concluded that "I very much doubt whether a methodology based on the intellectual practices of physicists and biologists (supposing that methodology to be sound) would be of any great use to sociologists" (Medawar 1969:8, 13). By that reckoning, Figure 1.1 should show the methodologies of the individual sciences dispersed, with no area in which they would all overlap.

Is it plausible that, contrary to Figure 1.1, the methodologies of the various branches of science have no overlap, no shared general principles? Asking a few concrete questions should clarify the issues. Do astronomers use deductive logic, but not microbiologists? Do psychologists use inductive logic (including statistics) to draw conclusions from data, but not geologists? Are probability concepts and calculations used in biology, but not in sociology? Do medical researchers care about parsimonious models and explanations, but not electrical engineers? Does physics have presuppositions about the existence and comprehensibility of the physical world, but not genetics? If the answers to such questions are no, then Figure 1.1 stands as a plausible picture of science's methodology.

The AAAS position on method

Beyond such brief and rudimentary reasoning about science's methodology, it merits mention that the thesis proposed here accords with the official position of the American Association for the Advancement of Science (AAAS). The AAAS is the world's largest scientific society, the umbrella organization for almost 300 scientific organizations and publisher of the prestigious journal *Science*. Accordingly, the AAAS position bids fair as an expression of mainstream science. The AAAS views scientific methodology as a combination of general principles and specialized techniques, as depicted in Figure 1.1.

> Scientists share certain basic beliefs and attitudes about what they do and how they view their work.... Fundamentally, the various scientific disciplines are alike in their reliance on evidence, the use of hypotheses and theories, the kinds of logic used, and much more. Nevertheless, scientists differ greatly from one another in what phenomena they investigate and in how they go about their work; in the reliance they place on historical data or on experimental findings and on qualitative or quantitative methods; in their recourse to fundamental principles; and in how much they draw on the

> findings of other sciences.... Organizationally, science can be thought of as the collection of all of the different scientific fields, or content disciplines. From anthropology through zoology, there are dozens of such disciplines.... With respect to purpose and philosophy, however, all are equally scientific and together make up the same scientific endeavor. (AAAS 1989:25–26, 29)

Regarding the general principles, "Some important themes pervade science, mathematics, and technology and appear over and over again, whether we are looking at an ancient civilization, the human body, or a comet. They are ideas that transcend disciplinary boundaries and prove fruitful in explanation, in theory, in observation, and in design" (AAAS 1989:123). Accordingly, "Students should have the opportunity to learn the nature of the 'scientific method'" (AAAS 1990:xii; also see AAAS 1993). That verdict is affirmed in official documents from the National Academy of Sciences (NAS 1995), the National Commission on Excellence in Education (NCEE 1983), the National Research Council of the NAS (NRC 1996, 1997, 1999, 2012), the National Science Foundation (NSF 1996), the National Science Teachers Association (NSTA 1995), and the counterparts of those organizations in many other nations (Matthews 2000:321–351). In all of these reports, scientific method holds a prominent position.

Science as a liberal art

An important difference between specialized techniques and general principles is that the former are discussed in essentially scientific and technical terms, whereas the latter inevitably involve a wider world of ideas. Accordingly, the central premise of the AAAS position paper on *The Liberal Art of Science* is extremely important: "Science is one of the liberal arts and... science must be taught as one of the liberal arts, which it unquestionably is" (AAAS 1990:xi).

Indeed, in antiquity, the liberal arts included some science. Grammar, logic, and rhetoric were in the lower division, the trivium; and arithmetic, geometry, astronomy, and music were in the higher division, the quadrivium. An early accretion was geology, and clearly the AAAS now includes all branches of contemporary science in the liberal art of science. A mosaic in a Cornell University chapel beautifully depicts the integration of all learning, with Philosophy the central figure flanked by Truth and Beauty (not shown) and the Arts and the Sciences to the right and left (Figure 1.4).

Many of the broad principles of scientific inquiry are not unique to science but also pervade rational inquiry more generally, as depicted in Figure 1.2. "All sciences share certain aspects of understanding—common perspectives that transcend disciplinary boundaries. Indeed, many of these fundamental values

Figure 1.4 The Arts and the Sciences. The Arts are represented by Literature, Architecture, and Music, and the Sciences by Biology, Astronomy, and Physics. These details are from the mosaic *The Realm of Learning*, in Sage Chapel, Cornell University, that was designed by Ella Condie Lamb. (These photographs by Robert Barker of Cornell University Photography are reproduced with his kind permission.)

and aspects are also the province of the humanities, the fine and practical arts, and the social sciences" (AAAS 1990:xii; also see p. 11).

Furthermore, the continuity between science and common sense is respected, which implies productive applicability of scientific attitudes and thinking in daily life. "Although all sorts of imagination and thought may be used in coming up with hypotheses and theories, sooner or later scientific arguments must conform to the principles of logical reasoning—that is, to testing the validity of arguments by applying certain criteria of inference, demonstration, and common sense" (AAAS 1989:27). "There are . . . certain features of science that give it a distinctive character as a mode of inquiry. Although those features are especially characteristic of the work of professional scientists, everyone can exercise them in thinking scientifically about many matters of interest in everyday life" (AAAS 1989:26; also see AAAS 1990:16).

Because the general principles of science involve a wider world of ideas, many vital aspects cannot be understood satisfactorily by looking at science in isolation. Rather, they can be mastered properly only by seeing science in context, especially in philosophical and historical context. Therefore, this book's pursuit of the principles of scientific method sometimes ranges into discourse that has a distinctively philosophical or historical or sociological character. There is a natural and synergistic traffic of great ideas among the liberal arts, including science. The AAAS suggested several practical advantages from placing science within the liberal-arts tradition.

> Without the study of science and its relationships to other domains of knowledge, neither the intrinsic value of liberal education nor the practical benefits deriving from it can be achieved. Science, like the other liberal arts, contributes to the satisfaction of the human desire to know and understand. Moreover, a liberal education is the most practical education because it develops habits of mind that are essential for the conduct of the examined life. Ideally, a liberal education produces persons who are openminded and free from provincialism, dogma, preconception, and ideology; conscious of their opinions and judgments; reflective of their actions; and aware of their place in the social and natural worlds. The experience of learning science as a liberal art must be extended to all young people so that they can discover the sheer pleasure and intellectual satisfaction of understanding science. In this way, they will be empowered to participate more fully and fruitfully in their chosen professions and in civic affairs. . . . Education in science is more than the transmission of factual information: it must provide students with a knowledge base that enables them to educate themselves about the scientific and technological issues of their times; it must provide students with an understanding of the nature of science and its place in society; and it must provide them with an understanding of the methods and processes of scientific inquiry. (AAAS1990:xi–xii)

Matthews (1994:2) agreed: "Contributors to the liberal tradition believe that science taught . . . and informed by the history and philosophy of the subject can engender understanding of nature, the appreciation of beauty in both nature and science, and the awareness of ethical issues unveiled by scientific knowledge and created by scientific practice." He offered a specific example: "To teach Boyle's Law without reflection on what 'law' means in science, without considering what constitutes evidence for a law in science, and without attention to who Boyle was, when he lived, and what he did, is to teach in a truncated way. More can be made of the educational moment than merely teaching, or assisting students to discover that for a given gas at a constant temperature, pressure times volume is a constant" (Matthews 1994:3).

Indeed, concepts that are rich in philosophical content and meaning pervade science, such as rationality, truth, evidence, and cause. And deductive logic, probability theory, and other relevant topics have been addressed by both scientists and philosophers. Accordingly, an adequate understanding of science, for science and nonscience majors alike, must see science as one of the liberal arts. A humanities-rich vision of science surpasses a humanities-poor vision.

Certainly, the depictions by the AAAS of productive interactions between science and the other liberal arts are decidedly convivial and promising. But it must be acknowledged that science's recommended partners, the humanities, currently are in a state of tremendous turmoil and controversy.

With keen insight, Matthews (1994:9) discerned that there are "two broad camps" in the history and philosophy of science (HPS) literature, "those who appeal to HPS to support the teaching of science, and those who appeal to HPS to puncture the perceived arrogance and authority of science." This second camp stresses "the human face of science" and argues for pervasive "skepticism about scientific knowledge claims." Matthews's sensible reaction was to "embrace a number of the positions of the second group: science does have a human, cultural, and historical dimension, it is closely connected with philosophy, interests and values, and its knowledge claims are frequently tentative," and yet, "none of these admissions need lead to skepticism about the cognitive claims of science."

Given the profound internal controversies of the humanities, to suggest that science can gain strength by partnering with the humanities might seem like suggesting that a sober person seek support from a staggering drunk! But that would be an unfortunate overreaction. True, there are enough troubles in the humanities that a wanton relationship could weaken science. But much more importantly, there are enough insights and glories in the humanities that a discerning relationship can greatly strengthen science.

For the present, however, the foregoing rather cheerful and innocent account of science as a liberal art provides a fitting point of departure. Unquestionably and wonderfully, science is a liberal art.

Benefits and challenges

The expected benefits from studying scientific method are increased productivity and enhanced perspective. But, regrettably, for most university students, the current situation is challenging. Few science majors ever take a course in scientific method, logic, or the history and philosophy of science. "The hapless student is inevitably left to his or her own devices to pick up casually and

Elementary Scientific Method

- Hypothesis formulation
- Hypothesis testing
- Deductive and inductive logic
- Controlled experiments; replication and repeatability
- Interactions between data and theory
- Limits to science's domain

Figure 1.5 Typical topics in an elementary presentation of scientific method intended for college freshmen and sophomores. Introductory science texts often start with several pages on scientific method, discussing the formulation and testing of hypotheses, collection of data from controlled and replicated experiments, and so on. They are unlikely, however, to include any discussion of parsimony or any exploration of the history of scientific method beyond a passing mention of Aristotle.

randomly, from here and there, unorganized bits of the scientific method, as well as bits of *un*scientific methods" (Theocharis and Psimopoulos 1987). And the same is true for most science professors and professionals: "Ask a scientist what he conceives the scientific method to be, and he will adopt an expression that is at once solemn and shifty-eyed: solemn, because he feels he ought to declare an opinion; shifty-eyed, because he is wondering how to conceal the fact that he has no opinion to declare" (Medawar 1969:11).

The exposure of university students to science's principles is usually limited to the occasional science textbook that begins with brief remarks on scientific method. Figure 1.5 lists typical contents. But such an elementary view of scientific method is wholly inadequate at the university level for science and nonscience students alike.

What are the benefits from studying scientific method? The best answers have not come from scientists or philosophers but rather from science educators. They have conducted hundreds of careful empirical studies to characterize and quantify and compare the specific benefits that can result from learning the scientific method. Many of those studies have involved impressive sample sizes and carefully controlled experiments to quantify educational outcomes for students who either have or else have not received instruction in science's general principles. Because Chapter 13 will review the literature in science education, here only brief remarks without documentation will be presented, by way of anticipation.

(1) Better Comprehension. The specialized techniques and subject knowledge that so obviously make for productive scientists are better comprehended when the underlying principles of scientific method are understood. Giving adequate attention to both specialized knowledge and general principles creates a win-win situation.

(2) **Greater Adaptability.** It is facility with the general principles of science that contributes the most to a scientist's ability to be adaptable and to transfer knowledge and strategies from a familiar context to new ones. Adaptability is crucial as science and technology experience increasingly rapid and pervasive changes.

(3) **Greater Interest.** Most people find a humanities-rich version of science, with its wider perspective and big picture, much more engaging and interesting than a humanities-poor version. Including science's method, history, and philosophy in the science curriculum increases retention rates of students in the sciences.

(4) **More Realism.** An understanding of the scientific method leads to a realistic perspective on science's powers and limits. It also promotes balanced views of the complementary roles of the sciences and the humanities.

(5) **Better Researchers.** Researchers who master science's general principles gain productivity because they can make better decisions about whether or not to question an earlier interpretation of their data as a result of new evidence, whether or not there is a need to repeat an experiment, and where to look for other scientific work related to their project. They better assess how certain or accurate their conclusions are.

(6) **Better Teachers.** Teachers and professors who master science's general principles prove to be better at communicating science content. They are better at detecting and correcting students' prior mistaken notions and logic, and hence such teachers can better equip the next generation of scientists to be productive.

The facts of the case are clear, having been established by hundreds of empirical studies involving various age groups, nations, and science subjects: understanding the principles of scientific method does increase productivity and enhance perspective. Why? The most plausible explanation is simply that the thesis of this book is true: it really is the case that scientific methodology has two components, the general principles of scientific method and the research techniques of a chosen specialty, and the winning combination is strength in both.

Personal experience

Thus far, this introductory chapter has drawn on the insights of others, especially those of the AAAS and science educators, to support this book's thesis. But perhaps some readers would be interested in the personal experience that has prompted my interest in the principles of scientific method.

Figure 1.6 A soybean yield trial conducted in Aurora, New York. The soybean varieties here varied in terms of numerous traits. For example, the variety in the center foreground matured more quickly than the varieties to its left and right, making its leaves light yellow rather than dark green as the end of the growing season approached. Yield is a particularly important trait. (Reprinted from Gauch, 1992:3, with kind permission from Elsevier Science.)

My research specialty at Cornell University from 1970 to 2010 has been the statistical analysis of ecological and agricultural data. A special focus in this work has been agricultural yield trials. Worldwide, billions of dollars are spent annually to test various cultivars, fertilizers, insecticides, and so on. For instance, Figure 1.6 shows a soybean yield trial conducted to determine which cultivars perform best in various locations throughout the state of New York. The main objective of yield-trial research is to increase crop yields.

From studying the philosophy and method of science, but not from reading the agricultural literature, I came to realize that a statistical model can provide greater accuracy than can its raw data. As will be explained in Chapter 10 – on parsimony, which also is called simplicity or Ockham's razor – often statistical modeling increases accuracy as much as would collecting much more data. But the modeling costs merely a few seconds of computer time, whereas expanding data collection costs tens to hundreds of thousands of dollars in various instances, so this statistical gain in accuracy is spectacularly cost-effective. And greater accuracy improves decisions, increases repeatability, and accelerates progress.

The salient feature of that story is that the requisite statistical analyses and theory had been developed by 1955 and computers had become widely available to agronomists and breeders by 1970. However, no one had capitalized on that opportunity until Gauch (1988).

What has been the opportunity cost? Standard practices in agricultural research today are increasing the yields for most of the world's major crops by about 0.5% to 1.5% per year. A conservative estimate is that statistical models of yield-trial data often can support an additional increment of about 0.4% per year. Hence, for a typical case, if ordinary data analysis supports an average annual yield increase of 1%, whereas aggressive analysis supports 1.4%, then progress can be accelerated by another 40% simply by putting statistics to work. Regrettably, the opportunity cost for delaying that annual yield increment for a couple of decades equates to losing enough food for several hundreds of millions of persons, more than the population of North America.

What caused this reduction in crop productivity? Recalling the resource inventory in Figure 1.3, it was neither the lack of specialized research techniques nor the ability to easily perform billions of arithmetic steps. Rather, it was lack of understanding of parsimony, one of the general principles of scientific method. What was missing was the last of the critical resources listed in Figure 1.3. Method matters.

The larger issue that this experience raises is that many other scientific and technological specialties present us with tremendous opportunities that cannot be realized until some specialist in a given discipline masters and applies a critical general principle. Precisely because these are *general* principles, my suspicion is that my own experience is representative of what can be encountered in countless other specialties (Gauch 1993, 2006).

Furthermore, my own experience resonates with the AAAS (1990:xi) expectation that a broad vision of science as a liberal art is worthwhile for "the sheer pleasure and intellectual satisfaction of understanding science." I had a restless curiosity and deep interest regarding the basic principles of scientific thinking. But that spark of curiosity had received no stimulus or encouragement whatsoever from the courses and ideas presented in my university education.

While a graduate student at Cornell, I stumbled across a book by Arthur Burks not long after it was first published in 1963, which is now available in a newer edition (Burks 1977). He was a professor of both philosophy and computer science. His book was quite long, about 700 pages, and frequently was repetitious and tedious. However, it had the content that I had been seeking and had not yet found anywhere else. There at last I had found an intellectually satisfying account of the underlying principles and rationality of scientific thinking. That book immediately became a great favorite of mine. Subsequently, I sought and occasionally found additional books to nourish my ongoing interest in the principles of scientific method, most notably that by Jeffreys (1983), first published in 1961, and more recently Howson and Urbach (2006).

Thus, my interest in science's principles dates to about 1965. My motivation for that interest was – to echo the AAAS – the "sheer pleasure" that accompanies "the human desire to know and understand" (AAAS 1990:xi). Grasping the big ideas that are woven throughout the fabric of the entire scientific enterprise generates delight and confidence. However, the idea that mastery of those principles could also promote productivity did not awaken in my mind until a couple of decades later (Gauch 1988). Since then, my interest in these principles has been motivated by desires for both intellectual perspective and scientific productivity. During the 2000s, my interest in the general principles of scientific method has been further stimulated by co-teaching a course on scientific method with a colleague and thereby enjoying the intriguing and creative thinking of Cornell graduate and undergraduate students in science and nonscience majors.

Seven streams

Scientific method, as explored in the 14 chapters of this book, involves numerous topics. Readers wanting more information on a given topic should find the citations helpful. It may also be helpful to have an overview of the literature that bears on scientific method. The relevant literature is widely scattered and it has seven streams. These seven streams are complementary, so all seven are needed for a rich understanding of scientific method.

(1) Books on scientific method by scientists are the most obvious and directly relevant literature, although most are several decades old.
(2) Statistics provides the principal literature on a crucial component of scientific method: inductive logic (including experimental design, parameter estimation, data summary, and hypothesis testing).
(3) Philosophy of science provides profound insights on science.
(4) History of science provides essential perspective on science.
(5) Sociology of science reveals the human context of the scientific community.
(6) Science education is essential for the existence and flourishing of the scientific enterprise and for improving pedagogy. However, as seems quite natural and appropriate, these literatures primarily address the distinctive interests and purposes of philosophers, historians, sociologists, and educators. Nevertheless, there is a fraction of these literatures that can serve a different purpose, helping scientists to become better scientists. That is the fraction selectively emphasized in my citations to and quotations from these four literature streams.
(7) Last and immensely valuable, there are position papers on science from the AAAS, NAS, NSF, and other leading scientific organizations, as well as their counterparts in many other nations.

The important roles of position papers and science education standards in this book merits explanation. The nature of scientific method and the reliability of scientific findings have been debated for centuries, including hot debates in recent decades, as Chapter 4 documents. More broadly, the status of human knowledge has been contested without interruption from antiquity to the present, as Chapters 2 and 3 explain. Multiple positions increase the complexity of a book on scientific method, unavoidably. But against this complex backdrop of multiple positions, these prominent position papers describe, distinguish, and privilege one particular position as being the mainstream position. As will be evident from the historical material in this book (as well as in these position papers), many of the main features of scientific method in its contemporary, mainstream manifestation have been stable features of science for several centuries, or even as long as two millennia for some key ideas.

My intention as an author on scientific method, which I want to make known to my readers explicitly and clearly in this first chapter, is to align with mainstream science. Not only does this represent my own personal convictions, it also best serves the needs of both the scientific community and the general public. Mainstream science is ideal for developing technology, appreciating nature, and interacting with the humanities in a fruitful manner.

The principal position papers and education standards engaged here are *Science for All Americans* (AAAS 1989); *The Liberal Art of Science* (AAAS 1990); *Reshaping the Graduate Education of Scientists and Engineers* (NAS 1995); *National Science Education Standards* (NRC 1996); *Shaping the Future: New Expectations for Undergraduate Education in Science, Mathematics, Engineering, and Technology* (NSF 1996); *Transforming Undergraduate Education in Science, Mathematics, Engineering, and Technology* (NRC 1999); *On Being a Scientist: A Guide to Responsible Conduct in Research* (NAS 2009); and *A Framework for K–12 Education: Practices, Crosscutting Concepts, and Core Ideas* (NRC 2012). These careful documents involved hundreds of outstanding scientists and scholars as contributors and reviewers, who worked through drafts over several years.

Listed here are several books from each of these seven streams that express a diversity of views on science and scientific method. Exemplary books on scientific method include Nash (1963), Burks (1977), and Derry (1999), with Carey (2012) an admirable book at a somewhat more elementary level. Exceptional books on statistics include Berger (1985), Gelman et al. (2004), Taper and Lele (2004), Howson and Urbach (2006), Robert (2007), and Hoff (2009). Philosophical perspectives on science are addressed by Trigg (1993), Godfrey-Smith (2003), Nola and Sankey (2007), Gimbel (2011), and Rosenberg (2012); historical perspectives by Gower (1997), Losee (2001), McClellan and Dorn (2006), and Lindberg (2007); sociological perspectives by Merton (1973), Merton and Sztompka (1996), and Stehr and Meja (2005); and educational perspectives by Matthews (1994), McComas (1998), Hodson (2009), and Niaz (2011). Finally, the scientific community would benefit from greater awareness of the position papers on science, already cited herein, from the AAAS, NAS, and NSF.

Historical and future outlook

Despite the unanimous recommendation from the AAAS and many other leading scientific organizations in many nations that scientific method be emphasized in the science curriculum, the current situation at the university level is one of pervasive neglect. Despite the AAAS verdict that science unquestionably is one of the liberal arts, in recent decades, this has not been generally and clearly appreciated. Consequently, turning the AAAS vision into reality will require some effort: "In spite of the importance of science and the ubiquity of its applications, science has not been integrated adequately into the totality of human experience.... Understanding science and its influence on society and the natural world will require a vast reform in science education from preschool to university" (AAAS 1990:xi). What went wrong?

To understand the huge discrepancy between the AAAS vision of humanities-rich science and the current reality of humanities-poor science, some rudimentary historical perspective is needed. By AD 500, the classical liberal arts had already become well codified in the trivium and quadrivium, which included some science. Then, around 1200, and coincident with the founding of the earliest universities, there was a great influx of knowledge into Western Europe.

Around 1850, at about the time when many of the great universities in the USA and elsewhere beyond Europe were being founded, there were two revealing developments: invention of the new word "scientist" and acquisition of a new meaning for the word "science." The term "scientist" was coined in 1834 by members of the British Association for the Advancement of Science to describe students of nature, by analogy with the previously existing term "artist." Subsequently, that new word was established securely in 1840 through William Whewell's popular writings. Somewhat later, in the 1860s, the *Oxford English Dictionary* (OED) recognized that "science" had come to have a new meaning as "physical and experimental science, to the exclusion of theological and metaphysical," and the 1987 supplement to the OED remarked that "this is now the dominant sense in ordinary use" (http://dictionary.oed.com). Those new or modified words certified science's coming of age, with its own independent intellectual identity. Increasingly since 1850, science has also had its own institutional identity.

The rift between the sciences and the humanities reached its peak around the 1920s and 1930s, with a prevailing conception of science that discounted human factors in science and intentionally disdained philosophy, especially metaphysics. A turning point came in 1959 with the publication of two books destined to have enormous influence: Sir Karl Popper's *The Logic of Scientific*

Discovery called for a human-sized account of science, with significant philosophical, historical, and sociological content (Popper 1968); and C. P. Snow's *The Two Cultures and the Scientific Revolution* drew attention to the divide between the sciences and the humanities with its resulting lamentable intellectual fragmentation (Snow 1993). So the connection between science and the humanities has varied somewhat during the twentieth century, but on the whole there has been a considerable rift between the two cultures.

For twenty-two centuries, from Aristotle until the twentieth century, it was the universal practice of the scientific community to produce scholars who understood both philosophy and science. To use a term fittingly applied to Einstein by Schilpp (1951), they were philosopher-scientists. Einstein rightly insisted that "Science without Epistemology is – in so far as it is thinkable at all – primitive and muddled" (Rosenthal-Schneider 1980:27). But most contemporary scientists receive meager training in the history and philosophy of science, epistemology, the principles of scientific method, and logic. It would be beneficial for the scientific community to return to the venerable tradition, which served previous generations well, of producing philosopher-scientists.

The twentieth century has been the one and only century in science's rich history to have produced mostly scientists rather than philosopher-scientists. Consequently, many contemporary scientists cannot defend science's rationality and credibility from various intellectual attacks and they cannot optimize the methods and productivity of their own research programs. Weakness in scientific method is costly, wasting research dollars, compromising competitive advantages, delaying scientific discoveries and technological advances, and reducing the sheer intellectual pleasure that could be derived from a humanities-rich version of science.

Summary

Science and technology have immense cultural and economic significance. Scientific method is the gateway into scientific discoveries that in turn prompt technological advances and cultural influences. Science is best understood in a humanities-rich version that perceives science as a liberal art.

The thesis of this book is that there exist general principles of scientific method that are applicable across all of the sciences, undergird science's rationality, and greatly influence science's productivity and perspective. These general methodological principles include deductive and inductive logic, probability, parsimony, and hypothesis testing as well as science's presuppositions, limitations, and ethics. The implicit contrast is with specialized techniques that occur only in some sciences or applications. The winning combination for scientists is strength in both. Neither basic principles nor research techniques can substitute for one another. This winning combination increases productivity and enhances perspective.

On five counts, this thesis merits serious consideration. First, that science has a scientific method with general principles and that these principles can benefit scientists is the official, considered view of the AAAS (and the NAS, NRC, NSF, and other major scientific organizations in the United States, as well as similar entities in numerous other nations). Second, science's basic concepts and methods – such as rationality, truth, deductive and inductive logic, and parsimony – interconnect with philosophy, history, and other humanities, so the official view of the AAAS is compelling that a humanities-rich version of science as a liberal art stimulates clarity and perspective. Third, science educators have demonstrated in hundreds of empirical studies, often involving sizable samples and controlled experiments, that learning science's general principles can benefit students and scientists in several specific, quantifiable, important respects. Fourth, my own research experience, primarily involving agricultural yield-trial experiments, demonstrates the practical value of a particular principle of scientific method, namely parsimony. Fifth and finally, prior to the twentieth century, for twenty-two centuries following Aristotle, the customary practice of the scientific community, which served it well, had been to produce philosopher-scientists.

This book's message is that a disproportionately large share of future advances in science and technology will come from those researchers who have mastered their specialties like everyone else but who also have mastered the basics of science's philosophy and method. Also, philosopher-scientists will be prominent among those scholars providing the best reflections on science's rationality, relationship with the humanities, powers and limits, and roles in culture and life.

Study questions

(1) Would you suspect that many other seemingly simple and ordinary products, besides a loaf of bread or cup of rice, incorporate extensive scientific research and technological development? Can you give a specific example or two?

(2) Characterize the distinction between specialized techniques and general principles of scientific method. Where does or should the latter appear in the science curriculum, particularly for undergraduates and graduates in science?

(3) How do you react to the AAAS position that science is one of the liberal arts, unquestionably? Is this position familiar or foreign in your own university education? What are some clear implications and potential benefits from a humanities-rich version of science?

(4) What benefits from studying scientific method in particular, or the nature of science more generally, have science educators demonstrated with hundreds of empirical studies?
(5) If you are in the sciences, which specific weaknesses within your own specialty might plausibly be attributed to an inadequate understanding of scientific method? Or if you are in the humanities, have you been satisfied with prevalent characterizations of science's role and significance?

2

Four bold claims

This is the first of five chapters (2–6) directed mainly at this book's purpose of cultivating a humanities-rich perspective on science. The following five chapters (7–11) are directed mainly at this book's other purpose of increasing scientific productivity.

Consider a familiar scientific fact: water is composed of hydrogen and oxygen, having the chemical formula H_2O. The objective of this and the following chapter is to comprehend exactly what claims science makes for such findings. Accordingly, this chapter explicates the concepts of rationality, truth, objectivity, and realism. Mainstream science uses these four concepts incessantly, although usually implicitly, so the philosophical literature on these concepts can enrich scientists' understanding of their own craft. The next chapter explores the historical development of the concept of truth as applied to knowledge about the physical world, from Aristotle to the present. Finally, toward the end of the next chapter, additional scientific information will be presented to complete this story about science's rational, true, objective, and realistic knowledge that water is H_2O. Science worthy of the name must attend not only to facts about electrons, bacteria, humans, and galaxies but also to concepts of rationality, truth, objectivity, and realism.

Rationality

Rationality is good reasoning. The traditional concept of rationality in philosophy, which is also singularly appropriate in science, is that reason holds a double office: regulating belief and guiding action. Rational beliefs have appropriate evidence and reasons that support their truth, and rational actions promote what is good. Rational persons seek true beliefs to guide good actions. "Pieces of behaviour, beliefs, arguments, policies, and other exercises of the human mind may all be described as rational. To accept something as rational is to accept it as making sense, as appropriate, or required, or in accordance with some acknowledged goal, such as aiming at truth or aiming at the good" (Blackburn 1994:319).

Scientific inquiry involves imagination, insight, creativity, and sometimes luck, but in no way does that negate science also involving good reasoning. "Although all sorts of imagination and thought may be used in coming up with hypotheses and theories, sooner or later scientific arguments must conform to the principles of logical reasoning—that is, to testing the validity of arguments by applying certain criteria of inference, demonstration, and common sense" (AAAS 1989:27).

Of science's four bold claims, rationality is discussed first because it is so integral to this book's topic, the scientific method. Although beliefs, persons, and other things can be the objects of a claim of rationality, the principal target here is method. Method precedes and produces results, so claims of rationality for science's conclusions are derivative from more strategic claims of rationality for science's method. Rational methods produce rational beliefs.

A claim of rational knowledge follows this formula: I hold belief *X* for reasons *R* with level of confidence *C*, where assertion of *X* is within the domain of competence of method *M* that accesses the relevant aspects of reality. The first-order belief *X* is accompanied by a second-order belief that assesses the strength of the reasons *R* and hence the appropriate level of confidence *C*, which may range from low probability to high probability to certainty. Besides supporting belief *X*, some effort may also be directed at discrediting various alternative beliefs, *Y* and *Z*. Lastly, the reasons and evidence have meaning and force from a third-order appeal to an appropriate method *M* that accesses the aspects of reality that are relevant for an inquiry into *X*. For example, the scientific method is directed at physical reality, and its domain of competence includes reaching a confident belief, based on compelling evidence, about the composition of table salt.

This business of giving reasons *R* for belief *X* must eventually stop somewhere, however, so not quite all knowledge claims can follow this formula. Rather, some must follow an alternative formula: I hold belief *X* because of presuppositions *P*. This is a story, however, that is better deferred to Chapter 5. The important story at present is just that methods underlie reasons, which in turn underlie beliefs and truth claims.

Reason's double office, of regulating belief and guiding action, means that true belief goes with good action. When belief and action do not agree, which is a moral problem rather than an intellectual problem, the result is insincerity and hypocrisy. When reason is wrongfully demoted to the single office of only regulating belief, thus severing belief from action, the inevitable consequence is sickly beliefs deliberately shielded from reality.

The traditional opponent of reason was passion, as in Plato's picture of reason as a charioteer commanding unruly passions as the horses. So a rational person is one who sincerely intends to believe the truth, even if occasionally strong desires go against reason's dictates.

The claim to be defended here, that science is rational, should not be misconstrued as the different and imperialistic claim that only science is rational. To the contrary, science is a form of rationality applied to physical objects, and science flourishes best when integrated with additional forms of rationality, including common sense and philosophy. "The method of natural science is not the sole and universal rational way of reaching truth; it is one version of rational method, adapted to a particular set of truths" (Caldin 1949:134).

Likewise, the claim to be defended here, that science is rational, should not be conflated with the different and indefensible claim that science is always beneficial. It is unfair to deem that atomic weapons and carcinogenic insecticides count against science's rationality. Obviously, the simple truth is that knowledge of physical reality can be used for good or for ill. Science in the mind is like a stick in the hand: it increases one's ability to work one's will, regardless of whether that will is good or bad, informed or careless.

Truth

Truth is a property of a statement, namely, that the statement corresponds with reality. This correspondence theory of truth goes back to Aristotle, who wrote that "To say of what is that it is not, or of what is not that it is, is false, while to say of what is that it is, and of what is not that it is not, is true" (McKeon 1941:749). This definition has three components: a statement declaring something about the world, the actual state of the world, and the relationship of correspondence between the statement and the world. For example, if I say "This glass contains orange juice" and the state of affairs is that this glass does contain orange juice, then this statement corresponds with the world and hence it is true. But if I say that it contains orange juice when it does not, or that it does not contain orange juice when it does, then such statements are false. Truth claims may be expressed with various levels of confidence, such as "I am certain that 'Table salt is sodium chloride' is true" or "The doctors believe that 'The tumor is not malignant' with 90% confidence" or "There is a 95% probability that the sample's true mass is within the interval 1,072 ± 3 grams." Figure 2.1 depicts Aristotle's correspondence concept of truth.

The correspondence theory of truth grants reality priority over beliefs: "the facts about the world determine the truth of statements, but the converse is not true," and this asymmetry is nothing less than "a defining feature of truth about objective reality" (Irwin 1988:5). "In claiming that truth is correspondence to the facts, Aristotle accepts a biconditional; it is true that *p* if and only if *p*. But he finds the mere biconditional inadequate for the asymmetry and natural priority he finds in the relation of correspondence; this asymmetry is to be captured

Figure 2.1 The correspondence concept of truth, with priority of nature over belief. Here the state of nature is a flower with five petals, and the person's belief is that the flower has five petals, so nature and belief correspond and, consequently this brilliant scientist's belief is true. It is the flower's petals, not the scientist's beliefs, that control the right answer. Beliefs corresponding with reality are true. (This drawing by Carl R. Whittaker is reproduced with his kind permission.)

in causal or explanatory terms" (Irwin 1988:5–6). Again, "Truth is accuracy or representation of an independent world – a world that, while it includes us and our acts of representing it, decides the accuracy of our representations of it and is not constructed by them" (Leplin 1997:29).

In the correspondence definition of truth, notice that the bearers of truth are statements, not persons. Persons are the bearers of statements, but statements are the bearers of truth. Accordingly, truth is not affected by who does or does not say it.

For better or for worse, philosophers have proposed numerous definitions of truth besides the correspondence theory advocated here. What is valid in those other definitions is best regarded as routine elaboration of the correspondence definition, which alone can serve science as the core concept of truth.

For example, the coherence theory says that truth consists in coherence (agreement) among a set of beliefs. The valid element here is that coherence is crucial. Thus, if I say that "Table salt is sodium chloride" and at the same time also blithely voice the contrary that "Table salt is not sodium chloride," then

I lose credit for this first statement because of the incoherence and insincerity caused by the second statement. Likewise, to be either true or false, a statement must at least make sense; "big it run brown" is neither true nor false, but nonsense.

For another example, the pragmatic theory of truth says that the truth is what works. The valid element here is that truth does have practical value for doing business with reality. Thus, if your doctor puts you on a low-sodium diet, then there is practical value in understanding the truth that table salt is sodium chloride. Again, reason holds the double office of regulating belief and guiding action. The danger here would be to let pragmatic actions replace true beliefs, rather than complement them, in a theory of truth.

When the correspondence, coherence, pragmatic, and other theories of truth are all considered seriously and respected equally, in practice none of them wins the day. Rather, the only winner would seem to be a "mystification theory" of truth, saying that it is beyond humans to understand or define truth. Is this *your* theory of truth? There is a simple test: your mother asks this question: "Did you eat the last cookie? Now tell the truth!" If you are capable of answering that question, then someone else may be mystified about what truth is, but you are not. The mystification theory of truth is just bad philosophy.

The definition of truth is one easy little bit of philosophy that scientists must get straight before their enterprise can make meaningful claims. A true statement corresponds with reality. A characteristic feature of antiscientific and postmodern views is to place the word "truth" in scare quotes, or else proudly to avoid this word altogether. Indeed, *every* kind and variety of antiscientific philosophy has, as an essential part of its machinery, a defective notion of truth that assists in the sad task of rendering truth elusive. Scientists must take warning from the words of Leplin (1997:28) that "All manner of truth-surrogates have been proposed" by some philosophers "as what science *really* aims for." Scientists must reject all substitutes.

The definition of truth plays the important role of making scientific hypotheses meaningful even before collecting and analyzing data to test it. For example, the hypothesis that "a carbon atom contains nine protons" is meaningful precisely because it is understood as an attempt at truth, although in this particular case, experimental data would result in rejection of that hypothesis.

Truth is guarded by science's insistent demand for evidence. "Sooner or later, the validity of scientific claims is settled by referring to observations of phenomena.... When faced with a claim that something is true, scientists respond by asking what evidence supports it" (AAAS 1989:26, 28).

Because true statements correspond with objective reality, a theory of truth should be complemented by theories of objectivity and realism. Accordingly, the next two sections discuss these two related concepts.

Objectivity

In its primary usage, the concept of objectivity often appears in adjectival form as objective belief, objective knowledge, or objective truth. This concept is complex and somewhat subtle, having three interrelated aspects. Objective knowledge is about an object, rather than a subject or knower; it is achievable by the exercise of ordinary endowments common to all humans, so agreement among persons is possible; and it is not subverted and undone by differences between persons in their worldview commitments, at least for nearly all worldviews.

The first of the three interrelated aspects of objectivity is that objective knowledge is about an object. The AAAS characterizes science beautifully in the simple words that "science is the art of interrogating nature" with "commitment to understanding the natural world" (AAAS 1990:17). For example, "Table salt is sodium chloride" expresses an objective claim about an object, table salt, while expressing nothing about persons who do or do not hold this belief. Because objective beliefs are about objects themselves, not the persons expressing beliefs, the truth or falsity of an objective belief is determined by the belief's object, such as table salt. This thinking reflects and respects the correspondence theory of truth and its priority of reality over beliefs.

In Aristotle's terms, an objective truth about nature is a truth "known by nature," meaning that it expresses a real feature of the physical world, not just an opinion suited to our cognitive capacities or our questionable theories (Irwin 1988:5). Indeed, "As one physicist remarked, physics is about how atoms appear to atoms," and "in science the ultimate dissenting voice is nature itself, and that is a voice which even an entrenched scientific establishment cannot silence for ever" (O'Hear 1989:229, 215). Science's goal is "observer-independent truths about a world independent of us," and "The truths science attempts to reveal about atoms and the solar system and even about microbes and bacteria would still be true even if human beings had never existed" (O'Hear 1989:231, 6).

The second aspect of objectivity is that objective knowledge is achievable by the exercise of ordinary endowments common to all humans, so agreement among persons is possible. Consequently, science's claims are public and verifiable. "Men and women of all ethnic and national backgrounds participate in science and its applications.... Because of the social nature of science, the dissemination of scientific information is crucial to its progress" (AAAS 1989:28–29). The link between objective truth and inter-subjective agreement is so strong that the former is difficult to defend when the latter fails.

The third and final aspect of objectivity is immunity to worldview differences. A major reason why science is respected is that it cuts across political, cultural, and religious divisions.

> The impartiality of nature to our feelings, beliefs, and desires means that the work of testing and developing scientific theories is insensitive to the ideological background of

individual scientists.... [Indeed,] science does cut through political ideology, because its theories are about nature, and made true or false by a nonpartisan nature, whatever the race or beliefs of their inventor, and however they conform or fail to conform to political or religious opinion.... There is no such thing as British science, or Catholic science, or Communist science, though there are Britons, Catholics, and Communists who are scientists, and who should, as scientists, be able to communicate fully with each other. (O'Hear 1989:6–7, 2, 8)

There is humility, openness, and generosity of spirit in realizing that not only your own worldview supports science, but also most other worldviews allow science to make sense. But having just emphasized that science rises above worldview divisions, on balance it must also be said that this immunity to worldview differences is substantial and satisfactory, but not total. Although held by only a small minority of the world's population, there are some worldviews that are so deeply skeptical or relativistic that they do not and cannot support anything recognizable as science's ordinary claims. And those worldview commitments have a deeper role and greater influence than any and all of science's evidence. But that is a story better told in Chapter 5, on science's presuppositions. For the present, it suffices to acknowledge that science is for almost everyone, but not quite everyone.

"Objectivity" also has a secondary usage that applies to persons rather than to beliefs. When formulating their beliefs, objective persons are willing to allow facts and truth to overrule prejudices and desires. Science "forbids a man to sink into himself and his selfish claims, and shifts the centre of interest from within himself to outside" (Caldin 1949:135–136). Objective inquirers welcome truth.

Furthermore, it must be emphasized that objective knowledge is claimed or possessed by human subjects, for otherwise, unrealistic and indefensible versions of objectivity would emerge. Scientists, as human beings, "must inevitably see the universe from a centre lying within ourselves and speak about it in terms of a human language shaped by the exigencies of human intercourse. Any attempt rigorously to eliminate our human perspective from our picture of the world must lead to absurdity" (Polanyi 1962:3). Objective knowledge that is shared among numerous persons gives science a convivial social aspect, the scientific community.

Articulate systems which foster and satisfy an intellectual passion can survive only with the support of a society which respects the values affirmed by these passions.... [Thus,] our adherence to the truth can be seen to imply our adherence to a society which respects the truth, and which we trust to respect it. Love of truth and of intellectual values in general will... reappear as the love of the kind of society which fosters these values, and submission to intellectual standards will be seen to imply participation in a society which accepts the cultural obligation to serve these standards. (Polanyi 1962:203)

But having acknowledged the subjective and social aspects of objectivity, a grave pathology develops if subjectivity supplants rather than complements objectivity. Such elevation of the knower over the known actually demeans the personal aspect of knowing because it leaves scientists with nothing for their beliefs to be about. That outcome illustrates the principle that every excess becomes its own punishment. Any attempt to eliminate physical objects from science's picture of the world and any attempt to eliminate human persons from science's picture of the world must alike lead to absurdity.

Realism

Realism, as regards the physical world, is the philosophical theory that both human thoughts and independent physical objects exist and that human endowments render the physical world substantially intelligible and reliably known. Scientific realism embodies the claim that the scientific method provides rational access to physical reality, generating much objective knowledge. Realistic beliefs correspond with reality. Realistic persons welcome reality.

> We are trying to refer to reality whenever we say what we think exists. Some may wish to talk of God, and others may think matter is the ultimate reality. Nevertheless, we all talk about tables and chairs, cats and rabbits. They exist, and are real, and do not just depend in some way on our thought for their existence. . . . Man himself is part of reality, and causally interacts with other segments of reality. He can change things, and even sometimes control them. He does not decide what is real and what is not, but he can make up his mind what he thinks real. This is the pursuit of truth. Man's attempt to make true assertions about the self-subsistent world of which he is a part may not always be successful, and may not always prove easy or straightforward. The repudiation of it as a goal would not only destroy science, but would make human intellectual activity totally pointless. (Trigg 1980:200)

Reality does not come in degrees because something either does or does not exist. Thus, one little potato is fully as real as is the entire universe. It is not as big, not as important, and not as enduring, but it is just as real. Likewise, one little potato that exists fleetingly now is completely real regardless of whatever ultimate reality may be invoked to explain or cause or sustain its existence. Science claims to deal with reality. But, clearly, some humility is in order regarding the extent of science's reach. Scientists can agree that a little potato is real even while there is disagreement, uncertainty, or even ignorance about the deep philosophical or physical explanation of its existence.

Common-sense belief in reality is practically universal. For example, a child may say "I am patting my cat." What does this mean? Manifestly, the philosophical story, too obvious to be elaborated in ordinary discourse, is that the child feels and sees and enjoys the cat by virtue of having hands and eyes and brain in close proximity to the furry quadruped. And science's realism is the same. “The simple and unscientific man’s belief in reality is fundamentally the same as that of the scientist” (Max Born, quoted by Nash 1963:29). On the basis of numerous conversations, Rosenthal-Schneider (1980:30) summarized Einstein’s view: “Correspondence to the real physical universe, to nature, was for him the essential feature, the only one which would give ‘truth-value’ to any theory.”

The opposite of realism is antirealism, in any of its many variants. Recall from this section’s opening definition that realism combines two tenets: the existence of objects and minds, and the intelligibility of objects to minds. Idealism denies the first tenet. It says that only minds exist and that “objects” are just illusions imagined by minds. Constructivism claims that the physical world is a projection of the mind, so we construct rather than discover reality. Instrumentalism denies that external physical objects should be the targets of our truth claims, substituting internal perceptions and thoughts as the material for analysis. Skepticism denies the second tenet. It does not deny that the physical world exists, but it denies that we do have or could have any reliable knowledge about the physical world. Relativism accepts personal truth-for-me but not public truth-for-everyone, so there is no objective and shared knowledge about the world such as the scientific community claims.

Ordinary science is so thoroughly tied to realism that realism’s competitors seem to scientists to be somewhat like the philosophical joke expressed well in a little story by Wittgenstein: “I am sitting with a philosopher in the garden; he says again and again ‘I know that that’s a tree’, pointing to a tree that is near us. Someone else arrives and hears this, and I tell him: ‘This fellow isn’t insane. We are only doing philosophy’” (Anscombe and von Wright 1969:61e). Without realism, ordinary science perishes.

The full force of science’s claims results from the joint assertion of all four: rationality, truth, objectivity, and realism. Science claims to have a rational method that provides humans with objective truth about physical reality. The meanings of science’s four claims are reviewed in Figure 2.2.

Science and common sense

The choice of a suitable strategy for defending science’s four bold claims in subsequent chapters is greatly affected by the relationship between science and common sense. However, for better or for worse, that relationship is highly contentious. There are two basic choices. Science can be seen as a refinement of common sense, so the defense of science’s four bold claims begins with an appeal to common sense. Or, science can be seen as an unnatural and counter-intuitive enterprise relative to simplistic common sense, so science’s defense must locate other resources. As exemplars of these two choices, this section considers Nash (1963) and Wolpert (1993).

Science's Four Claims

Rationality
Rational methods of inquiry use reason and evidence correctly to achieve substantial and specified success in finding truth, and rational actions use rational and true beliefs to guide good actions.

Truth
True statements correspond with reality:

	Correspondence	
External Physical World of Objects and Events	← →	Internal Mental World of Perceptions and Beliefs

Objectivity
Objective beliefs concern external physical objects; they can be tested and verified so that consensus will emerge among knowledgeable persons; and they do not depend on controversial presuppositions or special worldviews.

Realism
Realism is correspondence of human thoughts with an external and independent reality, including physical objects.

Figure 2.2 Science's claims of rationality, truth, objectivity, and realism.

The Nature of the Natural Sciences by Nash (1963) has a first chapter titled "Common Sense (and Science)" and a second chapter titled "Science (and Common Sense)." Nash began: "Science is a way of looking at the world. There are, of course, other ways. The man of common sense sees the world in his own way. So does the artist, the philosopher, the theologian. The view of the scientist, if at all unique, is characterized by its heavy involvement of elements drawn from all the others" (page 3). But, given those basic elements, the scientist then "seeks a higher unity, a deeper understanding, unknown to common sense" (page 3). He added: "Though between science and common sense there exist dissimilarities we must not (and will not) overlook, the strong similarities between them establish for us a point of departure. Seeking to understand science, we begin by trying to understand the nature of common sense" (page 4). Nash recommended that we follow Einstein, whom he quoted as saying:

> The whole of science is nothing more than a refinement of everyday thinking. It is for this reason that the critical thinking of the physicist cannot possibly be restricted to the examination of the concepts of his own specific field. He cannot proceed without considering critically a much more difficult problem, the problem of analyzing the nature of everyday thinking. (Albert Einstein, quoted in Nash 1963:4)

By contrast, *The Unnatural Nature of Science* by Wolpert (1993) has a first chapter titled "Unnatural Thoughts" that gives many examples of scientific discoveries that seem unnatural from the perspective of common sense. For instance, objects move in different paths than common sense leads most people to expect, white light is composed of a mixture of different colors, and correct probability judgments are often counter-intuitive. He expressed his perspective concisely:

> The central theme presented in this book is that many of the misunderstandings about the nature of science might be corrected once it is realized just how 'unnatural' science is. I will argue that science involves a special mode of thought and is unnatural for two main reasons.... Firstly, the world just is not constructed on a common-sensical basis. This means that 'natural' thinking – ordinary, day-to-day common sense – will never give an understanding about the nature of science. Scientific ideas are, with rare exceptions, counter-intuitive: they cannot be acquired by simple inspection of phenomena and are often outside everyday experience. Secondly, doing science requires a conscious awareness of the pitfalls of 'natural' thinking. For common sense is prone to error when applied to problems requiring rigorous and quantitative thinking; lay theories are highly unreliable. (Wolpert 1993:xi–xii)

Upon encountering these opposing views, the first necessity is to recognize that any two things are partly similar and partly dissimilar. For instance, a bird and a stone are similar with respect to being physical objects, but they are dissimilar with respect to being alive. The same holds for science and common sense, being similar in some respects and dissimilar in others. Given that simple insight, this book can accommodate Nash and Wolpert alike.

On the one hand, the similarity of science and common sense is asserted here with respect to two absolutely crucial matters. First, both have the same concept of truth. When a child says "I ate three cookies" and a scientist says "Table salt is NaCl," the same concept and criterion of truth is at work, correspondence of a statement with reality. The concept of truth did not originate with the emergence of science! Rather, all four of science's bold claims – rationality, truth, objectivity, and realism – have a continuity with those claims in common sense. Second, as will be elaborated in Chapter 5, science's presuppositions of a real and comprehensible world, which are indispensable for mainstream science, are best legitimated by an appeal to common sense. One might suppose that a viable or preferable alternative would be an appeal to philosophy, but philosophy is in the same position as science as regards its dependence on common sense for these presuppositions.

On the other hand, the dissimilarity of science and common sense is asserted here with respect to the more advanced and exacting methods of science and the frequently surprising and bizarre findings of science. That scientific method is demanding and sometimes even counter-intuitive is precisely why books like this are needed! And scientific findings in everything from quantum mechanics to biology to cosmology are amazing and decidedly beyond common sense's reach.

Accordingly, a proper understanding of science must hold in tension both the similarities and dissimilarities between science and common sense. For example, consider the common-sense view that time passes at a constant rate, which has been overturned by the surprising view in Einstein's relativity theory that time passes at different rates depending on an object's speed relative to a given observer. Those different rates can actually be measured for satellites and other fast-moving objects by using extremely accurate clocks, and these measurements agree precisely with theory. Nevertheless, the strange world of relativity theory or quantum mechanics is not detached from the humdrum world of common sense because a scientist looks at a clock or measures a speed or whatever, and those appeals to empirical evidence necessarily require presuppositions about a real and comprehensible world that were best legitimated by a previous appeal to common sense. Furthermore, within the common-sense realm of ordinary speeds and ordinary clocks, relativity confirms, rather than contradicts, the common-sense perception that time always passes at a constant rate. No relativistic corrections are needed in a baseball park.

Science and common sense are partly dissimilar and partly similar. Scientific method, as compared to common-sense thinking, has complicated evidence and advanced logic supporting remarkable conclusions, but it also has identical presuppositions and shared concepts such as rationality and truth.

Summary

Science's four traditional claims are rationality, truth, objectivity, and realism. This chapter explores these four bold claims, drawing on the relevant philosophical literature that tends to be unfamiliar to scientists.

Rationality is good reasoning. Reason holds the double office of regulating belief and guiding action. Rational methods of inquiry, including scientific method, use reason and evidence correctly to achieve substantial and specified success in finding truth, and rational actions use rational and true beliefs to guide good actions.

Truth consists of correspondence between a statement and the actual state of affairs. This correspondence theory of truth presumes and subsumes the coherence theory requiring agreement among a set of beliefs, and it implies and confirms the pragmatic theory saying that truth promotes business with reality in a manner unmatched by ignorance and error. Nevertheless, the principal and essential concept is that of correspondence. The definition of truth is one very simple bit of philosophy: true statements correspond with reality. The real challenge is not the definition of truth but rather the implementation of effective methods for sorting true from false statements.

Objectivity has three interrelated aspects. Principally, objective beliefs are about objects themselves, rather than persons expressing beliefs, so the truth or falsity of an objective belief is determined by the belief's object, such as table salt. Secondly, objective knowledge is attainable by the exercise of ordinary endowments common to all humans, so agreement among persons is possible. Thirdly, objectivity involves immunity to deep worldview differences or philosophical debates, thereby allowing a worldwide scientific community to exist and flourish. A major reason why science is respected is that it cuts across political, cultural, and religious divisions.

Realism, as regards the physical world, is the philosophical theory that both human thoughts and independent physical objects exist and that human endowments render the physical world substantially intelligible and reliably known. Scientific realism embodies the claim that the scientific method provides rational access to physical reality, generating much objective knowledge.

The full force of science's claims results from the joint assertion of all four: rationality, truth, objectivity, and realism. Science claims to have a rational method that provides humans with objective truth about physical reality.

Science is both similar to and different from common sense in various respects. The concepts of rationality, truth, objectivity, and realism are rich and meaningful precisely because they are not unique to science but rather are shared by common sense, philosophy, history, law, and so on. Also, science inherits indispensable presuppositions about the world being real and comprehensible from common sense. But scientific method is more exacting and unnatural than is common-sense thinking. Also, scientific findings are often surprising and even bizarre relative to common-sense beliefs.

Study questions

(1) Define rationality. In which academic disciplines besides science is rationality also applicable and important?
(2) Define truth. How would you compare and relate the correspondence, coherence, and pragmatic theories of truth?
(3) Define objectivity. What are the three interrelated aspects of objectivity and why is objectivity important in science?
(4) Define realism. What are the two basic tenets of realism, and what philosophical positions result from denying one or the other of those tenets?
(5) How would you relate science and common sense?

3

A brief history of truth

This chapter's history of the conceptions of truth covers 23 centuries in about as many pages. Such extreme brevity allows only four stops, each separated by several centuries: Aristotle around 350 BC, Augustine around AD 400, several scholars in the fledgling medieval universities of Paris and Oxford in the 1200s, and philosopher-scientists of the past several centuries until 1960. Subsequent developments are deferred to the next chapter. This history focuses specifically on truth about the physical world, that is, scientific truth.

For many scientists, their research frontiers are moving so rapidly that most relevant work comes from the past several years. However, this book's topic of scientific method is different from routine scientific research in having a far greater debt to history and benefit from history. Concepts of truth, objectivity, rationality, and method have been around for quite some time. Consequently, great minds from earlier times still offer us diverse perspectives and penetrating insights that can significantly improve our chances of arriving at rich and productive solutions. Also, current thinking and debates about scientific method can be better understood in the light of science's intellectual history.

The most elemental question

The most elemental question about scientific method concerns identifying its basic components: What inputs are required for us humans to reach true conclusions about the physical world? In other words, what must go in so that scientific conclusions can come out? After resolving that initial question, subsequent questions then concern how to secure and optimize these inputs.

The history of attempts to answer this elemental question can be comprehended better by alerting readers from the outset to this chapter's overarching theme. This theme is the subtle and indecisive struggle over the centuries among rationalism, empiricism, and skepticism, caused by an underlying confusion about how to integrate science's logic, evidence, and presuppositions.

Inputs Emphasized by Various Schools and Scholars

Inputs	Schools and Scholars
Logic/Reason	Rationalists *Aristotle (ideal), René Descartes, Gottfried Leibniz*
Evidence	Empiricists *Aristotle (actual), John Locke, George Berkeley*
Presuppositions (worried)	Skeptics *Pyrrho of Elis, Sextus Empiricus, David Hume*
Presuppositions (confident)	Mainstream Scholars *Albertus Magnus, Isaac Newton, Thomas Reid*
Logic + Evidence	Logical Empiricists of 1920–1960 *Rudolf Carnap, C. G. Hempel, W. V. Quine*

Presuppositions (confident version) + Evidence + Logic = PEL model in Gauch 2002, with precedents from Albertus Magnus, Robert Grosseteste, and Isaac Newton.

Figure 3.1 Inputs required to support scientific conclusions. Historically, various schools have emphasized different inputs. Logic or reason was emphasized by rationalists, whereas evidence or experience was emphasized by empiricists. Aristotle expressed both an ideal science that aligned with rationalism and an actual science that aligned with empiricism. Presuppositions have been formulated in two quite different versions: a worried version by skeptics and a confident version by mainstream scholars. Logic and evidence were combined by logical empiricists. All three inputs – presuppositions, evidence, and logic – are integrated in the PEL model.

Figure 3.1 lists the inputs emphasized by various schools and scholars. The rationalists expected logic or reason to generate scientific truth. By contrast, the empiricists saw evidence or experience as the touchstone of knowledge and truth. And the skeptics were so worried about science's presuppositions of a real and comprehensible world that they despaired of offering any truth claims, although mainstream scholars advocated a confident version of science's presuppositions. As troubles mounted over the centuries for both rationalism and empiricism, in part because of skeptical attacks, the logical empiricists realized that neither reason nor evidence is adequate separately, so their innovation was a scientific method that combined reason and evidence. But after only several decades, their short-lived project also encountered insurmountable troubles.

The resolution that I proposed in 2002 in my text, *Scientific Method in Practice*, reflects the scientific method of philosopher-scientists such as Albertus Magnus, Robert Grosseteste, and Isaac Newton. Presuppositions, evidence, and logic constitute the three inputs needed to support scientific conclusions. No subset of these three inputs is functional, but rather the combination of all three works.

The details that follow in this and subsequent chapters will make more sense for those readers who grasp two things at this point. First, the most elemental question about scientific method is: What inputs must go in so that scientific conclusions can come out? Second, a satisfactory account of scientific method that answers this question and supports mainstream science will necessarily involve securing and optimizing science's presuppositions, evidence, and logic.

Aristotle

Aristotle (384–322 BC) was enormously important in science's early development. He was a student of Plato (*c.* 429–347 BC), who was a student of Socrates (*c.* 470–399 BC), and he became the tutor of Alexander the Great. Aristotle established a school of philosophy called the Lyceum in Athens, Greece. He wrote more than 150 treatises, of which about 30 have survived.

Aristotle defined truth by the one and only definition that fits common sense and benefits science and technology: the correspondence concept of truth. A statement is true if it corresponds with reality; otherwise, it is false. This definition of truth is obvious and easy, despite the temptation to think that a philosophically respectable definition must be difficult, mysterious, and elusive. As Adler (1978:151) said, "The question 'What is truth?' is not a difficult question to answer. After you understand what truth is, the difficult question, as we shall see, is: How can we tell whether a particular statement is true or false?"

Aristotle had a deductivist vision of scientific method, at least for a mature or ideal science (Losee 2001:4–13). The implicit golden standard behind that vision was geometry. Ancient philosophers were quite taken with geometry's clear thinking and definitive proofs. Consequently, geometry became the standard of success against which all other kinds of knowledge were judged.

Naturally, despite that ideal of deductive certainty, Aristotle's actual method in the natural sciences featured careful observations of stars, plants, animals, and other objects, as well as inductive generalizations from the data. The axioms that seemed natural and powerful for geometry had, of course, no counterpart in the natural sciences. For example, no self-evident axioms could generate knowledge about a star's location, a plant's flowers, or an animal's teeth. Rather, one had to look at the world to see how things are. Accordingly, Aristotle devised an inductive–deductive method that used inductions from observations to infer general principles, deductions from those principles to check the principles against further observations, and additional cycles of induction and deduction to continue the advance of knowledge.

Aristotle gave natural science a tremendous boost. Ironically, his achievements are not often appreciated by contemporary scientists because he influenced certain raging debates about fundamentals that we now take for granted. It must be emphasized – even though modern readers can scarcely grasp how something so obvious to them could ever have been hotly debated – that Aristotle advanced science enormously and strategically simply by insisting that the physical world is real. Plato had diminished the reality or significance of the visible, physical world to an illusion – a derivative, fleeting shadow of the eternal, unreachable "Forms" that, Plato thought, composed true reality. But Aristotle rejected his teacher's theory of a dependent status for physical things, claiming rather an autonomous and real existence. "Moreover, the traits that give an individual object its character do not, Aristotle argued, have a prior and separate existence in a world of forms, but belong to the object itself. There is no perfect form of a dog, for example, existing independently in the world of forms and replicated imperfectly in individual dogs, imparting to them their attributes. For Aristotle, there were just individual dogs" (Lindberg 2007:46).

Another of Aristotle's immense contributions to science was to improve deductive logic. Aristotle's syllogistic logic was the first branch of mathematics to be based on axioms, pre-dating Euclid's geometry.

The greatest general deficiency of Aristotle's science was confusion about the integration and relative influences of philosophical presuppositions, empirical evidence, and deductive and inductive logic. How do all of these components fit together in a scientific method that can provide humans with considerable truth about the physical world? Aristotle's choice of geometry as the standard of success for the natural sciences amounts to asking deduction to do a job that can be done only by a scientific method that combines presuppositions, observational evidence, deduction, and induction. Aristotle never reconciled and integrated the deductivism in his ideal science and the empiricism in his actual science. Furthermore, the comforting notion that logic and geometry had special, self-evidently true axioms was destined to evaporate two millennia later with the discovery of nonstandard logics and non-Euclidean geometries. Inevitably, the natural sciences could not be just like geometry. The study of physical things and the study of abstract ideas could not proceed by identical methods.

The greatest specific deficiency of Aristotle's science was profound disinterest in manipulating nature to carry out experiments. For Aristotle, genuine science concerned undisturbed nature rather than dissected plants or manipulated rocks. Regrettably, his predilection to leave nature undisturbed greatly impeded the development of experimental science. Even in Aristotle's time, much about rudimentary experimental methods could have been learned from the simple trial-and-error procedures that had already been successful for improving agriculture and medicine. But for Aristotle, reflection on the practical arts was beneath the dignity of philosophers, so philosophy gained nothing from that prior experience with experimentation in other realms.

It is difficult to give a specific and meaningful number, but I would say that Aristotle got 70% of scientific method right. His contribution is impressive, especially for a philosopher-scientist living more than two millennia ago.

Augustine

Skipping forward seven centuries in this brief history of truth, from Aristotle to Augustine (AD 354–430), the standard of truth and grounds of truth had shifted considerably. Augustine is *the* towering intellect of Western civilization, the one and only individual whose influence dominated an entire millennium. He is remembered primarily as a theologian and philosopher – as a church father and saint. Yet his contribution to science was also substantial. Augustine's treatise on logic, *Principia dialecticae*, adopted Aristotle's logic (rather than its main competitor, Stoic logic), thereby ensuring great influence for Aristotle in subsequent medieval logic.

Lindberg (2007:150) has nicely summarized the relationship between science and the church in antiquity: "If we compare the early church with a modern research university or the National Science Foundation, the church will prove to have failed abysmally as a supporter of science and natural philosophy. But such a comparison is obviously unfair. If, instead, we compare the support given to the study of nature by the early church with the support available from any other contemporary social institution, it will become apparent that the church was the major patron of scientific learning."

For Augustine, the foremost standard of rationality and truth was not Euclid's geometry. Rather, it was Christian theology, revealed by God in Holy Scripture. Theology had the benefit of revelation from God, the All-Knowing Knower. Accordingly, theology replaced geometry as queen of the sciences and the standard of truth. But Augustine's view of how humans acquire even ordinary scientific knowledge relied heavily on divine illumination, particularly as set forth in *The Teacher* (King 1995). He "claimed that whatever one held to be true even in knowledge attained naturally – that is to say, without the special intervention of God as in prophecy or in glorification – one knew as such because God's light, the light of Truth, shone upon the mind" (Marrone 1983:5).

With beautiful simplicity and great enthusiasm, Augustine saw that truth is inherently objective, public, communal, and sharable: "We possess in the truth . . . what we all may enjoy, equally and in common; in it are no defects or limitations. For truth receives all its lovers without arousing their envy. It is open to all, yet it is always chaste. . . . The food of truth can never be stolen. There is nothing that you can drink of it which I cannot drink too. . . . Whatever you may take from truth and wisdom, they still remain complete for me" (Benjamin and Hackstaff 1964:69).

Augustine is also notable for his book against skepticism, *Against the Academicians.* He argued that skepticism was incoherent and that we can possess several kinds of knowledge impervious to skeptical doubts. Augustine defended the general reliability of sense perception. He appealed to common sense by asking if an influential skeptic, Carneades, knew whether he was a man or a bug! Conventional views of science's method and success continue to be challenged by skepticism and relativism, so Augustine's analysis remains relevant.

Medieval scholars

Moving forward another eight centuries in this brief history of truth, from Augustine, around AD 400, to the beginnings of medieval universities in the 1200s, the standard and grounds of scientific truth faced the most perplexing, exciting, and productive shift in the entire history of the philosophy of science. Some leading figures were Robert Grosseteste (*c.* 1168–1253) and later William of Ockham (*c.* 1285–1347) at the university in Oxford; William of Auvergne (*c.* 1180–1249), Albertus Magnus or Albert the Great (*c.* 1200–1280), and Thomas Aquinas (*c.* 1225–1274) at the university in Paris; and Roger Bacon (*c.* 1214–1294) and John Duns Scotus (*c.* 1265–1308) at both universities. The rise of universities happened to coincide with the rediscovery and wide circulation of Aristotle's books and their Arabic commentaries.

The immensely original contribution of those medieval scholars was to ask a new question about science's truth, a question that may seem ordinary now, but it had not previously been asked or answered. Indeed, after Augustine, eight centuries would pass before the question would be asked clearly, and still another century would pass before it would be answered satisfactorily. It is a slight variant on the most elemental question, placing emphasis on the human and social aspects of science. It can be expressed thus: What human-sized and public method can provide scientists with truth about the physical world? "Scholastics of the thirteenth and fourteenth centuries wanted to know how to identify that true knowledge which any intelligent person could have merely by exercising his or her natural intellectual capabilities" (Marrone 1983:3). They skillfully crafted a reinforced scientific method incorporating five great new ideas.

(1) Experimental Methods. Despite Aristotle's disinterest in manipulated nature, experimental methods were finally being developed in science, greatly expanding the opportunities to collect the specific data that could be used to discriminate effectively between competing hypotheses. Grosseteste was "the principal figure" in bringing about "a more adequate method of scientific inquiry" by which "medieval scientists were able eventually to outstrip their ancient European and Muslim teachers" (Dales 1973:62). He initiated a productive shift in science's emphasis, away from presuppositions and ancient authorities, and toward empirical evidence, controlled experiments, and mathematical descriptions. He combined the logic from philosophy and the empiricism from practical arts into a new scientific method. "He stands out from his contemporaries . . . because he, before anyone else, was able to see that the major problems to be investigated, if science was to progress, were those of scientific method. . . . He seems first to have worked out a methodology applicable to the physical world, and then to have applied it in the particular sciences" (A. C. Crombie, in Callus 1955:99, 101).

Roger Bacon, the Admirable Doctor, was influenced by Grosseteste. He expressed the heart of the new experimental science in terms of three great prerogatives. The first prerogative of experimental science was that conclusions reached by induction should be submitted to further experimental testing; the second prerogative was that experimental facts had priority over any initial presuppositions or reasons and could augment the factual basis of science; the third prerogative was that scientific research could be extended to entirely new problems, many with practical value. The Admirable Doctor conducted numerous experiments in optics. He was eloquent about science's power to benefit humanity.

(2) **Powerful Logic.** An army of brilliant medieval logicians greatly extended the deductive and inductive logic needed by science. That stronger logic, combined with the richer data coming from new experiments with manipulated objects, as well as traditional observations of unaltered nature, brought data to bear on theory choices with new rigor and power.

(3) **Theory Choice.** Medieval philosopher-scientists enriched science's criteria for choosing a theory. The most obvious criterion is that a theory must fit the data. Ordinarily, a theory is in trouble if it predicts or explains one thing but something else is observed. But awareness was growing that theories also had to satisfy additional criteria, such as parsimony.

William of Ockham, the Venerable Inceptor, is probably the medieval philosopher who is best known to contemporary scientists through the familiar principle of parsimony, often called "Ockham's razor." From Aristotle to Grosseteste, philosopher-scientists had valued parsimony, but Ockham advanced the discussion considerably. In essence, Ockham's razor advises scientists to prefer the simplest theory among those that fit the data equally well. Ockham's rejection, on grounds of parsimony, of Aristotle's theory of impetus paved the way for Newton's theory of inertia.

(4) **Science's Presuppositions.** Albertus Magnus, the Universal Doctor, handled science's presuppositions with exquisite finesse, as will be elaborated in Chapter 5. He gave Aristotle the most painstaking attention yet, writing more

than 8,000 pages of commentary. Albertus Magnus grounded science in common sense. For instance, seeing someone sitting justifies the belief that such is the truth. That common-sense grounding enabled the Universal Doctor to grant science considerable intellectual independence from worldview presuppositions and theological disputes.

Thomas Aquinas, the Angelic Doctor, was an enormously influential student of Albertus Magnus, and he accepted his teacher's view of science. Aquinas's support alone would have been sufficient to ensure widespread acceptance in all medieval universities of Albertus's approach for legitimating presuppositions and demarcating science. Although primarily a theologian, the Angelic Doctor also wrote extensive commentaries on several of Aristotle's books, including the *Physics.*

(5) Scientific Truth. Finally, medieval philosopher-scientists adopted a conception of scientific truth that was more broad, fitting, and attainable than had Aristotle. Ockham "made a distinction between the science of real entities (*scientia realis*), which was concerned with what was known by experience to exist and in which names stood for things existing in nature, and the science of logical entities (*scientia rationalis*), which was concerned with logical constructions and in which names stood merely for concepts" (Crombie 1962:172). Thus, the natural sciences had quit trying to be just like geometry.

Those five great ideas account for much of the medieval reinforcement of scientific method that vitalized science's pursuit of truth. The main deficiencies of Aristotelian science were remedied in the thirteenth century. Experiments with manipulated objects were seen to provide relevant data with which to test hypotheses. Also, a workable integration of presuppositions, evidence, and logic emerged that endowed scientific method with accessible truth. Medieval philosopher-scientists also demarcated science apart from philosophy and theology, thereby granting science substantial intellectual and even institutional independence.

The thirteenth century began with a scientific method that lacked experimental methods and lacked an approach to truth that applied naturally to physical things. It concluded with an essentially complete scientific method with a workable notion of truth. Because of Robert Grosseteste at Oxford, Albertus Magnus at Paris, and other medieval scholars, it was the golden age of scientific method. No other century has seen such a great advance in scientific method. The long struggle of sixteen centuries, from Aristotle to Aquinas, had succeeded at last in producing an articulated and workable scientific method with a viable conception of truth. Science had come of age. From the prestigious universities in Oxford and Paris, the new experimental science of Robert Grosseteste and Roger Bacon spread rapidly throughout the medieval universities: "And so it went to Galileo, William Gilbert, Francis Bacon, William Harvey, Descartes, Robert Hooke, Newton, Leibniz, and the world of the seventeenth century" (Crombie 1962:15). So it went to us also.

Modern scholars

Skipping forward a final time in this brief history of truth, science's method and concept of truth have been developed further in modern times beginning around 1500. Developments after 1960, which have been the primary determinants of the current scene, will be taken up in greater detail in the next chapter.

The development of increasingly powerful scientific instruments has been a prominent feature of scientific method during the modern era. An influential early example was the observatory of Tycho Brahe (1546–1601), with an unprecedented accuracy of four minutes of arc, nearly the limit possible without a telescope. Galileo Galilei (1564–1642) constructed an early telescope and invented the first thermometer. He carefully estimated measurement errors and took them into account when fitting models to his data. Blaise Pascal (1623–1662) invented the barometer and an early calculating machine.

Mathematical tools were also advanced. Pascal, Pierre de Fermat (1601–1665), Jacob Bernoulli (1654–1705), Thomas Bayes (1701–1761), and others developed probability theory and elementary statistics. Sir Isaac Newton (1642–1727) and Gottfried Leibniz (1646–1716) invented calculus. Thomas Reid (1710–1796) invented a non-Euclidean geometry in 1764. That discovery, that Euclid's axioms are not uniquely self-evident and true, further eroded the ancient reputation of geometry as the paradigmatic science. Although syllogistic logic was axiomatized by Aristotle, and geometry by Euclid, about 23 centuries ago, arithmetic was first axiomatized by Giuseppe Peano (1858–1932) a mere one century ago.

Sir Francis Bacon (1561–1626) popularized the application of science to the furtherance of mankind's estate with his enduring slogan, "Knowledge is power." His attempt to win financial support for science from the English crown failed in his own lifetime but bore fruit shortly thereafter.

In 1562, the French scholar Henri Etienne (1531–1598) first printed, in Latin translation, the *Outlines of Pyrrhonism*, by the ancient skeptic Sextus Empiricus (*fl.* AD 150). "It was the rediscovery of Sextus and of Greek scepticism which shaped the course of philosophy for the next three hundred years" (Annas and Barnes 1985:5). That the preceding millennium had struggled rather little with skepticism may have been due to the perception that Augustine's refutation sufficed. But René Descartes (1596–1650), George Berkeley (1685–1753), David Hume (1711–1776), Immanuel Kant (1724–1804), and other modern thinkers struggled mightily with Sextus' challenges.

"In his *Outlines of Pyrrhonism* Sextus defends the conclusions of Pyrrhonian scepticism, that our faculties are such that we ought to suspend judgement on all matters of reality and content ourselves with appearances" (Woolhouse 1988:4). The skeptics' opponents, to use their own term, were the "dogmatists" who believed that truth was attainable. Sextus observed that two criteria for discovering truth were offered: reason by the rationalists, and the senses by the empiricists. He argued that neither reason nor sense perception could guarantee truth. To a considerable extent, the philosophies of Descartes and Leibniz can be understood as attempts to make reason work despite Sextus' skeptical criticisms. Likewise, the philosophies of Francis Bacon and Locke attempt to make sense perception and empirical data work despite the ordeal by skepticism. So, although quite different, rationalism and empiricism had in common the same opponent, skepticism. Chatalian (1991) argued persuasively that the Greek skeptics, Pyrrho of Elis (*c.* 360–270 BC) and Sextus Empiricus, were often superficially studied and poorly understood. Nevertheless, rationalism and empiricism sought to guard truth from skepticism's attacks.

René Descartes exemplified rationalism, which emphasized philosophical reasoning as the surest source of truth rather than uncertain observations and risky inductions. Descartes agreed with Francis Bacon that science had both general principles and individual observations, but his progression was the reverse. The empiricist Bacon sought to collect empirical data and then progress inductively to general relations, whereas the rationalist Descartes sought to begin with general philosophical principles and then deduce the details of expected data. To obtain the needed stockpile of indubitable general principles, Descartes's method was to reject the unverified assumptions of ancient authorities and begin with universal doubt, starting afresh with that which is most certain.

His chosen starting point for indubitable truth was his famous "*Cogito ergo sum*," "I think, therefore I exist." He then moved on to establish the existence of God, whose goodness assured humans that their sense perceptions were not utterly deceptive, so they could conclude that the physical world exists.

George Berkeley was an empiricist. The battle cry of empiricists was back to experience. In essence, "an empiricist will seek to relate the contents of our minds, our knowledge and beliefs, and their acquisition, to sense-based experience and observation. He will hold that experience is the touchstone of truth and meaning, and that we cannot know, or even sensibly speak of, things which go beyond our experience" (Woolhouse 1988:2). Berkeley was also an idealist, believing that only minds and ideas exist, not the physical world.

Berkeley applauded Newton's careful distinction between mathematical axioms and empirical applications, in essence, between ideas and things. But Berkeley was concerned that such a distinction would invite a dreaded skepticism: "Once a distinction is made between our perceptions of material things and those things themselves, 'then are we involved all in *scepticism*'. For it follows from this distinction that we see only the appearances of things, images of them in our minds, not the things themselves, 'so that, for aught we know, all we see, hear, and feel, may be only phantom and vain chimera, and not at all agree with the real things'" (Woolhouse 1988:110). What was the solution? "Faced with the evidently troublesome distinction between things and ideas, Berkeley in effect collapses it; he concludes that *ideas are things*. As he explains, 'Those immediate objects of perception, which according to [some] . . . are only appearances of things, I take to be the real things themselves'" (Woolhouse 1988:113). Ideas and minds were all of reality; there were no such things as physical objects. Accordingly, science's proper goal was to account for the mind's experiences and perceptions, rather than an external physical reality.

Isaac Newton continued Aquinas's broad perspective on truth in science, in contrast to Aristotle's narrow vision. Newton believed that science could make valid assertions about unobservable entities and properties. For example, from the hardness of observable objects, one could infer the hardness of their constituent particles that were too small to be observed. He also believed that science should generally trust induction: "In experimental philosophy we are to look upon propositions inferred by general induction from phenomena as accurately or very nearly true, notwithstanding any contrary hypotheses that may be imagined, till such time as other phenomena occur, by which they may either be made more accurate, or liable to exceptions" (Cajori 1947:400). Also, Newton insisted, contrary to Leibniz, that science could claim legitimate knowledge even in the absence of deep explanation. Thus, the observed inverse-square law applying to gravitational attraction counted as real knowledge, even without any deep understanding of the nature or cause of gravity.

Newton's view of scientific method, which has influenced modern science so strongly, corresponded with that of Grosseteste: "Of his 'Rules of Reasoning in Philosophy' the first, second, and fourth were, respectively, the well-established principles of economy [parsimony], uniformity, and experimental verification and falsification, and the third was a derivative of these three. And when he came to describe his method in full, he described precisely the double procedure that had been worked out since Grosseteste in the thirteenth century," namely, induction of generalities from numerous observations, and deduction of specific predictions from generalities (Crombie 1962:317). "We reach the conclusion that despite the enormous increase in power that the new mathematics brought in the seventeenth century, the logical structure and problems of experimental science had remained basically the same since the beginning of its modern history some four centuries earlier" (Crombie 1962:318).

David Hume could be considered an empiricist or a skeptic. "Among all the philosophers who wrote before the twentieth century none is more important for the philosophy of science than David Hume. This is because Hume is widely recognized to have been the chief philosophical inspiration of the most important twentieth-century school in the philosophy of science – the so-called logical positivists," also called logical empiricists (Alexander Rosenberg, in Norton 1993:64). Hume admired Francis Bacon and greatly admired Newton, "the greatest and rarest genius that ever rose for the ornament and instruction of the species" (Woolhouse 1988:135). Hume took himself to be discovering a science of man, or principles of human understanding more specifically, that was akin to Newton's science of mechanics in its method and rigor.

Hume's analysis began with two fundamental moves. First, he insisted that the objective was *human* understanding, so he examined human nature to assess our mental capacities and limitations. "There is no question of importance, whose decision is not compriz'd in the science of man; and there is none, which can be decided with any certainty, before we become acquainted with that science" (John Biro, in Norton 1993:34). Second, Hume rigorously adopted an empiricist theory of meaning, requiring statements to be grounded in experience, that is, in sense perceptions and ideas based on them. "As to those *impressions*, which arise from the *senses*, their ultimate cause is, in my opinion, perfectly inexplicable by human reason, and 'twill always be impossible to decide with certainty, whether they arise immediately from the object, or are produc'd by the creative power of the mind, or are deriv'd from the author of our being. Nor is such a question any way material to our present purpose. We may draw inferences from the coherence of our perceptions, whether they be true or false; whether they represent nature justly, or be mere illusions of the senses" (David F. Norton, in Norton 1993:6–7).

It is difficult to induce contemporary scientists, who think that rocks and trees are real and knowable, to grasp the earnestness of Hume's empiricism. Hume's empiricist science concerned mental perceptions, not physical things. His concern was with "our perceptions, qua perceptions, with perceptions as, simply, the *elements or objects of the mind* and not as *representations* of external existences" (David F. Norton, in Norton 1993:8). For example, he was concerned with our mental perceptions and ideas of trees, not with trees as external physical objects. Accordingly, to report that "I see a tree" was, for Hume, a philosophical blunder, because this "I see" posits a mental perception, while this "tree" posits a corresponding physical object. He called that blunder the "double existence" (or "representational realism") theory – "the theory that while we experience only impressions and ideas, there is also another set of existences, namely objects" (Alexander Rosenberg, in Norton 1993:69). Of course, earlier thinkers, like Aristotle, had a more flattering name for that theory, the correspondence theory of truth. Anyway, for Hume, the corrected report would read something like "I am being appeared to treely," which skillfully avoids the double existence of perceptions and objects and instead confines itself to the single existence of perceptions.

So although Hume's avowed hero was Newton, their philosophies of science were strikingly different because Newton's science concerned truth about a knowable physical world. Hume and Newton could agree on the truism that science was done by scientists – by humans. But Hume's "humans" were post-skeptical philosophers, whereas Newton's "humans" were common-sensical scientists. Likewise, Hume's "observations" were strictly mental perceptions, whereas Newton's "observations" were sensory responses corresponding reliably to external physical objects. Hume says, "I am being appeared to treely," but Newton says "I see a tree."

Thomas Reid, quite in contrast to his fellow Scot David Hume, grounded philosophy in an initial appeal to common sense, as in this quotation from Hamilton's edition of Reid's work:

Philosophy... has no other root but the principles of Common Sense; it grows out of them, and draws its nourishment from them. Severed from this root, its honours wither, its sap is dried up, it dies and rots.... It is a bold philosophy that rejects, without ceremony, principles which irresistibly govern the belief and the conduct of all mankind in the common concerns of life: and to which the philosopher himself must yield, after he imagines he hath confuted them. Such principles [of common sense] are older, and of more authority, than Philosophy: she rests upon them as her basis, not they upon her. (Hamilton 1872:101–102)

Wolterstorff offers an insightful commentary on this passage from Reid:

The philosopher has no option but to join with the rest of humanity in conducting his thinking within the confines of common sense. He cannot lift himself above the herd....

Alternatively, philosophers sometimes insist that it is the calling of the philosopher to *justify* the principles of common sense – not to reject them but to ground them. Close scrutiny shows that this too is a vain attempt; all justification takes for granted one or more of the principles. Philosophical thought, like all thought and practice, rests at bottom not on grounding but on trust. (Nicholas Wolterstorff, in Cuneo and van Woudenberg 2004:77–78)

Reid avoided the hopeless attempt to make natural science just like geometry by accepting both the deductions of geometry and the reliability of observation:

That there is such a city as Rome, I am as certain as of any proposition in Euclid; but the evidence is not demonstrative, but of that kind which philosophers call probable. Yet, in common language, it would sound oddly to say, it is probable there is such a city as Rome, because it would imply some degree of doubt or uncertainty. (Hamilton 1872:482)

Representing common sense as eyes and philosophy as a telescope, Reid offered the analogy that a telescope can help a man see farther if he has eyes, but will show nothing to a man without eyes (Hamilton 1872:130). Accordingly, to the partial skeptic, Reid commended a dose of common sense as the best remedy; but to the total skeptic, Reid had nothing to say. Reid could give a man a telescope but not eyes.

Exactly what does Reid mean by common sense? He listed 12 principles as a sampling from the totality of such principles (Nicholas Wolterstorff, in Cuneo and van Woudenberg 2004:78–79). For example, these principles include "that the thoughts of which I am conscious are the thoughts of a being which I call myself" and "that there is life and intelligence in our fellow men with whom we converse" and "that those things do really exist which we distinctly perceive by our senses." The general reliability of sense perception looms large in Reid's writings (James van Cleve, in Cuneo and van Woudenberg 2004:101–133).

Immanuel Kant devised a new variant of rationalism intended to divert Hume's skepticism and to support a thoroughly subjective, human-sized version of scientific truth. His influential *Critique of Pure Reason* (1781, revised 1787) was followed by a popularization, the *Prolegomena to any Future Metaphysics That Shall Come Forth as Scientific* (1783). He was not happy with his predecessors. Against Descartes, Hume, Berkeley, and Reid and their failed metaphysics, Kant promised us a keen pilot that can steer our metaphysical ship safely. But Kant's thinking is remarkably complex and subtle.

Fortunately, however, the opening pages of his *Prolegomena* lead us quickly into the very heart of enduring themes in his philosophy of science. The centerpiece is his response to Hume's problem of causality. In the entire history of metaphysics, "nothing has ever happened which was more decisive to its fate than the attack made upon it by David Hume," specifically the attack upon "a single but important concept in Metaphysics, viz., that of Cause and Effect" (Carus 1902:3–4).

The problem with causality, or any other general law of nature, was that such laws made claims that went beyond any possible empirical support. Empirical evidence for a causal law could only be of the form "All instances of *A* observed in the past were followed by *B*," whereas the law asserted the far grander claim that "All instances of *A*, observed or not and past or future, are followed by *B*." But that extension was inductive, excessive, and uncertain, exceeding its evidence. Consequently, something else had to be added to secure such a law.

Accordingly, Kant's solution combined two resources: a general philosophical principle of causality asserted by *a priori* reasoning, and specific causal laws discovered by *a posteriori* empirical observation and induction. By that combination, "particular empirical laws or uniformities are subsumed under the *a priori* concept of causality in such a way that they thereby become necessary and acquire a more than merely inductive status" (Michael Friedman, in Guyer 1992:173). For example, "The rule of uniformity according to which illuminated bodies happen to become warm is at first merely empirical and inductive; if it is to count as a genuine law of nature, however, this same empirical uniformity must be subsumed under the *a priori* concept of causality, whereupon it then becomes necessary and strictly universal" (Michael Friedman, in Guyer 1992:173). Thus, a general principle of causality upgraded the evidence for particular causal laws.

Moving forward about a century after Kant to almost a century ago, the period around 1920 was pivotal for the philosophy and method of science. Although the current scene is one of vigorous debate among several sizable schools, for a few decades following 1920, a single school dominated, logical empiricism (also called logical positivism, just positivism, and the Vienna Circle). Some of the leading members, associates, visitors, and collaborators were A. J. Ayer, Rudolf

Carnap, Albert Einstein, Herbert Feigl, Philip Frank, Kurt Gödel, Hans Hahn, C. G. Hempel, Ernest Nagel, Otto Neurath, W. V. Quine, Hans Reichenbach, Moritz Schlick, and Richard von Mises. Sir Karl Popper, who would become the circle's most influential critic, often attended but was not a member or associate. "Almost all work, foundational or applied, in English-language philosophy of science during the present century has either been produced within the tradition of logical empiricism or has been written in response to it. Indeed it is arguable that philosophy of science as an academic discipline is essentially a creation of logical empiricists and (derivatively) of the philosophical controversies that they sparked" (Richard Boyd, in Boyd, Gasper, and Trout 1991:3).

As its apt name suggests, "logical empiricism" combines logic and empiricism. "Logical empiricism arose in the twentieth century as a result of efforts by scientifically inclined philosophers to articulate the insights of traditional empiricism, especially the views of Hume, using newer developments in mathematical logic" (Richard Boyd, in Boyd et al. 1991:5). The central idea was to limit meaningful scientific statements to sensory-experience reports and logical inferences based on those reports. Considered separately, the rationalist tradition with its logic and the empiricist tradition with its sensory experience were deemed inadequate for science, but a clever integration of logic and experience was expected to work.

However, presuppositions were not part of logical empiricism. Indeed, "the fundamental motivation for logical empiricism" was "the elimination of metaphysics," including "doctrines about the fundamental nature of substances," "theological matters," and "our relation to external objects" (Richard Boyd, in Boyd et al. 1991:6). The perceived problem with metaphysical presuppositions was that they were not truths demonstrable by logic, and neither were they demonstrable by observational data, so for a logical empiricist, such ideas were just nonsense. Accordingly, science and philosophy parted ways. "The Circle rejected the need for a specifically philosophical epistemology that bestowed justification on knowledge claims from beyond science itself" (Thomas Uebel, in Audi 1999:956).

Clearly, the motivation of logical empiricism was to create a purified, hard, no-nonsense version of science based on solid data and avoiding philosophical speculation. Yet serious problems emerged that eroded its credibility by 1960.

Regrettably, logical empiricism rejected two medieval insights that have since been restored to their vital roles in philosophy of science. First, the innovation of the logical empiricists was not their combining of logic and empirical evidence, for their medieval predecessors had already done that several centuries earlier, but rather was in their rejection of presuppositions, especially metaphysical presuppositions about what exists. By dismissing presuppositions, science parted ways not only with philosophy but also with common sense. Even the primitive theory, for instance, that a person's perception of a cat results from the eyes seeing an actual physical cat *is* a metaphysical theory about what exists. "Given such a view" as logical empiricism, "difficult epistemological gaps arise between available evidence and the commonsense conclusions we want to reach about the world around us," including "enormous difficulty explaining how what we know about sensations could confirm for us assertions about an objective physical world" (Richard A. Fumerton, in Audi 1999:515).

Second, medieval scholars had engaged the practical question: What human-sized and public method can provide scientists with truth about the physical world? But logical empiricism's stringent science used logic and data in a rather mechanical fashion, guaranteed to be scientific and to guard truth, while largely disregarding human factors. The rapid dismantling of logical empiricism around 1960 was a reaction against this science lacking a human face.

Water

An objective announced at the beginning of the previous chapter is to enable readers to comprehend a statement such as "Water is H_2O" in its full philosophical and scientific richness. That chapter explained the meanings of science's four bold claims: rationality, truth, objectivity, and realism. This chapter has presented the intellectual history of the concept of truth as applied to the physical world. With this philosophical and historical background in place, the additional scientific information can now be added to complete the story about water.

What things are made of was one of the principal scientific questions that began to be asked in antiquity:

> Thales of Miletos, who lived in about 600 BC, was the first we know of who tried to explain the world not in terms of myths but in more concrete terms, terms that might be subject to verification. What, he wondered might the world be made of? His unexpected answer was: water. Water could clearly change its form from solid to liquid to gas and back again; clouds and rivers were in essence watery; and water was essential for life. His suggestion was fantastical perhaps, but such unnatural thoughts – contrary to common sense – are often the essence of science. But more important than his answer was his explicit attempt to find a fundamental unity in nature. (Wolpert 1993:35)

Thales of Miletos (*c.* 625–546 BC) got the wrong answer about water, but Wolpert credited him for being in essence the first scientist because he was asking the right question. And that was quite an innovation indeed! But about two and a half millennia later, the right answer about water's composition has finally emerged.

In 1800, William Nicholson decomposed water into H_2 and O_2 by electrolysis, but it remained until 1805 for Joseph Louis Gay-Lussac and Alexander von Humboldt to discover the proper ratio of two parts H_2 and one part O_2, and hence the chemical formula H_2O. Furthermore, these two gases can be ignited and thereby recombined to reconstitute the water. These simple experiments are easily replicated in high school or college chemistry classes (Eggen et al. 2012).

That table salt is NaCl was discovered a few years later. The element sodium was discovered in 1807 by Sir Humphry Davy by electrolysis of molten sodium hydroxide and the element chlorine in 1810 by Davy (by repeating an earlier experiment of 1774 by Karl Wilhelm Scheele whose reaction of MnO_2 and HCl produced chlorine gas but without Scheele understanding that chlorine is an element). Hence, table salt is a compound of a caustic metal and a poisonous gas.

The nature of a chemical element, such as hydrogen or chlorine, was illuminated substantially by the invention of the periodic table of the chemical elements by Dmitri Mendeleev in 1869. Ernest Rutherford discovered the atomic nucleus in 1911, and that same year Robert Millikan and Harvey Fletcher published an accurate measurement of an electron's charge. Within a decade, chemists understood that the place of each element in the periodic table is determined by the number of protons in its nucleus. In another decade, they discovered neutrons, which are also in atomic nuclei and have a mass nearly the same as protons. At long last, there was a rather satisfactory understanding of a chemical element. Hydrogen and oxygen are elements 1 and 8, and sodium and chlorine are elements 11 and 17 in the periodic table.

Chemists further discovered that a given element can have several isotopes due to its atoms having the same number of protons but different numbers of neutrons. For instance, hydrogen has three naturally occurring isotopes with zero to two neutrons denoted by ^{1}H to ^{3}H, and oxygen has three naturally occurring isotopes with eight to ten neutrons denoted by ^{16}O to ^{18}O. Accordingly, pure water has 18 distinguishable kinds of H_2O molecules – and about 2 per billion of these molecules are dissociated into H^+ ions of 3 kinds and OH^- ions of 9 kinds, for a total of 30 constituents (although more than 99% of water is a single constituent, H_2O molecules composed of the most common isotopes, ^{1}H and ^{16}O). Even deeper understanding of matter continues with the discovery that protons and neutrons are composed of quarks and gluons, but that goes beyond what needs to be discussed here.

To recapitulate the story of water, it began with Thales asking what things are made of. It progressed with Aristotle who, unlike Plato, insisted that physical objects are thoroughly real. It advanced with medieval philosopher-scientists finally asking and answering the most elemental question about scientific method: What inputs are required for us humans to reach true conclusions about the physical world? It further advanced with the scientific revolution in the 1600s and 1700s. Finally, scientific discoveries from about 1800 to 1930 clarified the atomic makeup of the elements hydrogen and oxygen that combine to form water. To properly comprehend that "Water is H_2O," one must understand not only the relevant scientific discoveries since 1800 but also the indispensable philosophical and historical background beginning around 600 BC that gives meaning and credence to a scientific claim of rationality, truth, objectivity, and realism.

Summary

To understand science's method and claims in historical perspective, this brief history of truth has examined the standards and evidence expected for truth claims during the past 23 centuries, from Aristotle to 1960. The most elemental question remains: What inputs must go in so that scientific conclusions can come out? Aristotle got much of scientific method right, but he disregarded experimental methods and had a somewhat confused expectation that a mature version of the natural sciences should be much like geometry in its method and certainty. Those deficiencies were remedied in the fledgling medieval universities in the 1200s. From 1500 to the present, tremendous advances have been made, especially regarding deductive and inductive logic, instruments for collecting data, and computers for analyzing data.

History reveals a tremendous diversity of views on science. Rationalists emphasized reason and logic; empiricists emphasized sensory experience and empirical evidence; and logical empiricists combined logic and empirical evidence while attempting to avoid presuppositions. Science's presupposition of a real and comprehensible world has had two versions: the worried version of skeptics such as Pyrrho of Elis and Sextus Empiricus, and the confident version of Albertus Magnus, Isaac Newton, and Thomas Reid. At this time in history, the way ahead for science's general methodological principles will require a deep integration of these three inputs: presuppositions, evidence, and logic. This is necessary to support science's four bold claims: rationality, truth, objectivity, and realism.

Study questions

(1) The most elemental question about scientific method concerns the inputs required for us humans to reach true conclusions about the physical world. What are the three inputs identified in this chapter, and which of these inputs were emphasized by rationalists, empiricists, and skeptics?

(2) What aspects of scientific method do you think Aristotle got right, and what other important aspects remained to be clarified by later philosopher-scientists?

(3) What were Augustine's contributions to science?

(4) What were the five great ideas of medieval philosopher-scientists that advanced science greatly?

(5) Recall the diverse views on science of Descartes, Berkeley, Newton, Hume, Reid, and Kant, and then select one who you find particularly interesting. Which of his ideas most intrigue you, and do you think that those particular ideas have stood the test of time as indicated by their still being accepted as important ideas in contemporary science?

4

Science's contested rationality

Does science have a rational method of inquiry that provides humans with a considerable amount of objective truth about physical reality? Certainly, a reply of "yes" represents the traditional claims of mainstream science, as delineated in Chapter 2. Furthermore, anyone who confidently believed the scientific story in Chapter 3 that water is H_2O has given every appearance of being in the camp that replies "yes" to this question.

Nevertheless, a controversy has raged over science's claims of rationality and truth, especially in the 1990s, although with roots going back to the 1960s and even back into antiquity. This controversy had such intensity in the 1990s that it went by the name of the "science wars" and even made the front pages of the world's leading newspapers.

Views that directly contradict and intentionally erode mainstream science are this chapter's topic. In this book that explicitly and repeatedly aligns with mainstream science, why devote a whole chapter to these contrary positions? Two reasons may be suggested.

First, for better or for worse, attacks on science's rationality have substantial cultural influence. The specific arguments and inflammatory rhetoric of "science wars" quickly came and went in a mere decade, which is quite ephemeral in the grand sweep of history, but skeptical and relativistic attacks on truth are perennial features of intellectual history. So, attacks on science's rationality are too influential and persistent to be ignored.

Second and more important, "those who know only their side of a case know very little of that" (Susan Haack, in Gross, Levitt, and Lewis 1996:57). What hinders scientists the most from mastering scientific method in order to enhance perspective and increase productivity is not their opponents' attacks but rather their own complacency – assuming that they already know scientific method well, and hence no further effort or study is required. Exposure to the other side attacking science can press the incisive questions that disturb insidious complacency and thereby prompt rigorous answers.

This chapter begins by considering who has legitimate rights to be auditors of science: scientists only, or else additional scholars also. It then examines four deadly threats to science's rationality: elusive truth, theory-laden data, incommensurable paradigms, and empty consensus. Reactions to these woes are reviewed, emphasizing articles in *Nature* and other scientific journals that are especially visible to scientists. The posture of the American Association for the Advancement of Science (AAAS) is also noted. Finally, a suggestion is offered for discerning the principal action in this complex debate, namely, whether an attack on rationality targets science alone, or else both science and common sense. This chapter introduces debates over science's rationality, leaving resolution to subsequent chapters, especially Chapter 14.

Science's auditors

Businesses have long been accustomed to having external auditors check their financial assets, liabilities, and ratings. But, increasingly, scientists have to get used to facing an accounting, although in their case the account is intellectual rather than financial. Many philosophers, historians, sociologists, and others have become external auditors of science. Their ratings of science's claims of rationality and truth are becoming increasingly influential, strongly affecting public perceptions of science.

This situation raises questions. Do philosophers and other nonscientists have a right to check science's claims? Or should scientists have the prerogative of setting their own standards for their truth claims without interference from anyone else?

Precisely because science is one of the liberal arts and because such fundamental intellectual notions as rationality and truth pervade the liberal arts, certainly it is within the purview of philosophy and history and other disciplines to have a voice in the weighing of science's intellectual claims. Every scientific claim of truth, expressed either as a certainty or as a probability, has both scientific and philosophical dimensions. Especially the general principles of scientific method, as contrasted with specialized techniques, have strong connections with many disciplines across the humanities.

Much, and perhaps most, of the probing of science's method and rationality by philosophers, historians, and sociologists is simply an earnest attempt to determine exactly what science's actual methods imply for science's legitimate claims. It would be decidedly unrealistic, however, not to recognize that some of this probing constitutes a militant call for scientists to promise less and for the public to expect less – much less! Scholars with relativistic, skeptical, and postmodern leanings routinely reach verdicts on science that are much more negative than what even the most cautious scientists reach.

Figure 4.1 An important philosopher of science, Sir Karl Popper. (This photograph by David Levenson is reproduced with kind permission of Black Star.)

During the second half of the twentieth century, four philosophers of science have been especially prominent: Sir Karl Popper (1902–1994), Imre Lakatos (1922–1974), Thomas Kuhn (1922–1996), and Paul Feyerabend (1924–1994). They are the four irrationalists of Stove (1982) and the four villains of Theocharis and Psimopoulos (1987). Among philosophers, numerous philosophers of science are well known; but among scientists, Popper and Kuhn probably are better known than all the others combined.

The earliest of these four philosophers is Popper, shown in Figure 4.1. His reassessment of science's claims began with the publication of *Logik der Forschung* in 1934, which appeared in English as *The Logic of Scientific Discovery* in 1959 (second edition 1968). "Popper is far and away the most influential philosopher of modern science – among scientists if not other philosophers. He is best known for his assertion that scientific theories can never be proved through experimental tests but only disproved, or 'falsified'" (Horgan 1992). He had two particularly influential students, Imre Lakatos and Paul Feyerabend. An especially important contribution has been *The Structure of Scientific Revolutions* by Kuhn (1962, second edition 1970). It has sold more than a million copies in 20 languages and is commonly considered "the most influential treatise ever written on how science does (or does not) proceed" (Horgan 1991).

The incisive thinking and penetrating analyses of Popper, Kuhn, and other scholars have had many positive effects. Especially valuable are their effective criticisms of the logical empiricism that had preceded their generation,

Popper's insistence on falsifiability, and Kuhn's recognition of the human and historical elements in science. Nevertheless, their writings have also mounted a sustained and influential attack on science's rationality, even though scientists often fail to recognize that. These critics have challenged science with four deadly woes: elusive truth, theory-laden data, incommensurable paradigms, and empty consensus.

Elusive truth

The first of the four deadly woes is elusive truth. What criterion can demarcate science from nonscience? In 1919, that question triggered Popper's interest in the philosophy of science (Popper 1974:33). He clearly distinguished that question, of whether or not a theory is scientific, from the different question of whether or not a theory is true. His question was occasioned by various claims that Einstein's physics, Marx's history, Freud's psychology, and Adler's psychology were all scientific theories, whereas Popper suspected that only the first of those claims was legitimate. Exactly what was the difference?

The received answer, from Francis Bacon several centuries earlier and from logical empiricists more recently, was that science is distinguished from pseudoscience and philosophy (especially metaphysics) by its empirical method, proceeding from observations and experiments to theories by means of inductive generalizations. But that answer did not satisfy Popper because admirers of Marx, Freud, and Adler also claimed an incessant stream of confirmatory observations to support their theories. Whatever happened was always and readily explained by their theories. What did that confirm? "No more than that a case could be interpreted in the light of the theory" (Popper 1974:35).

By sharp contrast, Einstein's theory of relativity could not sit easily with any and all outcomes. Rather, it made specific and bold predictions that put the theory at risk of disconfirmation if observations should turn out to be contrary to expectation. Einstein's theory claimed that gravity attracts light just as it attracts physical objects. Accordingly, starlight passing near the sun would be bent measurably, making a star's location appear to shift outward from the sun. Several years later, in 1919, a total eclipse of the sun afforded an opportunity to test that theory. An expedition led by Sir Arthur Eddington made the observations, clearly showing the apparent shift in stars' positions and thus confirming Einstein's theory. What impressed Popper most of all was the risk that relativity theory took, because an observation of no shift would have proved the theory false.

So, comparing supposedly "scientific" theories that are ready to explain anything with genuinely scientific theories that predict specific outcomes and thereby risk disconfirmation, Popper latched on to falsifiability as the essential criterion that demarcates science from nonscience. "Irrefutability is not a virtue of a theory (as people often think) but a vice. . . . [The] criterion of the scientific status of a theory is its falsifiability" (Popper 1974:36–37).

Notice that science was distinguished by falsification, not verification. Popper insisted that conjectures or theories could be proved false but no theory could ever be proved true. Why? Because he respected deductive logic but agreed with David Hume that inductive logic is a failure (Popper 1974:42). No quantity of observations can possibly allow induction to verify a general theory because further observations might bring surprises.

Hence, the best that science can do is to offer numerous conjectures, refute the worst with contrary data, and accept the survivors in a tentative manner. Conjecture followed by refutation was the scientific method, by Popper's account. But that implies that although "we search for truth . . . we can never be sure we have found it" (Popper 1974:56). Truth is forever elusive. So, Popper offered his demarcation criterion of falsifiability to separate science from nonscience, but at the cost of separating science from truth.

Theory-laden data

The second of four deadly woes is theory-laden data. A prominent feature of the logical empiricism that dominated the philosophy of science preceding Popper and Kuhn was a sharp boundary between data and theory. According to that view, true and scientific statements were based on empirical observations and their deductive logical consequences – hence the name, logical empiricism. By contrast, Popper insisted that observations are deeply theory-laden: "But sense-data, untheoretical items of observation, simply do not exist. . . . We can never free observation from the theoretical elements of interpretation" (Karl Popper, in Lakatos and Musgrave 1968:163). Why? This claim that data are theory-laden has many facets, but here it must suffice to mention three principal arguments.

First, in order to make any observations at all, scientists must be driven by a theoretical framework that raises specific questions and generates specific interests. Popper (1974:46) explained the point nicely: "But in fact the belief that we can start with pure observations alone, without anything in the nature of a theory, is absurd. . . . I tried to bring home the same point to a group of physics students in Vienna by beginning a lecture with the following instructions: 'Take a pencil and paper; carefully observe, and write down what you have observed!' They asked, of course, *what* I wanted them to observe. Clearly the instruction, 'Observe!' is absurd. . . . Observation is always selective. It needs a chosen object, a definite task, an interest, a point of view, a problem."

Second, what may seem to be a simple observation statement, put to work to advance one hypothesis or to deny another, actually has meaning and force only within an involved context of theory. For example, a pH meter may give a reading of 6.42, but the interpretation of that datum depends on the validity of a host of chemical and electronic theories involved in the design and operation of that instrument.

Third, theory choice involves numerous criteria that entail subtle trade-offs and subjective judgments. For example, scientists want theories to fit the observational data accurately and also want theories to be simple or parsimonious. But if one theory fits the data more accurately whereas another theory is more parsimonious, which theory accords better with the data? Clearly, theory choice is guided not only by the observational data but also by some deep theories about scientific criteria and method.

This problem that data are theory-laden is related to a similar problem, the underdetermination of theory by data. For any given set of observations, it is always possible to construct many different and incompatible theories that will fit the data equally well. Consequently, no amount of data is ever adequate to determine that one theory is better than its numerous equal alternatives.

What do such problems mean for science? "But if observations are theory laden, this means that observations are simply theories, and then how can one theory falsify (never mind verify) another theory? Curiously, the full implications of this little complication were not fully grasped by Popper, but by Imre Lakatos: not only are scientific theories not verifiable, they are not falsifiable either" (Theocharis and Psimopoulos 1987).

So the first woe was that science could not verify truths. Now this second woe is that science cannot falsify errors either. Science cannot declare any theory either true or false! "So back to square one: if verifiability and falsifiability are not the criteria, then what makes a proposition scientific?" (Theocharis and Psimopoulos 1987). These are huge problems, and yet there follows a third woe.

Incommensurable paradigms

The third of four deadly woes is incommensurable paradigms. "Thomas S. Kuhn unleashed 'paradigm' on the world," reads the subtitle of an interview with him in *Scientific American* (Horgan 1991). It was reported that "Kuhn . . . traces his view of science to a single 'Eureka!' moment in 1947. . . . Searching for a simple case history that could illuminate the roots of Newtonian mechanics, Kuhn opened Aristotle's *Physics* and was astonished at how 'wrong' it was. How could someone so brilliant on other topics be so misguided in physics? Kuhn was pondering this mystery, staring out of the window of his dormitory room ('I can still see the vines and the shade two thirds of the way down'), when suddenly Aristotle 'made sense.' . . . Understood on its own terms, Aristotle's physics 'wasn't just bad Newton,' Kuhn says; it was just different. . . . He wrestled with the ideas awakened in him by Aristotle for 15 years. . . . 'I sweated blood and blood and blood,' he says, 'and finally I had a breakthrough.' The breakthrough was the concept of paradigm."

Just what is a paradigm? The meaning of Kuhn's key concept is disturbingly elusive. Margaret Masterman (in Lakatos and Musgrave 1970:59–89) counted 21 different meanings, and later in that book, Kuhn himself admitted that the concept was "badly confused" (p. 234). In response to criticisms, Kuhn clarified two main meanings: a paradigm is an exemplar of a past scientific success, or is the broad common ground and disciplinary matrix that unites particular groups of scientists at particular times.

The latter, broad sense is most relevant here. A paradigm is a "strong network of commitments – conceptual, theoretical, instrumental, and methodological" (Kuhn 1970:42). Scientific ideas do not have clear meanings and evidential support in isolation but rather within the broad matrix of a paradigm. For example: "The earth orbits around the sun" is meaningless apart from concepts of space and time, theories of motion and gravity, observations with the unaided eye and with various instruments, and a scientific methodology for comparing theories and weighing evidence.

The history of science, in Kuhn's view, has alternating episodes of normal science, which refine and apply an accepted paradigm, and episodes of revolutionary science, which switch to a new paradigm because anomalies proliferate and unsettle the old paradigm. A favorite example is Newton's mechanics giving way to Einstein's relativity when experiments of many kinds piled up facts that falsified the former but fit the latter theory.

But why did Kuhn speak of revolutions in science when others have been content to speak of progress? The problem is that different paradigms, before and after a paradigm shift, are incommensurable. This term means that no common measure or criterion can be applied to competing paradigms to make a rational, objective choice between them. "In learning a paradigm the scientist acquires theory, methods, and standards together, usually in an inextricable mixture. Therefore, when paradigms change, there are usually significant shifts in the criteria determining the legitimacy both of problems and of proposed solutions" (Kuhn 1970:109). Also, in a paradigm shift, the very meanings of key terms shift, so scientists are not talking about the same thing before and after the shift, even if some words are the same. For example, "Kuhn realized that Aristotle's view of such basic concepts as motion and matter were totally unlike Newton's" (Horgan 1991).

Well, if successive paradigms are incommensurable, what does that imply for science's rationality? In his interview in *Scientific American*, Kuhn remarked, "with no trace of a smile," that science is "arational" (Horgan 1991). To say that science is arational is to say that science is among those things, like cabbage, that have neither the property of being rational nor the property of being irrational. Consequently, saying that science is arational is an even stronger attack on science's rationality than saying that science is irrational.

What happens to realism? The interview by Horgan (1991) says that Kuhn's "most profound argument" is that "scientists can never fully understand the 'real world,'" with the real world sequestered here in scare quotes. "There is, I think, no theory independent way to reconstruct phrases like 'really there'; the notion of a match between the ontology of a theory and its 'real' counterpart in nature now seems to me illusive in principle" (Kuhn 1970:206). This detachment of science from nature is expressed with complete finality by these strong words, "illusive in principle."

What happens to truth? Science has no truth. Indeed, what should be said is "Not that scientists discover truth about nature, nor that they approach ever closer to the truth," because "we cannot recognize progress towards that goal" of truth (Thomas Kuhn, in Lakatos and Musgrave 1970:20). Likewise, the back cover of the current edition (1970) of Kuhn's book quotes, unashamedly and approvingly, the review in *Science* by Wade (1977) saying that "Kuhn does not permit truth to be a criterion of scientific theories."

Obviously scandalized, Theocharis and Psimopoulos (1987) observed that "according to Kuhn, the business of science is not about truth and reality; rather, it is about transient vogues – ephemeral and disposable paradigms. In fact three pages from the end of his book *The Structure of Scientific Revolutions*, Kuhn himself drew attention to the fact that up to that point he had not once used the term 'truth'. And when he used it, it was to dismiss it: 'We may have to relinquish the notion that changes of paradigm carry scientists . . . closer and closer to the truth.'" With rationality, realism, and truth gone, there follows yet another woe for science.

Empty consensus

The fourth and final deadly woe is empty consensus. For more than two millennia since Aristotle, and preeminently in the fledgling universities in Oxford and Paris during the 1200s, philosopher-scientists labored to develop, refine, and establish scientific method. The intention was for scientific method to embody and support science's four traditional claims of rationality, truth, objectivity, and realism. But various arguments developed during the past century, including the problems discussed in the preceding three sections, have led some scholars to abandon science's traditional claims and substitute mere consensus among scientists.

"According to the common-sense view, of course, the assent of the [scientific] community is dictated by certain agreed standards, enabling us to say that the preferred theory is the better one. But Kuhn turns this upside down. It is not a higher standard which determines the community's assent, but the community's assent which dictates what is to count as the highest standard" (Banner 1990:12). What makes a statement scientific is that scientists say it; nothing more. Sociology replaces method in that account of what is scientific.

A particularly radical reinterpretation of science came from Paul Feyerabend. In an interview with Feyerabend in *Science*, "Equal weight, he says, should be given to competing avenues of knowledge such as astrology, acupuncture, and witchcraft. . . . 'Respect for all traditions,' he writes, 'will gradually erode the narrow and self-serving "rationalism" of those [scientists] who are now using tax money to destroy the traditions of the taxpayers, to ruin their minds, to rape their environment, and quite generally to turn living human beings into well-trained slaves.' . . . Feyerabend is dead set against what has been called 'scientism' – the faith in the existence of a unique [scientific] 'method' whose application leads to exclusive 'truths' about the world" (Broad 1979).

Similarly, a more recent interview with Feyerabend in *Scientific American* says that "For decades, . . . Feyerabend . . . has waged war against what he calls 'the tyranny of truth.' . . . According to Feyerabend, there are no objective standards by which to establish truth. 'Anything goes,' he says. . . . 'Leading intellectuals with their zeal for objectivity . . . are criminals, not the liberators of mankind.' . . . Jutting out his chin, he intones mockingly, 'I am searching for the truth. Oh boy, what a great person.' . . . Feyerabend contends that the very notion of 'this one-day fly, a human being, this little bit of nothing' discovering the secret of existence is 'crazy.' . . . The unknowability of reality is one theme of . . . Feyerabend" (Horgan 1993).

Ironically, these four woes bring the status of science full circle. Popper started with the problem of demarcating science from nonscience in order to grant credibility to science and to withhold credibility from nonscience. Despite considerable limitations, science was something special. But a mere generation later, his student Feyerabend followed his teacher's ideas to their logical conclusion by judging that science is neither different from nor superior to any other way of knowing. Science started out superior to astrology; it ended equivalent. For millennia, science involved methodology for finding objective truth; it ended with sociology for explaining empty consensus.

Finally, so what? After having told us that science is arational, that science finds no truth, that reality is eternally illusive in principle, and that science's supposed claims are to be explained away in sociological terms, a calm Kuhn tells us that "I no longer feel that anything is lost, least of all the ability to explain scientific progress, by taking this position" (Thomas Kuhn, in Lakatos and Musgrave 1970:26). For Kuhn, the rationality, truth, objectivity, and realism that scientists are accustomed to, just do not matter.

Reactions from scientists

The preceding four sections discussed four deadly woes: elusive truth, theory-laden data, incommensurable paradigms, and empty consensus. How do scientists react? Do they think that these philosophical criticisms are valid, forcing honest scientists to adjust and downgrade their claims, or not?

The following account of scientists' reactions focuses on material that is readily seen by scientists, especially articles in *Nature* and *Science*. The first notable exchange began with the provocative commentary by Theocharis and Psimopoulos (1987) in *Nature.* It stimulated a lively correspondence, from which *Nature* published 18 letters, until the editor closed the correspondence and gave the authors an opportunity to reply (*Nature* 1987, 330:308, 689–690; 1988, 331:129–130, 204, 384, 558; reply 1988, 333:389).

The dominant tenor of the 18 letters to the editor was that of numerous scientists rushing in to defend Popper and Kuhn from what they perceived as an unreasonable or even malicious attack by Theocharis and Psimopoulos. One letter even recommended that Theocharis and Psimopoulos (and perhaps also the journal *Nature*) offer Sir Karl Popper a public apology. Some letters rejected the claim that objective truth was important for science. One letter claimed that the "most basic truth is that there can be no objective truth." Another reader replaced truth with prediction: "This process of making ever better prediction *is* scientific progress, and it circumvents entirely the problem of defining scientific truth."

One of the strongest letters was from sociologist Harry Collins. Apparently considering himself to have ascended to high moral ground indeed, he suggested that "The only thing that makes clear good sense in Theocharis and Psimopoulos is the claim that the privileged image of science has been diminished by the philosophical, historical and sociological work of past decades. One hopes this is the case. Grasping for special privilege above and beyond the world we make for ourselves – the new fundamentalism that Theocharis and Psimopoulos press upon us – indicates bankruptcy of spirit luckily not yet widespread in the scientific community." So pursuit of truth had been transmuted into bankruptcy of spirit!

Some other reactions, however, were favorable. One responder wrote sympathetically that "Philosophical complacency . . . will not do; contrary to what both sceptics and conservatives often seem to believe, philosophical questions do matter." Another responded that "There are very good reasons why twentieth century philosophy of science, under the malign influence of Popper through to Feyerabend, is profoundly hostile to science itself. . . . It is indeed unfortunate that many scientists, through ignorance, quote these philosophers approvingly. The most effective victories are those in which the losers unwittingly assist their opponents."

In their reply to these 18 letters, Theocharis and Psimopoulos offered a poignant and intriguing remark about the uniqueness of the contemporary scene: "Natural philosophy has had enemies throughout its 2,600 or so years of recorded history. But the present era is unique in that it is the first civilized society in which an effective antiscience movement flourishes contemporaneously with the unprecedently magnificent technological and medical applications of modern science. This is a curious paradox which cries out for clarification."

Another interesting exchange was precipitated by the publication of a book by Collins and Pinch (1993). It was reviewed and criticized by N. David Mermin in

Physics Today [1996, 49(3):11, 13; 49(4):11, 13], with further exchanges [1996, 49(7):11, 13, 15; 1997, 50(1):11, 13, 15, 92, 94–95]. Mermin strongly rejected Collins and Pinch's conclusion that "Science works the way it does not because of any absolute constraint from Nature, but because we make our science the way that we do." The collective import of many such declarations was that science does not discover objective truths about physical reality but rather just constructs consensus among scientists. Obviously, Mermin said, scientists are humans involved in a social structure that is a real and integral aspect of our science, but "Agreement is reached not just because scientists are so very good at agreeing to agree." Mermin suggested that a crucial feature of science that Collins and Pinch overlooked in their denigrating account was the role of interlocking evidence: "an enormous multiplicity of strands of evidence, many of them weak and ambiguous, can make a coherent logical bond whose strength is enormous."

The letters to the editor were generally sympathetic to Mermin's defense of science's rationality. One letter offered the perceptive remark that "No modern-day consensus on the nature of science will be reached until we agree that what we are talking about is neither sociology nor science, but philosophy." Another letter declared quite simply that "There really are results and facts."

Another important exchange began with a commentary on science wars in *Nature* by Gottfried and Wilson (1997). Subsequently, Colin Macilwain and David Dickerson continued the discussion in *Nature* (1997, 387:331–334); readers provided four letters, and Gottfried and Wilson replied (1997, 387:543–546), and then two more letters appeared (1997, 388:13; 389:538).

The main concern in Gottfried and Wilson's 1997 commentary was with attacks on science from sociologists, in contrast to Theocharis and Psimopoulos's 1987 commentary in the same journal a decade earlier that had focused on attacks from philosophers. A school of sociology, so-called Science Studies, had vigorously attacked science's traditional claims. Variants of that movement went under several names, such as the "strong program" or the Edinburgh school of sociology, but here the brief name "constructivism" suffices. Gottfried and Wilson got quickly to the very heart of the debate over science's status: "Scientists eventually settle on one theory on the basis of imperfect data, whereas logicians have shown that a finite body of data cannot uniquely determine a single theory. Among scientists this rarely causes insomnia, but it has tormented many a philosopher."

Seven lines of evidence were cited to show that science has a strong grip on reality: (1) steadily improving predictions, often unambiguous, precise, diverse, and even surprising; (2) increasingly accurate and extensive data; (3) increasingly specific and comprehensive theories; (4) interlocking evidence of diverse sorts; (5) progress over time in describing and explaining nature; (6) reproducible experiments; and (7) science-based technology that works. Of those seven witnesses to science's success, the first, "predictive power," was "the strongest evidence that the natural sciences have an objective grip on reality."

A particularly notorious episode in the science wars, mentioned by Gottfried and Wilson (1997), was the so-called Sokal affair. To spoof postmodern and constructivist views of science, Sokal (1996) published an article in *Social Text*, only later to expose it as a hoax (Alan Sokal, in Koertge 1998:9–22; Sokal 2008). "'It took me a lot of writing and rewriting and rewriting before the article reached the desired level of unclarity,' he chuckles," in an interesting interview in *Scientific American* (Mukerjee 1998). That hoax provoked front-page articles in the *New York Times*, the *International Herald Tribune*, the London *Observer*, and *Le Monde*.

A recent essay in *Nature* by sociologist Collins (2009) had the byline, "Scientists have been too dogmatic about scientific truth and sociologists have fostered too much scepticism – social scientists must now elect to put science back at the core of society." This two-page essay provides an ideal summary for Collins's work over the preceding couple of decades. He characterized the sociology of science (or "science studies") as having three waves. The first wave "coincided with post-war confidence in science" after the Second World War, during which "social scientists took science to be the ultimate form of knowledge." The second wave began in the 1960s and culminated in the science wars that earnestly began around 1990 and essentially ended around 2000, and it "was characterized by scepticism about science." Collins (2009) proposed a new third wave "to counter the scepticism that threatens to swamp us all" and "to put the values that underpin scientific thinking back in the centre of our world." His main suggestion for implementing this third wave was "to analyse and classify the nature of expertise to provide the tools for an initial weighting of opinion," as explained more fully in a book that he and his colleagues had recently authored. He also suggested that "scientists must think of themselves as moral leaders" promoting "the good society."

His critique of this renounced skepticism had the crucial insight – which every science student and professional should fully appreciate and which Collins expressed skillfully and concisely – that skepticism is utterly and irremediably unfalsifiable. "By definition, the logic of a sceptical argument defeats any amount of evidence" because the skeptic can always appeal to several potential philosophical problems that cut deeper than any empirical evidence, such as that "one cannot be sure that the future will be like the past." And the troubling consequence rapidly follows that "One can justify anything with scepticism" since it recognizes no real knowledge to constrain belief or guide action.

Nevertheless, this essay's renunciation of skepticism about science left intact the chief tenet of this skepticism, namely, that science finds no settled truth. Scientists are "reaching towards universal truths but inevitably falling short" and they "must teach fallibility, not absolute truth." Again, "Science's findings . . . are not certain. They are a better grounding for society precisely, and only, because they are provisional." Consequently, "when we outsiders judge scientists, we must do it not to the standard of truth, but to the much softer standard of expertise."

However, one need not search very far to see that this insistent and repeated rejection of truth claims, which has been a persistent theme in Collins's publications for two decades, is wholly uncharacteristic of actual science. For instance, turn the pages of the 2009 issue of *Nature* containing Collins's essay to glance at its other articles. They concern the molecular basis of transport and regulation in the Na^{+}/betaine symporter BetP, a candidate sub-parsec supermassive binary black hole system, the electronic acceleration of atomic motions and disordering in bismuth, the innate immune recognition of infected apoptotic cells directing $T_{H}17$ cell differentiation, transcriptome sequencing for detecting gene fusions in cancer, determining protein structure in living cells by in-cell NMR spectroscopy, and such. The rest of this issue of *Nature* is immersed and soaked in truth claims, of which some are probable to a specified degree and some are certain, based on extensive evidence and careful reasoning that has been checked by competent colleagues and peer reviewers. Doubtless, someone somewhere is exaggerating science's role, but such nonsense is not characteristic of the mainstream science that readers encounter in the pages of *Nature* and *Science*, and neither is such nonsense best refuted by the opposite exaggerating of science's complete inability to find any settled truth.

Collins's essay prompted four replies. The title of the first reply effectively captured the tenor of this correspondence: "Let's not reignite an unproductive controversy." In my view, Collins's essay denounces exaggerated claims of science's powers, only to substitute exaggerated claims of science's limits, exaggerated fears of skepticism's threat to swamp us all, exaggerated pronouncements (in a science journal!) on religion's demerits, exaggerated expectations for science's role in politics and society, and exaggerated plaudits for scientific expertise after detaching expertise from truth.

Popper and Kuhn appear not only in contests over science's rationality but also when scientists publish routine research papers citing them. Such citations are moderately common, with Popper's *The Logic of Scientific Discovery* having more than 12,000 citations and Kuhn's *The Structure of Scientific Revolutions* having more than 44,000 citations. Most scientists have at least a passing familiarity with those influential intellectuals. What concepts are scientists drawing from such philosophers in routine scientific publications?

For the most part, despite occasional noteworthy exceptions, it must be said that the routine use of Popper and Kuhn's ideas by scientists is rather selective and superficial. For instance, the first of the 18 replies to Theocharis and Psimopoulos (1987), written by a physiologist who also taught a course in the philosophy of science, was quite revealing. "Popper began it all by his concern to distinguish good science from bad. He identified Einstein as good and Adler as bad by characterizing Einstein's predictions as *falsifiable* but not as false. . . . I share the view that the next stage of Popper's thought sees him following up certain ideas to unbalanced and therefore somewhat antirational conclusions, but this first, key perception [regarding falsifiability] is firmly on the side of objectivity and truth. This is the one bit of Popper that I teach." His sensible posture toward the philosophers was that "If their accounts are unbalanced, it is up to us [scientists] to balance, not dismiss them."

The one bit of Popper that does show up frequently in scientists' research papers is that a proposed hypothesis must make testable predictions that render the hypothesis falsifiable. And, more pointedly, a scientist should give his or her own favored hypothesis a trial by fire, deliberately looking for potentially disconfirming instances, not just instances that are likely to be confirming. Doubtless, this is wholesome advice. But is this some fancy, new insight? Hardly! It is as old as ancient *modus tollens* arguments (not *B*; *A* implies *B*; therefore not *A*) and the medieval Method of Falsification of Robert Grosseteste. Looking for potentially contradictory evidence seems more in the province of simple honesty than fancy philosophy.

Similarly, the one bit of Kuhn that does show up frequently in scientists' writings is the dramatic idea of a paradigm shift. Needless to say, this idea is particularly popular among those scientists who take themselves to be the innovators who are precipitating some big paradigm shifts in their own disciplines. Of course, the standard claim in scientists' papers is that their shiny new paradigms are a whole lot better than their predecessors, and even are true or at least approximately true. But unwittingly such scientists are bad disciples of their presumed master, Kuhn. His own view was that successive paradigms are incommensurable, so it makes no sense whatsoever to say that one paradigm is better than another. Of course, Kuhn's own view takes all the fun and prestige out of coming up with a slick new paradigm! So it is not too surprising that scientists have generally failed to get that discouraging bit of Kuhn.

What is the bottom line? The bad news is a recent history of scientists mostly citing skeptical philosophers of science whose actual views undermine science's traditional claims but evading potential harm by selective and superficial use of the occasional bits that are sensible for science practitioners. The good news is that nothing is keeping scientists from a future history of using many bits of great ideas from mainstream philosophers of science for the practical purposes of increasing productivity and enhancing perspective.

The AAAS posture

For several decades, particularly since the books by Popper and Kuhn appeared in 1959 and 1962, science's traditional claims of rational realism and objective truth have been under significant reappraisal and even sustained attack. Does the AAAS rebut these new ideas about science, accept them, or just ignore them?

Their most general statements about science reveal most incisively just what the AAAS takes science to be. For instance, "science is the art of interrogating nature" (AAAS 1990:17). That simple but profound remark claims that science is objective in the fundamental sense of being about an object with its own independent existence and properties. Also, truth is sought: "When faced with a claim that something is true, scientists respond by asking what evidence supports it" (AAAS 1989:28). The basic elements in scientific method are observation and evidence, controlled experiments, and logical thought (AAAS 1989:25–28). Such remarks presume and express science's traditional claims of rationality, truth, objectivity, and realism.

The AAAS also expresses some ideas associated with Popper and Kuhn. For instance, scientific thinking demands falsifiability, as Popper insisted. "A hypothesis that cannot in principle be put to the test of evidence may be interesting, but it is not scientifically useful" (AAAS 1989:27; also see AAAS 1990:xiii). And data are theory-laden because theory guides the choice, organization, and interpretation of the data (AAAS 1989:27, 1990:17–18). Likewise, scientific research is guided by paradigms that are "metaphorical or analogical abstractions" that "dictate research questions and methodology," as Kuhn emphasized (AAAS 1990:21, 24). "Because paradigms or theories are products of the human mind, they are constrained by attitudes, beliefs, and historical conditions" (AAAS 1990:21).

Science has a decidedly human face, quite unlike its earlier images offered by Francis Bacon in the 1600s or the logical empiricists in the early 1900s. Indeed, "human aspects of inquiry . . . are involved in every step of the scientific process from the initial questioning of nature through final interpretation" (AAAS 1990:18). "Science as an enterprise has individual, social, and institutional dimensions" (AAAS 1989:28). This humanity brings risks and biases: "Scientists' nationality, sex, ethnic origin, age, political convictions, and so on may incline them to look for or emphasize one or another kind of evidence or interpretation" (AAAS 1989:28). But, on balance, scientists do attempt to identify and reduce biases: "One safeguard against undetected bias in an area of study is to have many different investigators or groups of investigators working in it" (AAAS 1989:28).

How much success does science enjoy in getting at the truth? "Scientific knowledge is not absolute; rather, it is tentative, approximate, and subject to revision" (AAAS 1990:20), and "scientists reject the notion of attaining absolute truth and accept some uncertainty as part of nature" (AAAS 1989:26). "Current theories are taken to be 'true,' the way the world is believed to be, according to the scientific thinking of the day" (AAAS 1990:21). Note that "true," here sequestered in scare quotes, is equated to nothing more real or enduring than "the scientific thinking of the day." Furthermore, science's checkered history "underscores the tentativeness of scientific knowledge" (AAAS 1990:24).

Nevertheless, "most scientific knowledge is durable," and "even if there is no way to secure complete and absolute truth, increasingly accurate approximations can be made" (AAAS 1989:26). However, deeming durability to be something good or admirable presumes that durability is serving as some sort of truth surrogate, because otherwise the durability or persistence of a false idea is bad. Hence, switching from "true" to "durable" is not a successful escape from the issue of truth. One of the most positive remarks is that "the growing ability of scientists to make accurate predictions about natural phenomena provides convincing evidence that we really are gaining in our understanding of how the world works" (AAAS 1989:26).

All in all, the AAAS verdict is a nuanced mix of positives and negatives: "Continuity and stability are as characteristic of science as change is, and confidence is as prevalent as tentativeness. . . . Moreover, although there may be at any one time a broad consensus on the bulk of scientific knowledge, the agreement does not extend to all scientific issues, let alone to all science related social issues" (AAAS 1989:26, 30).

One must also observe, however, that the pages of AAAS (1989) catalogue literally hundreds of facts about the universe, the earth, cells, germs, heredity, human reproduction and health, culture and society, agriculture, manufacturing, communications, and other matters. Unquestionably, the vast majority of these facts are presented with every appearance of truth and certainty and without even a trace of revisability or tentativeness. For instance, science has declared that the earth moves around the sun (and around our galaxy), and the former theory that the earth is the unmoving center of the universe is not expected to make a stunning comeback because of some new data or theory!

Admittedly, it is awkward that some AAAS declarations sound as though they reflect the concept that all scientific knowledge is tentative, whereas other statements apparently present numerous settled certainties. Perhaps the AAAS verdict on science and truth is rather unclear, or perhaps some isolated statements lend themselves to an unbalanced or unfair reading relative to the overall message. Anyway, it may be suggested that, given a charitable reading, the AAAS position papers say that some scientific knowledge is true and certain, some is probable, and some is tentative or even speculative and that scientists usually have good reasons that support legitimate consensus about which level of certainty is justified for a given knowledge claim.

Although the AAAS acknowledges revolutionary changes in paradigms, they explicitly deny that successive paradigms are incommensurable or fail to move closer to the truth: "Albert Einstein's theories of relativity—revolutionary in their own right—did not overthrow the world of Newton, but modified some of its most fundamental concepts" (AAAS 1989:113; also see p. 26). Newton's mechanics and Einstein's relativity lead to different predictions about motions that can be observed, so they are commensurable; and relativity's predictions have been more accurate, so it is the better theory (AAAS 1989:114). And yet, looking to the future, physicists pursue "a more [nearly] complete theory still, one that will link general relativity to the quantum theory of atomic behavior" (AAAS 1989:114).

Likewise, the AAAS repeatedly discusses the logic of falsifiability. But, unlike Popper, they also repeatedly discuss the logic of confirmation, including sophisticated remarks about the criteria for theory choice (AAAS 1989:27–28, 113–115, 135). Also, the acknowledgement of science's human face causes no despair about humans rationally investigating an objective reality with considerable success.

With admirable candor, the AAAS (1990:26) recognizes that even their careful position papers "are set in a historical context and that all the issues addressed will and should continue to be debated." One curious feature of AAAS (1989, 1990) is that despite the numerous unmistakable allusions to Popper and Kuhn's influential ideas, those figures are neither named nor cited. In the future, it will be interesting to see whether the AAAS decides to engage science's external auditors more directly.

Clear targets

When scientists encounter philosophers and others in grand discussions of science's rationality, just one simple question is surpassingly most essential. What are the targets of an argument against rationality: science only, or else both science and common sense? These two options are depicted in Figure 4.2.

The attacks on science's rationality involve a thousand complicated technicalities, but the principal action concerns this one simple matter of the scope of an attack. If the target is science only, then the argument presents a challenge that scientists really need to answer. But, if the targets are science and common sense, then the argument is merely some variant of radical skepticism, and the scientific community is under no obligation to find it of any interest. Science *begins* with the presupposition that the physical world is comprehensible to us (AAAS 1989:25, 1990:16). Therefore, a legitimate attack on science's rationality must target science alone, not both science and common sense. Why this is so can be illuminated by an analogy.

People often have the perception that many disease organisms are difficult to kill because numerous terrible diseases still ravage millions of suffering persons, and scientists have no satisfactory cure. But, in fact, all of these viruses, bacteria, and other microbes are easy to kill – every last one of them. A strong dose of arsenic or cyanide could kill them all, not to mention the even easier expedient of merely heating them to 500°C. It is easy to kill any pathogen. The trick is not to kill the host at the same time! Medicine's challenge is to kill the pathogen *and* not kill the host. So a strong dose of arsenic fails to qualify as a medicine, not because it kills too little, but because it kills too much.

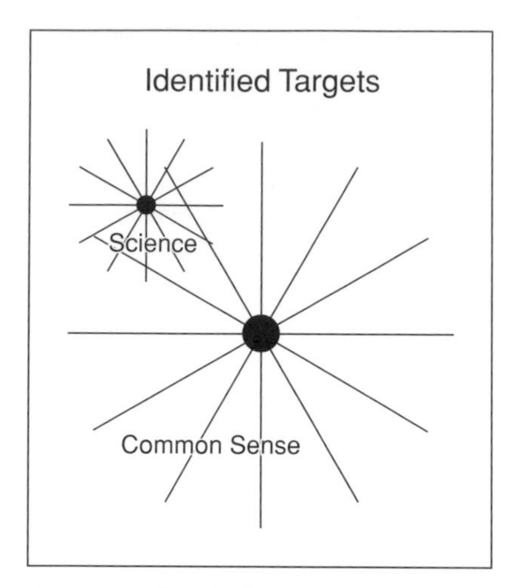

Figure 4.2 Identified targets for arguments against science's rationality. Arguments vary in their explosive force and intended targets. Some attack science only, whereas others attack both science and common sense. Because these two options call for very different analyses and responses, the targets of a given argument must be clearly identified.

Now, the same situation applies to science and common sense. Countless philosophical arguments and intellectual blunders kill common sense, which obviously kills science also. It is easy to kill both common sense and science. Indeed, a lackluster high school student can easily learn five skeptical objections in as many minutes – maybe our sense perceptions are unreliable, maybe some demon is deceiving us, maybe the physical world is only an illusion, maybe the future will be unlike the past, and so on. But, for any discipline such as science that begins with a nonnegotiable conviction that common sense delivers much truth because the world is comprehensible, a philosophical argument that kills science and common sense is unimpressive because it kills too much.

Consequently, the first and greatest burden placed on scientists when they read antiscientific arguments is to ask this discerning question: Does this philosophical argument kill science alone, or does it kill both science and common sense? Any argument that, if understood clearly and applied consistently, would imply that we cannot really know trifling trinkets of common-sense knowledge is just plain ridiculous, whether or not the scientist has enough philosophical training and acumen to spot and refute the specific steps at which the argument goes awry.

For example, Popper insisted that "We *cannot* justify our knowledge of the external world; *all* our knowledge, even our observational knowledge, is theoretical, corrigible, and fallible" (Karl Popper, in Lakatos and Musgrave 1968:164). Hence, in Popper's own estimation, problems extend to "*all* our knowledge" of any kind, including both science and common sense. Accordingly, this attack is suspect because it unsettles too much.

The defense of science's rationality begins with this insistence on clear targets for arguments against rationality. But this defense matures with painstaking, methodical development of the components of scientific thinking: presuppositions, evidence, and logic. That ensues in the following Chapters 5 to 11, after which my response to the four woes is given in the first section of the final Chapter 14.

Summary

Opinions about science's rationality and objective truth have always been strongly influenced by the claims of scientists themselves. But the assessments and judgments of philosophers, historians, sociologists, and others are becoming increasingly influential. External auditors are legitimate and beneficial because science is a liberal art and there is a rich traffic of fruitful ideas among the sciences and the humanities. Even highly critical views can be helpful, disturbing complacency and prompting scientists to think things through carefully.

Science's four traditional claims are rationality, truth, objectivity, and realism. But those claims have been under heavy attack, especially in terms of four intellectual problems: (1) Karl Popper claimed that empirical data could falsify a theory but never prove it, so science could never find truth. (2) Popper and others claimed that observations are theory-laden and that data underdetermined theory choice. From that, Imre Lakatos drew the implication that scientific theories cannot be falsified either, so science cannot declare any theory either true or false. (3) Thomas Kuhn said that paradigms are incommensurable, so science is arational. (4) Kuhn also said that what makes a statement scientific is merely that scientists say it. Accordingly, Paul Feyerabend concluded that there is nothing special about science.

The first notable exchange among scientists, philosophers, and others regarding science's rationality began with the commentary by Theocharis and Psimopoulos (1987) in *Nature*. Those scientists felt that the critiques by Popper and Kuhn and other philosophers were unjustified and exaggerated, but nevertheless quite influential, so it was incumbent upon the scientific community to give a satisfying defense of science's rationality, objectivity, and truth. More recent exchanges were prompted by a commentary by Gottfried and Wilson (1997) and an essay by Collins (2009) in *Nature*. Position papers from the AAAS have provided a mainstream institutional perspective.

Identifying the target of an argument against science's rationality is essential. An argument that attacks both science and common sense is simply some variant of radical skepticism, so the scientific community is not obliged to respond.

However, a full explanation and defense of science's rationality necessarily takes the form of an account of the presuppositions, evidence, and logic that together undergird science's rationality.

Study questions

(1) Have you encountered attacks on science's rationality? If so, are they coming from scientists or from others? How prevalent or influential do you perceive these attacks to be?
(2) In your own estimation, who are the legitimate auditors of science's claims?
(3) Recall the four deadly woes: elusive truth, theory-laden data, incommensurable paradigms, and empty consensus. Which one do you consider to be the most serious threat and why? How would you answer that threat in order to preserve science's rationality?
(4) Consider the reactions from scientists to the so-called science wars. Which two or three of the scientists' arguments do you think are the strongest? Can you think of any additional strong arguments that were not already mentioned in this chapter?
(5) Doubtless, the complexity of the four deadly woes means that any adequate response must take the form of a collection of numerous arguments. But such a collection needs to start somewhere. What do you regard as the first and most important clarification or argument?

5

Science's presuppositions

Essentially, a presupposition is a belief that is required to reach a particular conclusion, and yet it cannot possibly be proved. A presupposition cannot be proved in the ordinary sense of marshaling definitive evidence because presuppositions precede and empower evidence. But that does not necessarily mean that presuppositions are arbitrary and shaky. Rather, presuppositions should be chosen carefully, disclosed, and then legitimated. Because presuppositions are just as necessary as evidence for science to reach any conclusions, a reflective account of science must discuss them.

Although presuppositions and evidence are equally essential, in ordinary scientific discourse, the presuppositions are ignored, whereas the evidence is marshaled. Why? Within the context of ordinary science, the presuppositions needed in science are sensible and unproblematic and are taken for granted. Nevertheless, "Our presuppositions are always with us, never more so than when we think we are doing without them" (O'Hear 1989:54). Again, "Most scientists take for granted their metaphysical assumptions, but they are none the less necessary logically to the conclusions of science" (Caldin 1949:176).

This chapter's topic of presuppositions bears primarily on this book's purpose of enhancing perspective. To the extent that defending science's rationality is important for science's long-term health, however, this chapter is crucial. Primarily science's presuppositions, rather than its evidence or logic, prompt positive or negative assessments of science's rationality.

This chapter argues that science requires more logic, more evidence, more instrumentation, more education, and more work than does common sense, but nothing more by way of presuppositions. Presuppositions cannot be proved by logic or established by evidence; rather, they can be disclosed by philosophy and accepted by faith.

Historical perspective on presuppositions

Because presuppositions are the most subtle components of scientific method, some historical perspective is invaluable. Especially for scientists who have given science's presuppositions little thought, a brief historical survey provides a window into the various options and their implications.

Albertus Magnus (*c.* 1200–1280) handled science's presuppositions by an appeal to conditional necessity, a concept that ranks among the most important notions in the philosophy of science. He was building on the concept of suppositional reasoning that had been explained by Aristotle (384–332 BC). Albertus rendered the opening sentence of Book 2 of Chapter 9 of Aristotle's *Physics* as follows, with Aristotle's text in italic type and Albertus's amplification in roman type.

> "We ask therefore first *whether the necessity of physical things is a necessity simply or is a necessity 'ex suppositione'* and on the condition of some end that is presupposed. For example, a simple necessity is such that it is necessary that the heavy go down and the light go up, for it is not necessary that anything be presupposed for this for it to be necessary. Necessity 'ex conditione,' however, is that for whose necessity it is necessary to presuppose something, nor is it in itself necessary except 'ex suppositione'; and so it is necessary for you to sit if I see you sitting." (William A.Wallace, in Weisheipl 1980:116)

In Albertus's view, biology *could* be as certain as geometry: "Surely Albert entertained no doubts that . . . one could have certain and apodictic demonstrations even when treating of animals, provided the proper norms of *ex suppositione* reasoning were observed" (William A. Wallace, in Weisheipl 1980:127–128). For example, that (normal adult) horses eat grass and that Euclidean triangles have interior angles totaling 180 degrees have equal degrees of certainty, even though they have different grounds of certainty.

Suppositional reasoning became a device for demarcating a human-sized and public science apart from philosophical differences and theological debates. Bear in mind that in referring to philosophy, Albertus included natural philosophy or what we would now most commonly call science. Albertus "proposed to distinguish between philosophy and theology on methodological grounds, and to find out what philosophy alone, without any help from theology, could demonstrate about reality. . . . He acknowledged (with every other medieval thinker) that God is ultimately the cause of everything, but he argued that God customarily works through natural causes and that the natural philosopher's obligation was to take the latter to their limit" (Lindberg 2007:240–241).

In the subsequent formulations of Thomas Aquinas (*c.* 1225–1274), John Duns Scotus (*c.* 1265–1308), Jean Buridan (*c.* 1295–1358), and others, that notion of conditional necessity was gradually shifted in two significant ways. First, Aristotle and Albertus emphasized purpose in nature, but later views of scientific explanation gave this diminishing attention. Second, concern with necessity gave way to interest in certainty. Often, the motive for demonstrating that something is necessary had been to establish that it is certain. But necessity is a much richer concept than certainty, implicating a much larger and potentially more controversial story about how the world works.

Sir Francis Bacon (1561–1626) was an influential proponent of science. His *Novum Organum* claimed to offer a better scientific method than had his predecessors, especially Aristotle. It emphasized empirical evidence untainted by presuppositions. Bacon charged that someone led captive by presuppositions would reach a conclusion before doing proper experiments. Doubtless, Bacon's view of science, supposedly based on presuppositionless evidence, still typifies the way many scientists think about science.

David Hume (1711–1776) was a great skeptic. He thought that science's ambitions must be limited to describing our perceptions, avoiding philosophical speculations about some external physical world. But, despite his philosophical convictions, Hume conceded that common sense must rule in life's ordinary dealings. His writings reflect an awkward tension between common sense and philosophy that never gets resolved. Another challenging view was that of George Berkeley (1685–1753), who believed that only minds and ideas exist, and not physical objects.

Thomas Reid (1710–1796) was the great protagonist of common sense as the only secure foundation for philosophy and science, in marked contrast to Hume. Of course, previous ancient and medieval thinkers had developed the scientific method within a common-sensical framework. But subsequent challenges, especially from Berkeley and Hume, had necessitated exposing science's common-sense roots with greater clarity and force. According to Reid,

> Hume's error was to suppose that it made sense to justify first principles of our faculties by appeal to [philosophical] reason. It does not. . . . To attempt to justify the first principles of our faculties by reasoning is to attempt to justify what is the most evident by appeal to less evident premises, those of philosophers. . . . Philosophy, properly understood, does not justify these principles of common sense but grows from them as a tree grows from its roots. . . . The attempt to justify a conclusion that is evident to begin with, such as that I see a cat, by appeal to premises that are philosophically controversial is doomed to absurdity. When the conclusion of an argument is more evident to begin with than it could be shown to be by a philosophical argument, the latter is useless as the justification of the conclusion. . . . No such [philosophical] argument has the evidential potency of innate [common-sense] principles of the mind. (Lehrer 1989:19, 294)

Reid's conception of science based on common sense had five main elements.

(1) The Symmetry Thesis. An influential eighteenth-century science of the human mind, originating from John Locke (1632–1704), Berkeley, and Hume, said that real knowledge could be achieved only for our sensations and relations among sensations, not for objects supposedly causing our sensations, so science must settle for appearances rather than realities. As a corrective, Reid adopted a symmetry thesis that gave the internal world of sensations and external world of objects equal priority and status, with both taken as starting points for philosophical reflection.

(2) Harmonious Faculties. Hume and other skeptics granted philosophical reasoning priority over scientific observation. But Reid endorsed the basic reliability of all of our faculties, both sensory and mental, saying that "He must be either a fool or want to make a fool of me, that would reason me out of my reason and senses" (Hamilton 1872:104). Indeed, "Scepticism about the soundness of the sceptic's arguments is at least as justified as the scepticism which he urges upon us" (Dennis C. Holt, in Dalgarno and Matthews 1989:149).

(3) Parity among Presuppositions. Reid claimed a parity between realist and skeptical presuppositions. Reid noted that we have two choices: to trust our faculties as common sense enjoins, or not to trust our faculties and become skeptics. A critic may complain that Reid's appeal to common sense is dogmatic or circular: "Because propositions of common sense are foundational, it is not possible to provide constructive, independent grounds for their acceptance. The propositions of common sense constitute the final court of appeal; they cannot themselves be justified, at least in the manner appropriate to derivative propositions. For that reason it must seem to the committed idealist or sceptic that the defender of common sense begs the question" (Dennis C. Holt, in Dalgarno and Matthews 1989:147). But Reid's reply was that such exactly is the nature of a foundational presupposition: it can only be insisted upon. The realist presupposes that the world is real and comprehensible, whereas the skeptic presupposes that it is not. Therefore, the contest between realism and skepticism has to turn on considerations other than the choice or role of presuppositions.

(4) Asking Once or Twice. What is the basis for science's presuppositions? Reid's reply depended on whether that question was asked once or twice. Asked once, Reid supported science's presuppositions by an appeal to common sense. But if asked twice, the deeper issue became why the world was so constituted as common sense supposed. For instance, why does the physical world exist, rather than nothing? And why are we so constituted that the world is comprehensible to us?

Clearly, those deeper questions cannot be answered satisfactorily by a mere appeal to common sense but rather require the greater resources of some worldview. Regarding that deeper appeal to a worldview, Reid had two things to say.

First, Reid said that his own worldview, Christianity, explained and supported science's common-sense presuppositions. That worldview says that God made the physical world and made our senses reliable. "In Reid's doctrine the existence of common sense has theistic presuppositions; its truths are 'the inspiration of the Almighty.' Reid did not maintain that belief in them depends upon belief in

God; they are imposed upon us by the constitution of our nature, whatever our other beliefs. His implication is that we have to go behind common sense, if we are to explain its competence, to the fact that our nature has been constituted by God" [Selwyn A. Grave, in Edwards 1967(7):120–121].

Second, Reid claimed that virtually all other worldviews also respected the rudimentary common sense that provided science's presuppositions. Common sense was imposed on us "by the constitution of our nature," and that human nature was shared by all humans, regardless of whatever a person happened to believe or not believe about God.

Reid's strategy for supporting science's presuppositions had a wonderful clarity and balance. Worldview-independent, common-sense presuppositions preserved science's credibility. At the same time, there was no confusion or pretense that mere common sense provided a deep or ultimate explanation of why the world is as it is. That job has to be done by some worldview. Fortunately, although worldviews differ on many other points, they do not challenge each other over rudiments of common sense such as "The earth exists" or "I have two eyes." By seeing common sense as a penultimate rather than an ultimate defense of science, Reid invited the humanities to complement science's picture of the world.

(5) Reason's Double Office. Reid maintained that reason holds the traditional "double office" of "regulating our belief and our conduct" [Selwyn A. Grave, in Edwards 1967(7):121]. Belief and action should match. If not, the diagnosis is not the logical problem of incoherence between one belief and another contrary belief but rather the moral problem of insincerity or hypocrisy shown by mismatch between belief and action. As for the world of human actions, common sense was the only game in town. For example, a skeptic's mouth may say that we cannot be sure that a car is a real or hard object, but at a car's rapid approach, the skeptic's feet had better move!

Reid happily quoted Hume's own admission that a skeptic "finds himself absolutely and necessarily determined, to live and talk and act like all other people in the common affairs of life" (Hamilton 1872:485). Nevertheless, Hume went on to remark that "reason is incapable of dispelling these clouds" of skepticism. But if that was the case, it must be that Hume's version of reason held only the single office of regulating belief, rather than Reid's traditional reason that held the double office of regulating belief and action. In other words, only after first having adopted an impoverished notion of reason that pertains to belief but not to action is it possible for someone to regard as reasonable a skeptical philosophy that could not possibly be acted upon and lived out without jeopardizing the skeptic's survival. Greco provided a penetrating analysis of the key ideas in Reid's reply to the skeptics, and judged it extremely effective (John Greco, in Cuneo and van Woudenberg 2004:134–155).

Although regarded principally as a philosopher, Reid was also an accomplished scientist. He wrote and lectured on mathematics, optics, electricity, chemistry, astronomy, and natural history (Paul Wood, in Cuneo and van Woudenberg 2004:53–76).

Position papers from the American Association for the Advancement of Science (AAAS) provide a contemporary, mainstream expression of science's presuppositions. "Science presumes that the things and events in the universe occur in consistent patterns that are comprehensible through careful, systematic study. Scientists believe that through the use of the intellect, and with the aid of instruments that extend the senses, people can discover patterns in all of nature. Science also assumes that the universe is, as its name implies, a vast single system in which the basic rules are everywhere the same" (AAAS 1989:25). "All intellectual endeavors share a common purpose—making sense of the bewildering diversity of experience. The natural sciences search for regularity in the natural world. The search is predicated on the assumption that the natural world is orderly and can be comprehended and explained" (AAAS 1990:16).

Furthermore, careful scientific argumentation should disclose all premises. Indeed, the AAAS (1989:139) lists several "signs of weak arguments" that are useful for checking both others' and one's own arguments. One sign of a "shoddy" argument is that "The premises of the argument are not made explicit." Likewise, "Inquiry requires identification of assumptions" (NRC 1996:23). Therefore, science's presuppositions should be explicitly and fully disclosed.

The PEL model of full disclosure

A given scientific argument may be good or bad, and its conclusion may be true or false. But, in any case, the first step in assessing a scientific conclusion is merely to disclose the argument fully. Then, each and every piece of the argument can be inspected carefully and weighed intelligently, and every participant in the inquiry can enjoy clear communication with colleagues.

It is intellectually satisfying to be able, when need be, to present a scientific argument or conclusion with full disclosure. Also recall from Chapter 3 that the question "What goes in so that scientific conclusions can come out?" was asked by Aristotle and became the central question for the philosophy of science for two millennia. But, unfortunately, precious few scientists are trained to be able to answer that, the most elemental question that could possibly be asked about scientific inquiry.

What does it take to present a scientific conclusion with full disclosure? The basic model of scientific method presented in this book, named by the acronym the PEL model, says that presuppositions (P), evidence (E), and logic (L) combine to support scientific conclusions.

These three components interact so deeply that they must be understood and defined together. The situation is analogous to the three concepts of mothers, fathers, and children. It is easy to explain all three concepts together, but it would be impossible to give a nice explanation of mothers while saying nothing about fathers and children.

Remarkably, a simple example suffices to reveal the general structure of scientific reasoning, no matter how complex. Consider the following experiment, which you may either just imagine or else actually perform, as you prefer. Either envision or get an opaque cup, an opaque lid for the cup, and a coin. Ask someone else to flip the coin, without your observing the outcome or the subsequent setup. If the flip gives heads, place the coin in the cup and cover the cup with the lid. If the flip gives tails, hide the coin elsewhere and cover the cup with the lid. Now that the setup is completed, ask this question: "Is there a coin in the cup?"

The present assignment is to give a complete, fully disclosed argument with the conclusion that there is or is not a coin in the cup, as the case may be. This means that *all* premises needed to reach the conclusion must be stated explicitly, with nothing lacking or implicit. Before reading further in this section, you might find it quite instructive to write down your current answer to this problem for comparison with your response after studying this section.

To simplify the remaining discussion, the assumption is made that the actual state of affairs, to be discovered in due course through exemplary scientific experimentation and reasoning, is that "There is a coin in the cup." Those readers with this physical experiment before them may wish to make that so before proceeding with the assignment. Nevertheless, for purposes of the following story, we shall pretend that we do not yet know, and still need to discover, whether or not the cup contains a coin.

The question "Is there a coin in the cup?" can be expressed with scientific precision by stating its hypothesis set – the list of all possible answers. From the foregoing setup, particularly the coin flip, there are exactly two hypotheses:

H_1. There is a coin in the cup.

H_2. There is not a coin in the cup.

These two hypotheses are mutually exclusive, meaning that the truth of either implies the falsity of the other. They are also jointly exhaustive, meaning that they cover all of the possibilities. Consequently, exactly one hypothesis must be true.

How can we determine which hypothesis is true? The answer we seek is a contingent fact about the world. Thus, no armchair philosophizing can give the answer because nothing in the principles of logic or philosophy can imply that the cup does or does not contain a coin. Rather, to get an answer, we must look at the world to discover the actual state of nature. We must perform an experiment.

Various satisfactory experiments could be proposed. We could shake the cup and listen for the telltale clicking of a coin. We could take an X-ray photograph of the cup. But the easiest experiment is to lift the lid and look inside. Here, we presume a particular outcome, that we look and see a coin. That experimental outcome motivates the following argument and conclusion:

Premise. We see a coin in the cup.

Conclusion. There is a coin in the cup.

As a common-sense reply, this argument is superb, and its conclusion is certain that H_1 is true. Nevertheless, as a philosophical reply, this argument is incomplete and defective. Symbolize seeing the coin by "*S*" and the coin's existence in the cup by "*E*." Then this argument has the form "*S*; therefore *E*." It is a *non sequitur*, meaning that the conclusion does not follow from the premise. Something is missing, so let us try to complete this argument.

Another required premise is that "Seeing implies existence," or "*S* implies *E*," specifically for objects such as coins and cups. From the perspective of common sense, this is simply the presupposition that seeing is believing. In slightly greater philosophical detail, this premise incorporates several specific presuppositions, including that the physical world exists, our sense perceptions are generally reliable, human language is meaningful and adequate for discussing such matters, all humans share a common human nature with its various capabilities, and so on. The story of this premise can be told in versions as short or as long as desired.

With the addition of this second premise, the argument now runs as follows: "*S*; *S* implies *E*; therefore *E*." This is much better, following the valid argument form *modus ponens*. However, to achieve full disclosure, the logic used here must itself be disclosed by means of a third premise declaring that "*modus ponens* is a correct rule for deduction." Incidentally, to avoid a potential problem with infinite regress that philosophers have recognized for more than a century (Jeffreys 1973:198–200), note that here *modus ponens* is not being implemented in a formal system of logic but rather is merely being disclosed as a simple element in ordinary scientific reasoning.

Finally, a fourth premise is required. The "archive" is used as a technical philosophical term denoting all of a person's beliefs that are wholly irrelevant to a given inquiry. For example, given the current inquiry about a coin in a cup, my beliefs about the price of tea in China may be safely relegated to the archive. The archive serves the philosophical role, relative to a given inquiry, of providing for a complete partitioning of a person's beliefs. It also serves the necessary and practical role of dismissing irrelevant knowledge from consideration so that a finite analysis of the relevant material can yield a conclusion (whereas if one had to consider everything one knows before reaching a conclusion about the coin, no conclusion could ever be reached). For a particular scientific argument for a given person, each of that person's beliefs is one of the following: the argument's conclusion itself, or a presupposition, or an item of evidence, or a rule of logic, or an inert item in the archive.

Of course, to reside legitimately in the archive, a belief must be genuinely irrelevant and inert. Sometimes progress in science results from showing that a belief accidentally relegated to the archive is, in fact, relevant and must be exhibited as a presupposition, item of evidence, or logic rule. Anyway, a final premise is required here, saying that "the archive dismisses only irrelevant beliefs." It contains nothing with the power to unsettle or overturn the current conclusion.

Rearranging the preceding four premises in a convenient order, they can now be collected in one place to exhibit the argument entirely, with full disclosure:

Premise 1 [Presupposition].	Seeing implies existence.
Premise 2 [Evidence].	We see a coin in the cup.
Premise 3 [Logic].	*Modus ponens* is a correct rule for deduction.
Premise 4 [Archive].	The archive dismisses only irrelevant beliefs.
Conclusion.	There is a coin in the cup.

This elementary argument exemplifies full disclosure according to the PEL model. It could be called the PELA model to recognize all four inputs, including the archive, but because the archive is essentially inert, I prefer the briefer acronym PEL that focuses on just the three active components. The formula of the PEL model is that presuppositions, evidence, and logic give the conclusion. This structure of a rational argument, flushed out by this simple coin example, pervades all scientific claims of knowledge about the world, regardless of how elementary or advanced. Figure 5.1 summarizes the components of the PEL model.

With this model, the basic nature of presuppositions can be understood clearly. A presupposition is a belief that is necessary in order for any of the hypotheses to be meaningful and true but that is nondifferential regarding the credibilities of the individual hypotheses. The hypotheses originate from the question being asked that is the ultimate starting point of an inquiry, and then presuppositions emerge from comparing the hypotheses to see what they all have in common. For example, in order to declare either H_1 or H_2 to be true, it must be the case that the physical world exists and that human sense perceptions are generally reliable. But these presuppositions are completely nondifferential, making H_1 neither more nor less credible than H_2.

Presuppositions also serve another role, limiting the hypothesis set to a finite roster of sensible hypotheses. Were common-sense presuppositions ignored, the foregoing hypothesis set with only two hypotheses, H_1 and H_2, might not be jointly exhaustive. Instead, it could be expanded to include countless wild possibilities such as H_3, that "We are butterflies dreaming that we are humans looking at a cup containing a coin." But no empirical evidence could possibly discriminate among those three hypotheses, so this expanded hypothesis set would prevent science from reaching any conclusion. Numerous wild hypotheses, due to abandoning common-sense presuppositions, can undo science.

The PEL Model of Full Disclosure

Presuppositions + Evidence + Logic → Conclusions

Presuppositions are beliefs that are absolutely necessary in order for any of the hypotheses under consideration to be meaningful and true but that are completely non-differential regarding the credibilities of the individual hypotheses. Science requires several common-sense presuppositions, including that the physical world exists and that our sense perceptions are generally reliable. These presuppositions also serve to exclude wild ideas from inclusion among the sensible hypotheses under serious consideration.

Evidence is data that bear differentially on the credibilities of the hypotheses under consideration. Evidence must be admissible, being meaningful in view of the available presuppositions, and it must also be relevant, bearing differentially on the hypotheses.

Logic combines the presuppositional and evidential premises, using valid reasoning, to reach a conclusion. Science uses deductive and inductive logic.

A complete partitioning of a person's beliefs results from also recognizing an archive containing all beliefs that are irrelevant for a given inquiry, that is, beliefs that are not presuppositions or evidence or logic rules or conclusions. Irrelevant material must be ignored to avoid infinite and impossible mental processing. But the archive has no active role and hence is not indicated in the acronym for the PEL model.

Figure 5.1 Scientific conclusions emerge from three inputs: presuppositions, evidence, and logic.

Evidence has a dual nature, admissible and relevant. First, evidence is *admissible* relative to the available presuppositions. Hence, given common-sense presuppositions about the existence of the physical world and the general reliability of sense perceptions, it is admissible to cite the seeing of a coin; whereas without such presuppositions, such a claim would not be meaningful or admissible. Second, evidence is *relevant* relative to the stated hypotheses, bearing differentially on their credibilities. Hence, seeing a coin is relevant testimony because it bears powerfully on the hypotheses, making H_1 credible and H_2 incredible.

To avoid a possible embarrassment of riches, evidence can be further partitioned into two subsets: tendered evidence that is actually supplied, and reserved evidence that could be gathered or presented but is not because it would be superfluous. For example, before gathering any evidence whatsoever, the credibilities of hypotheses H_1 and H_2, that the cup does or does not contain a coin, can be represented by probabilities of 0.5 and 0.5. But, after tendering the evidence that "We see a coin in the cup," those probabilities become 1 and 0. After citing the additional evidence that "Shaking the cup causes a telltale clicking sound," those probabilities remain 1 and 0, as is still the case after also observing that "An X-ray photograph shows a coin inside the cup."

So, after initial evidence has already established a definitive conclusion, additional evidence has no further effect on the hypotheses' credibilities. At this point, wisdom directs us to close the current inquiry and move on to other pressing questions that are not yet resolved. Likewise, sometimes a conclusion will have reached a high probability of truth that may be considered adequate, even though more effort and evidence potentially could further strengthen the conclusion.

Comparing briefly, presuppositions answer the question: How can we reach any conclusion to an inquiry? But evidence answers the question: How can we assert one particular conclusion rather than another? For example, presuppositions about the existence of the physical world and the reliability of our sense perceptions are needed to reach any conclusion about a coin in the cup, whereas the evidence of seeing a coin in the cup supports the particular conclusion that there is a coin in the cup.

Logic serves to combine the premises to reach the conclusion. For example, the foregoing argument has the form "*S*; *S* implies *E*; therefore *E*," which follows the valid rule *modus ponens*. Finally, the archive serves to avoid infinite mental processing but does merit a check that its contents are truly irrelevant.

Note that the PEL model closely interlinks the concepts of presupposition, evidence, and logic. For example, half of the concept of evidence involves admissibility, which is determined by the presuppositions. Consequently, if one's concept of presuppositions is fuzzy, inexorably the concept of evidence will also be fuzzy, which will be disastrous. When presuppositions are not rightly understood, they become inordinately influential, suppressing the proper influence of evidence. Inquiry using the PEL model is depicted in Figure 5.2.

AAAS statements about the basic components of scientific thinking correspond with the PEL model proposed here. Evidence and logic are the most evident components: "The process [of scientific thinking] depends both on making careful observations of phenomena and on inventing theories for making sense out of those observations" (AAAS 1989:26; also see pp. 27–28 and AAAS 1990:16). Furthermore, the three inputs of the PEL model are brought together as the basis for scientific conclusions in the statement that "the principles of logical reasoning . . . connect evidence and assumptions with conclusions" (AAAS 1989:27), where "assumptions" here may be taken as a synonym for "presuppositions."

Finally, at most, a scientific argument may be correct; at the least, it should be fully disclosed. Full disclosure is the first and minimal requirement for clear scientific reasoning. Hence, when weighing scientific arguments and claims, it helps considerably to understand that when fully disclosed, *every* scientific conclusion emerges from exactly three inputs: presuppositions, evidence, and logic.

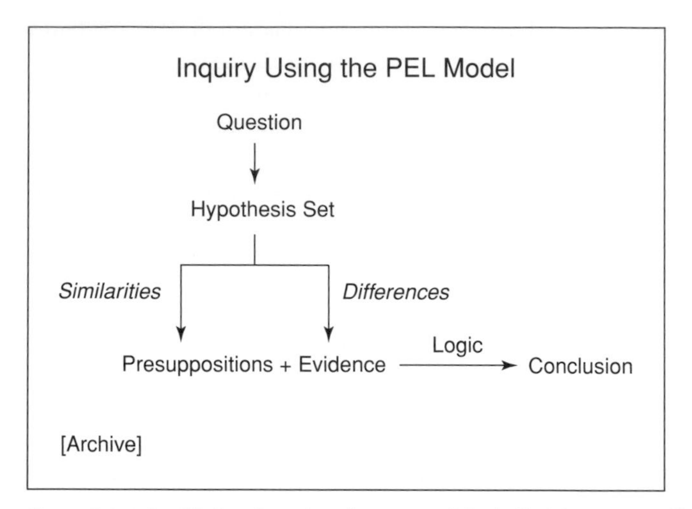

Figure 5.2 Scientific inquiry using the PEL model. Similarities among all of the hypotheses support presuppositions, whereas differences suggest potential evidence. Logic combines the presuppositions and evidence to reach the conclusion. Irrelevant knowledge is relegated to an inert archive.

Implementation of presuppositions

The method used here for implementing science's presuppositions proceeds in two steps. First, a little exemplar of common-sense knowledge about the world, called a "reality check," is selected that is as certain and universally known as is anything that could be mentioned. Second, philosophical reflection on this exemplar flushes out its presuppositions and reveals that they also suffice for scientific thinking.

The reason for choosing this particular method for implementing presuppositions is that it renders science's presuppositions as unimpeachable as our most certain knowledge. Science's presuppositions cannot be made unnecessary, but they can be made unproblematic. Also, this strategy fits historically with the thinking of many prominent scientists and mainstream philosophers. The text for the reality check, complete with its preamble, reads as follows:

> ***Reality Check***
> It is rational, true, objective, realistic, and certain that "Moving cars are hazardous to pedestrians."

To serve as a suitable object for philosophical analysis, however, it is essential that this text stand as common ground, believed by author and reader alike. So, do you believe this: that it is dangerous for pedestrians to step into the pathway of oncoming cars? I trust that this is the case. Indeed, readers who happen to have been star pupils in kindergarten will recognize this reality check as a sophisticated version of the command "Look both ways before crossing a road."

The choice, for or against accepting this reality check, is primordial and pretheoretical in that it is a common-sense conviction logically prior to all subsequent choices about the claims and methods of science. Common sense precedes science. Recall Reid's sentiment that the principles of common sense are older and of more authority than philosophy. Likewise, Wittgenstein insisted that rudimentary common-sense beliefs are oblivious to evidence because no evidence is more certain than such beliefs themselves: "my not having been on the moon is as sure a thing for me as any grounds I could give for it" (Ludwig Wittgenstein, in Anscombe and von Wright 1969:17e).

An appeal to common sense might be discredited or dismissed by a quick remark such as "Common sense isn't so common." Surely, this means that people sometimes spend more money than they earn, neglect the upkeep that could prevent costly repairs, and so on. But, clearly, my chosen exemplar of common sense is not gathered from the glorious heights of common sense, with offerings such as "Spend less than you earn" or "A stitch in time saves nine." Rather, "Moving cars are hazardous to pedestrians" is an exemplar of rudimentary common sense. So my appeal to rudimentary common sense should not be misinterpreted or dismissed as the unrealistic assumption that everyone is a paragon of good sense. Rather, it should be interpreted and taken seriously as the claim that all normal humans living on this one earth know some basics about physical reality.

In this academic book on scientific method, why fight for a meager scrap of common sense, that "Moving cars are hazardous to pedestrians"? What is the dreaded contrary of common sense? What is the threat?

Presumably, common sense's opponent is skepticism. But sincere skepticism is extremely rare. I, for one, have never met a single person who doubted, in any sense that could be taken as sincere, that "Moving cars are hazardous to pedestrians." Nor is it easy to imagine that such a person could survive apart from institutional care. Rather, the real opponent of common sense and science is ambivalent skepticism, which is common, just as is any other kind of inconsistency or insincerity.

The skeptical tradition, from start to finish, has been characteristically ambivalent. The founding figure of ancient Greek skepticism, Pyrrho of Elis, claimed not to trust his senses and once essayed to walk over a cliff, as if it could not matter. That sounds like gratifying, serious skepticism! But Pyrrho did that in the presence of his disciples, who kept their master from harm, and he lived to the ripe old age of ninety. He traveled to India with Alexander the Great.

The attempted coherence of the Pyrrhonistic skeptics is quite charming. Against the dogmatic Academic skeptics such as Sextus Empiricus, who claimed to show that knowledge was impossible, the Pyrrhonists claimed that they did not even know that they could not know. They were skeptical about whether or not they were skeptics! Much later, David Hume would continue that tradition of ambivalence, saying that the dictates of common sense must regulate ordinary daily life, even though they are not philosophically respectable.

A more recent example of that ambivalent tradition is Sir Karl Popper. On the one hand, some passages by Popper are reassuring to a common-sensical reader. He wrote of his "love" of common sense and said that "I am a great admirer of common sense" (Popper 1974:43, 1979:viii). Likewise, in his autobiography, in Schilpp (1974:71), Popper said that common-sense knowledge, such as "that the cat was on the mat; that Julius Caesar had been assassinated; that grass was green," is all "incredibly uninteresting" for his work because he focuses instead on genuinely "problematic knowledge" involving difficult scientific discoveries.

On the other hand, Popper wrote elsewhere that "The statement, 'Here is a glass of water' cannot be verified by any observational evidence" because of philosophical problems with induction and related matters that grip everyone, you and me included (Popper 1968:95). Now to say that you cannot know that "Here is a glass of water" is as plainly spoken a denial of common sense as to say that you cannot know my reality check that "Moving cars are hazardous to pedestrians." Clearly, common sense is under attack. But this attack is not consistent or sustained – nor could it be. Although Popper (1945:283) waxed eloquent about "the standards of intellectual honesty, a respect for truth, and . . . modest intellectual virtues," regrettably the force of such fine rhetoric is undercut by his saying elsewhere that trifling truths like "Here is a glass of water" are beyond a human's reach.

A discerning reader may detect the deep irony in Popper's declaration that *we* cannot know that "Here is a glass of water." His assumption that this incapacity afflicts all humans, rather than just him, requires knowing that all humans share similar endowments because of our common human nature. But with induction bankrupt according to Popper, how could he know this? Why is "Here is a glass of water" beyond reach, whereas "All humans share a common human nature" is within reach? The big "we" word is precisely what mainstream epistemology *is* entitled to, but skeptical epistemology *is not* entitled to.

The remedy for a disappointing, insincere skepticism is the sincere and cheerful acceptance of just one little scrap of common sense, such as that "Moving cars are hazardous to pedestrians." Reason must be restored to its double office of regulating both beliefs and actions. The insincere skeptic has a mouth whose words say that "Maybe cars are hazardous to pedestrians, and maybe not," but feet whose actions always say that "Moving cars are hazardous to pedestrians."

Like the declaration "I love you," the reality check can be voiced with varying degrees of conviction. Because of the frequent problems with superficial or ambivalent skepticism, the degree of conviction intended here must be made clear. Exactly which plaudits attend this reality check? To voice the reality check with clear conviction and to connect it with science's ambitions announced in Chapter 2, the reality check is proclaimed here with a preamble listing science's four bold claims: rationality, truth, objectivity, and realism. Furthermore, because this particular item of common-sense knowledge is so easy for all persons to learn and is absolutely exempt from sincere controversy, it is proclaimed here with one additional plaudit: certainty. It is voiced quite cheerfully with absolute confidence and unlimited boldness. It is voiced with no ambivalence, no superficiality, and no insincerity.

Given a nonnegotiable conviction that the reality check is true, what does this conviction imply for intellectual attacks on science's realism? It has a decisive implication, namely, any attack on science that also takes down common sense is simply incredible. Any attack that also targets common sense, including denying that we can know the reality check, fails because it does too much and thereby it loses credibility. Consequently, a legitimate attack on science's rationality must target science alone, not science and common sense both, as was emphasized in Chapter 4.

Having selected a little exemplar of knowledge about the world that is shared by author and reader alike, that "Moving cars are hazardous to pedestrians," the second and final step is philosophical reflection to disclose the presuppositions that are necessary for this reality check to be meaningful and true. "The logical premisses of factuality are not known to us or believed by us *before* we start establishing facts, but are recognized on the contrary *by reflecting on the way we establish facts*" (Polanyi 1962:162).

The presuppositions underlying the reality check can be organized in three broad groups: ontological, epistemological, and logical presuppositions. First, the ontological or metaphysical presuppositions include that physical reality has multiple things that are not all the same, such as cars that differ from pedestrians, or moving cars that differ from stationary cars. Because the universe is not merely one undifferentiated blob of being, there exists something to be comprehended. It also presumes that reality has natural kinds, to use the philosophical term. This means that multiple objects can be of the same kind (at a given level of description), such as numerous cats each being a cat. Human artifacts can also be of a given kind, such as numerous cars each being a car. One particularly important natural kind, from our perspective anyway, is the human being. The pedestrians or humans mentioned in the reality check share numerous properties, such as being soft and therefore vulnerable to strong impact from a large and hard object. It is not the case that car accidents are hazardous for some humans, whereas others are invincible. The reality check also presumes that physical reality is predictable. Its implicit advice not to step in front of a rapidly moving car obviously agrees with past experience regarding car accidents. But, equally, this advice is predictive, directed at preventing more accidents in the future.

Second, the epistemological presuppositions include that a human can know that an object is a rapidly approaching car and can act to move out of harm's way. This presupposes that our eyes, ears, and other sensory organs provide generally reliable information about the external world and that our brains can process and comprehend these sensory inputs. Furthermore, our brains can also direct our feet to move purposefully. Merely knowing without acting would not promote survival, so reason's double office of regulating belief and guiding action is evident. Another epistemological presupposition is that human language is meaningful. The reality check is expressed by several words in English. Humans have abilities of language and communication.

Third and finally, the logical presuppositions include coherence. To assert that "Moving cars are hazardous to pedestrians" legitimately, coherence demands that we not also assert the negation that "Moving cars are not hazardous to pedestrians." There is no credit for asserting the reality check if its opposite is also asserted. The reality check also presumes deductive and inductive logic. Logic is required to take a general principle and apply it to specific episodes of being near moving cars. Deduction is active in handling probability concepts, such as the idea of something being hazardous, meaning that harm is likely even if not certain. Induction is active in recognizing objects and in learning and using language.

The foregoing account of the reality check's presuppositions is not exhaustive. But, for most purposes, any further analysis would become technical and tedious. Furthermore, the presuppositions flushed out by analyzing this one representative little scrap of common sense, the reality check, pervade common sense. That is, philosophical analysis of "This cup contains a coin" or "Here is a glass of water" would evince the same presuppositions as this analysis of "Moving cars are hazardous to pedestrians." For any of these statements to be meaningful and true, the general makeup of ourselves and our world cannot follow just any conceivable story, but rather the world must be along the lines indicated by common-sense presuppositions. The following statement offers a concise expression of science's basic presuppositions:

(Mainstream) science's basic presuppositions

The physical world is real and orderly and we humans find it substantially comprehensible.

The presuppositions pervading science cannot be less than those encountered in one little scrap of common sense. "Although through our [scientific] theories, and the instrument-aided observations they lead to, we can go beyond and correct some of the pretheoretical picture of the world we have by virtue of our being human, there is always going to be a sense in which all our knowledge and theory is based on elements in that [common-sense] picture.... More theoretical knowledge of the world is always going to have some connection, however remote, with the humdrum level if it is to count as science fact rather than science fiction" (O'Hear 1989:95–96).

On the other hand, some have argued that the presuppositions of science must be more than those of common sense, or else at least partially different. They claim that these extra presuppositions provide science with esoteric content that is wholly absent from common-sense knowledge and reasoning. For instance, science has overturned common sense with certain surprises: that the objects around us are mostly empty space and that the rate at which time passes by is not constant but depends on an object's speed relative to the observer. However, those surprises are conclusions of science, not presuppositions. Indeed, those discoveries were eligible to become conclusions precisely because they never were presuppositions. Those surprises were established by empirical evidence that counted as evidence precisely because common-sense presuppositions were in effect. Science can overturn common-sense expectations and beliefs, but not common-sense presuppositions.

Likewise, some have argued for additional presuppositions drawn from a particular worldview to really explain why the world is as it is. However, the position taken here follows the mainstream scientific tradition of seven centuries, which began with Albertus Magnus, of distinguishing penultimate and ultimate accounts of science's presuppositions. A penultimate account must be included in science's own business because these presuppositions are necessary to reach any conclusions. By contrast, pursuit of an ultimate account obliges science to enlist support from the humanities, so it seems wrongheaded to expect such an account from only science itself. Whereas there are many worldviews, there is only one common sense shared by all persons, including the ubiquitous belief that "Moving cars are hazardous to pedestrians." Therefore, invoking science's presuppositions by a penultimate appeal to common sense preserves science's objective and public character.

In conclusion, if you believe the reality check, that "Moving cars are hazardous to pedestrians," then you have already adopted all of the presuppositions needed for science to flourish. You have already delivered science from the specter of skepticism. Compared with common sense, science requires more experimentation, data, reasoning, and work but absolutely nothing more by way of presuppositions. Building science on a base of common sense is a plausible and respected tradition (Nash 1963:3–62). According to Einstein, "The whole of science is nothing more than a refinement of everyday thinking" (Einstein 1954:290).

Science's worldview forum

If this book were claiming to address proponents of *every* worldview, including skepticism, then it would not be fair or correct to pretend that the preceding reality check is shared knowledge. Skepticism and the reality check are incompatible. If skepticism is true, then the reality check with its grand preamble is definitely unwarranted; but if the reality check is true, then skepticism is false. So something must go. My choice is to hold on to the reality check, accepting the consequence that this limits the range of worldviews engaged here.

This book's project is to presuppose common sense and then build scientific method, not to refute the skeptic and thereby establish common sense. The skeptic is unanswered here, not because of ill will on my part, but simply because I do not know what a skeptic wants from a nonskeptic or realist such as myself.

Fortunately, real skeptics are quite rare. My university work, including professional meetings of scientific and philosophical societies, causes me to meet many people from many nations. However, I have not yet personally met one real skeptic. Or, to be a little more accurate, I have met some people with skeptics' mouths but have not encountered any skeptics' feet. Their mouths may counter the reality check, saying it is uncertain; but their feet obey it with all diligence, as their survival attests. "Skeptics are like dragons. You never actually meet one, but keep on running across heroes who have just fought with them, and won" (Palmer 1985:14).

Realists may be perplexed that skepticism tends to be such an extreme position as to reject even the simple reality check. Recall, for instance, that Popper (1968:95) judged that even the simple common-sense belief that "Here is a glass of water" lies outside the bounds of human competence. But that extremism has a logical explanation. Imagine that you and I are enjoying lunch and beer at a pub. Suddenly, I am struck with remorse that I have never experienced being a skeptic and forthwith give it my best attempt. After struggling manfully for an hour, I proudly exclaim, "I've got it; I doubt that this salt shaker exists! Everything else still exists, but this salt shaker is gone – clean gone!" Understanding that I have lived for decades without the slightest inclination toward skepticism, doubtless your charity will move you to praise my fledgling skepticism. Nevertheless, you might be sorely tempted to say something like, "Well, let me move this pepper shaker that you can see right next to the salt shaker that you cannot see. Now can you see them both?" The same embarrassment would attend any other modest version of skepticism, such as doubting my beer but not yours, or doubting one chair in the pub but not anything else. Only the radical doubt of everything leaves no easy refutation close at hand.

The dismissal of skepticism has implications for worldviews. A worldview is a person's beliefs about the basic makeup of the world and life. Depending on a person's intellectual maturity, a worldview may be more or less explicit, articulate, and coherent. But everyone has a worldview. It supplies answers to life's big questions, such as: What exists? What can we know? What is good, true, and beautiful? What is the purpose of human life? What happens after death? Many worldviews are rooted in a religion or a philosophical position, but some persons hold views not affiliated with any widespread movement. There are many minor worldviews but relatively few major ones. The world's population, currently more than 7 billion persons, is approximately 32% Christians,

19% Muslims, 19% atheists, 14% Hindus, 9% tribal or animist religions, 6% Buddhists, and 1% other, which includes 0.3% Jews.

The most significant question that can be asked about worldviews is: Which one is true? However, that is not this book's question. Rather, this book is about scientific method, so its question is: How much does worldview pluralism affect science's claims and fortunes? It is simply a fact of life that historically there have been diverse views and that worldview pluralism is likely to continue for the foreseeable future. Is this a problem for science, or not? Does worldview diversity present insurmountable problems, motivating separate versions of science for each worldview, or even rendering science invalid for adherents of some worldviews? Or can science, preferably in one single version, work for essentially everyone, including atheists, Hindus, Buddhists, Jews, Christians, Muslims, and others?

The answer offered here is that a single version of science works fine for nearly every worldview, but not quite all. Science works in all worldviews that cheerfully assert the reality check, but it fails in all those that reject the reality check. Accordingly, science's worldview forum is comprised of all worldviews that assert the reality check.

Science could not be objective and public if science needed to depend on controversial philosophies or specific cultures. Fortunately, underneath these philosophical and cultural differences, there exists on this one earth a single human species with a shared human nature, and that commonality provides adequate resources for science's common-sense presuppositions. "Cultures may appear to differ, but they are all rooted in the same soil. . . . Human nature precedes culture and explains many of its features" (Roger Trigg, in Brown 1984:97).

Justification of knowledge claims

Besides the reality check itself, additional beliefs have also been deemed equally certain here, including the experimental result that "There is a coin in the cup." Likewise, Wittgenstein judged "the existence of the apparatus before my eyes" to have the same certainty as "my never having been on the moon" (Ludwig Wittgenstein, in Anscombe and von Wright 1969:43e). Such thinking – that various beliefs are equally certain – is intuitively appealing and undoubtedly right. This section formalizes that intuition.

Precisely what philosophical reasoning can be given for deeming various beliefs to be certain? How does the reality check's assumed certainty extend to other beliefs' demonstrated certainty? To make these questions more concrete, consider the belief that "There are elephants in Africa." What formal, philosophical account could be given for judging this belief to be certain? Prior to reading the rest of this section, I recommend that the reader try on his or her

A Model of Justification

Reality Check. Adopt as realistic, true, and certain the reality check's belief, denoted by A, that "Moving cars are hazardous to pedestrians."

Certainty Equivalence. Demonstrate that the belief, denoted by B, has certainty equivalence with the beliefs in the reality check, denoted by A.

This demonstration requires:

(1) the same presuppositions,
(2) equally admissible, relevant, and weighty evidence,
(3) equally valid or correct logic, and
(4) an equally inert and dismissible archive.

Rule for Justification. From the above beliefs, that A is certain and that A has certainty equivalence with B, infer that B is certain.

Figure 5.3 A model for justifying scientific beliefs based on the reality check, certainty equivalence, and rule for justification.

own to construct a philosophically rigorous proof that "There are elephants in Africa" is true and certain. The method of justification offered here has three steps, as depicted in Figure 5.3.

The first step merely reasserts the reality check's claim of certainty, denoted by belief *A*. The second step draws upon the PEL model to demonstrate that another belief *B* has the same certainty as belief *A*. This demonstration requires (1) the same presuppositions; (2) equally admissible, relevant, and weighty evidence; (3) equally valid or correct logic; and (4) an equally inert and dismissible archive. Then, the third and final step is a rule for justification that infers from the above premises, that *A* is certain and that *A* has certainty equivalence with *B*, that the conclusion *B* is certain.

For example, this model of justification can be applied to Africa's elephants as follows. Denote "Moving cars are hazardous to pedestrians" by *A*, and "There are elephants in Africa" by *B*. First, the reality check is asserted to be certain: *A* is certain. Second, the same presuppositions are necessary and sufficient to believe *A* or *B*; recent sightings or photographs of elephants in Africa are as admissible, relevant, and weighty as any evidence from sightings or photographs of car accidents with pedestrians that could be adduced for the reality check; equally valid or correct logic works in both cases; and both cases generate equally inert and dismissible archives. Thus, *A* has certainty equivalence with *B*. Third and finally, applying the rule for justification to these two premises produces the conclusion, "It is certain that 'There are elephants in Africa.'" The only way to unsettle the conviction that “There are elephants in Africa” would be to embrace such profound skepticism as would also unsettle the reality check that “Moving cars are hazardous to pedestrians.”

This model of justification explains the simple case of deeming other beliefs to be certain. However, much knowledge is probabilistic, such as a forecast that rain is likely tomorrow. This model is easily extended to justify probabilistic as well as certain conclusions, but probabilistic reasoning is better left to Chapters 8 and 9 on probability and statistics. The crucial move for justifying a probabilistic conclusion is to accept common-sense presuppositions so that the conclusion faces only the ordinary and workable challenge of imperfect evidence but not the insurmountable and debilitating challenge of skeptical presuppositions.

Review of functions

The role of presuppositions in scientific thinking is difficult to fully grasp because presuppositions serve so many functions. This chapter reveals at least these seven functions: Presuppositions are essential for reaching any scientific conclusions, for full disclosure of arguments, for rendering evidence admissible, for interconnecting science and common sense, for defending science’s rationality, for framing sensible questions that eliminate wild hypotheses, and for demarcating science’s worldview forum.

Yet the most vital function of presuppositions is to specify science’s referents, that is, what science is referring to or talking about. Consider the following thought experiment. Imagine that the contemporaries Berkeley, Hume, and Reid were brought together and were all patting a single horse. All three would report the same experience of a big furry animal, but their interpretations of that experience would differ. Berkeley would say that the physical horse does not exist but only the mind’s idea of a horse. Hume would say that science should concern our experience of the horse but would not say that the horse does or does not exist. Reid would say that philosophy and science should follow common sense with a confident and cheerful certainty that the physical horse does exist. Likewise, imagine stepping further back in the history of this debate and seeing the contemporaries Plato and Aristotle patting a single dog. For Plato, the dog would be but an illusory and fleeting shadow of its inaccessible but thoroughly real Form. But, for Aristotle, the dog itself would be accessible to our sensory experience and would be completely real. Clearly grasp that this perennial debate is not about the sensory data as such but rather is about the metaphysical interpretation of that data. This debate is altogether about philosophical presuppositions concerning what is real and knowable and is altogether not about scientific evidence.

Mainstream science follows common sense in presupposing that the physical world is real and orderly and we humans find it substantially comprehensible.

And mainstream science follows mainstream philosophy in granting reason the double office of regulating belief and action, thereby fostering sincerity and confidence.

Summary

By perceiving science's presuppositions, scientists can understand their discipline in greater depth and offer scientific arguments with full disclosure. An inquiry's presuppositions are those beliefs held in common by all of the hypotheses in the inquiry's hypothesis set. Analysis of a single scrap of common sense, such as the reality check that "Moving cars are hazardous to pedestrians," suffices to flush out the presuppositions that pervade common sense and science alike. Mainstream science's basic presuppositions are that the physical world is real and orderly and we humans find it substantially comprehensible.

Aristotle clearly accepted science's ordinary, common-sense presuppositions, but his deductivist vision worked better for mathematical sciences than for natural sciences. Albertus Magnus resolved that deficiency by conditional or suppositional reasoning, granting the natural sciences definitive empirical evidence on the supposition of common-sense presuppositions. That device also granted science substantial independence from philosophy and theology, a view subsequently endorsed by Aquinas, Duns Scotus, Buridan, and others. A tremendous diversity of views on science's presuppositions and claims emerged from the work of Francis Bacon, Berkeley, Hume, and Reid. The AAAS affirms science's common-sense presuppositions.

The PEL model of full disclosure shows that scientific method amounts to disclosing and securing the presuppositions, evidence, and logic needed to support scientific conclusions. Presuppositions are disclosed and legitimated by a procedure with two steps. First, a reality check is adopted by faith and with sincerity and confidence, that "Moving cars are hazardous to pedestrians." Second, philosophical reflection on this reality check reveals its ontological, epistemological, and logical presuppositions. The most obvious presuppositions are that the physical world exists and that our sense perceptions are generally reliable.

Insistence on a nonnegotiable reality check causes science's worldview forum to include all worldviews accepting this reality check, but to dismiss radical skepticism. This book's project is to presuppose common sense and then build scientific method, not to refute the skeptic and thereby establish common sense.

Knowledge claims are justified by asserting the reality check's certainty and then, in light of the PEL model, demonstrating that other beliefs have (1) the same presuppositions; (2) equally admissible, relevant, and weighty evidence; and (3) equally valid or correct logic (and an equally inert and dismissible archive). An example was given of justifying the common-sense belief that there are elephants in Africa. This model of justification for certain conclusions is readily extended to probable conclusions by adding probability theory and statistical analysis.

Presuppositions serve many functions in scientific thinking, being absolutely indispensable for reaching any conclusions whatsoever. The most vital function of presuppositions is to specify what science is talking about, namely the real, orderly, and comprehensible world engaged by mainstream science.

Study questions

(1) What are science's presuppositions? The text argues that science needs just common-sense presuppositions, nothing less and nothing more. What would be your best arguments for a different position, contrary to the text?

(2) The text implements science's presuppositions by a two-step procedure: adoption of a common-sense reality check, followed by philosophical reflection on its content. But often there are many ways to get a job done. Can you suggest an alternative implementation? What advantages or disadvantages does that alternative have over the recommended implementation?

(3) Presuppositions have the crucial role of supplying a necessary input for reaching any conclusions whatsoever. What other roles do presuppositions have?

(4) Explain the distinction between an ultimate and penultimate account of science's presuppositions. The text claims that a penultimate account, operating by an appeal to a worldview-independent and shared common sense, is the proper and sufficient business of science itself; whereas an ultimate account requires additional resources from the humanities, and hence is outside the purview of a book or course on scientific method. Do you agree or disagree? How would you argue for your position?

(5) Consider the scientific conclusion: The sun's mass is about 2×10^{30} kilograms. How would you apply the model of justification to this conclusion?

6

Science's powers and limits

What are science's powers and limits? That is, where is the boundary between what science is and is not able to discover? The American Association for the Advancement of Science has identified that issue as a critical component of science literacy: "Being liberally educated requires an awareness not only of the power of scientific knowledge but also of its limitations," and learning science's limits "should be a goal in all science courses" (AAAS 1990:20–21). The National Research Council concurs: "Students should develop an understanding of what science is, what science is not, what science can and cannot do, and how science contributes to culture" (NRC 1996:21).

People's motivations for exploring the limits of science can easily be misconstrued, so they should be made clear from the outset. Unfortunately, for some authors writing about science's limits, the motivation has been to exaggerate the limitations in order to cut science down, support antiscientific sentiments, or make more room for philosophy or religion. For others, the motivation has been to downplay science's limitations in order to enthrone science as the one and only source of real knowledge and truth. Neither of those excesses represents my intentions. I do not intend to fabricate specious problems to shrink science's domain, nor do I intend to ignore actual limitations to aggrandize science's claims. Rather, the motivation here is to characterize the actual boundary between what science can do and cannot do. One of the principal determinants of that boundary is the topic of this book, the scientific method.

Rather obvious limitations

Several limitations of science are rather obvious and hence are not controversial. The most obvious limitation is that scientists will never observe, know, and explain everything about even science's own domain, the physical world. The Heisenberg uncertainty principle, Gödel's theorem, and chaos theory set fundamental limits.

Besides these fundamental limits, there are also practical and financial limits. "Today, the costs of doing scientific work are met by public and corporate funds. Often, major areas of scientific endeavor are determined by the mission-oriented goals of government, industry, and the corporations that provide funds, which differ from the goals of science" (AAAS 1990:21).

The most striking limitation of science, already discussed in Chapter 5, is that science cannot prove its presuppositions. Nor can science appeal to philosophy to do this job on its behalf. Rather, science's presuppositions of a real and comprehensible world – as well as philosophy's presuppositions of the same – are legitimated by an appeal to rudimentary common sense followed by philosophical reflection.

However, the remainder of this chapter explores the powers and limits of science that are not especially obvious. Science's capacity to address big worldview questions is important but controversial. And an integrally related matter is the role of the humanities and the influence of individual experience on worldview convictions. A neglected topic meriting attention is science's power to enhance personal character and experiences of life.

The sciences and worldviews

Can science reach farther than its ordinary investigations of galaxies, flowers, bacteria, electrons, and such? Can science also tackle life's big questions, such as whether God exists and whether the universe is purposeful? This is the most complex – and perhaps the most significant – aspect of the boundary between science's powers and limits.

Life's grand questions could be termed religious or philosophical or worldview questions. But a single principal term is convenient and the rather broad term *worldview* is chosen here. A worldview sums up a person's basic beliefs about the world and life. The following account draws heavily from Gauch (2009a, 2009b).

Whether worldview implications are part of science's legitimate business is controversial. Nevertheless, the mainstream view, as represented by the AAAS, is that one of science's important ambitions is contributing to a meaningful worldview. "Science is one of the liberal arts" and "the ultimate goal of liberal education" is the "lifelong quest for knowledge of self and nature," including the quest "to seek meaning in life" and to achieve a "unity of knowledge" (AAAS 1990:xi, 12, 21). AAAS position papers offer numerous, mostly helpful perspectives on religion, God, the Bible, clergy, prayer, and miracles. The Dialogue on Science, Ethics, and Religion (DoSER) program of the AAAS offers ongoing events and publications.

The AAAS regards science's influence on worldviews not only as a desirable quest but also a historical reality. "The knowledge it [science] generates sometimes forces us to change—even discard—beliefs we have long held about ourselves and our significance in the grand scheme of things. The revolutions that we associate with Newton, Darwin, and Lyell have had as much to do with our sense of humanity as they do with our knowledge of the earth and its inhabitants.... Becoming aware of the impact of scientific and technological developments on human beliefs and feelings should be part of everyone's science education" (AAAS 1989:134). Likewise, "Scientific ideas not only influence the nature of scientific research, but also influence—and are influenced by—the wider world of ideas as well. For example, the scientific ideas of Copernicus, Newton, and Darwin... both altered the direction of scientific inquiry and influenced religious, philosophical, and social thought" (AAAS 1990:24).

But, unfortunately, on the specific worldview question of life's purposes, AAAS position papers are inconsistent. On the one hand, they say that science *does not* answer the big question about purposes: "There are many matters that cannot usefully be examined in a scientific way. There are, for instance, beliefs that—by their very nature—cannot be proved or disproved (such as the existence of supernatural powers and beings, or the true purposes of life)" (AAAS 1989:26). On the other hand, it is most perplexing that another AAAS position paper claims that science *does* answer this question: "There can be no understanding of science without understanding change and the fact that we live in a directional, though not teleological, universe" (AAAS 1990:xiii; also see p. 24). Now "teleological" just means "purposeful," so here the AAAS is boldly declaring, without any argumentation or evidence, that we live in a purposeless universe. Consequently, this is one of those rare instances in which AAAS statements have not provided reliable guidance because they are contradictory.

Science's powers and limits as regards ambitious worldview inquiries depend not only on science's method but also on social conventions that define science's boundaries and interests. A social convention prevalent in contemporary science, *methodological naturalism*, limits science's interests and explanations to natural things and events, not supernatural entities such as God or angels. Methodological naturalism has roots in antiquity with Thales (*c.* 624–546 BC) and others. Subsequently, medieval scholars emphasized pushing their understanding of natural causes to its limits (Lindberg 2007:240–241; Ronald L. Numbers, in Lindberg and Numbers 2003:265–285). But the name "methodological naturalism" is of recent origin, only three decades ago.

Methodological naturalism contrasts with metaphysical or *ontological naturalism* that asserts natural entities exist but nothing is supernatural, as claimed by atheists. Hence, methodological naturalism does not deny that the supernatural exists but rather stipulates that it is outside science's purview. Unfortunately, methodological naturalism is sometimes confused with ontological naturalism. To insist that science obeys methodological naturalism *and* that science supports atheism is to get high marks for enthusiasm but low marks for logic.

Many worldview matters might seem to reside within science's limits, rather than its powers, given that methodological naturalism excludes the supernatural. Indeed, questions such as whether God exists and whether the universe is purposeful, which inherently involve the supernatural, are precisely the kinds of questions that are foremost in worldview inquiries.

However, to be realistic, contemporary science is replete with vigorous discussions of worldview matters. For starters, consider the exceptional science books that manage to become bestsellers. The great majority of them are extremely popular *precisely because* they have tremendous worldview import, such as Collins (2006) and Dawkins (2006). Less popular but more academic books also concern science and worldviews, such as Ecklund (2010).

Furthermore, interest in science's worldview import is a minor but consistent element in mainstream science journals. For instance, religious experience provides one of the standard arguments for theism, but in *American Scientist*, psychologist Jesse Bering (2006) attempted to explain away belief in a deity or an afterlife as a spurious evolutionary by-product of our useful abilities to reason about the minds of others. Likewise, Michael Shermer, the editor of *Skeptic*, has a monthly column in *Scientific American* with provocative items such as "God's number is up" (Shermer 2004). Also, survey results on the religious convictions of scientists were published in *Science* (Easterbrook 1997), and significant commentary on science and religion was provided in *Nature* (Turner 2010; Grayling 2011; Waldrop 2011). To gauge the extent of worldview interests in mainstream science, an interesting little exercise is to visit the websites of journals such as *Nature* and *Science* and search for "religion" to see how many thousands of hits result.

Hence, contemporary scientific practice is far from a consistent and convincing implementation of methodological naturalism. Nor is the present scene uncharacteristic, given the broad interests of natural philosophers (now known as "scientists" since around 1850) in ancient, medieval, and modern times. Of course, methodological naturalism is characteristic of routine scientific investigations, such as sequencing the genome of the virus that causes the common cold, but that does not necessarily mean that it extends to every last scientific interest or publication.

Whereas mainstream science can and does have some worldview import, prominent variants of fringe science are problematic, particularly scientism and skepticism. They are opposite errors. At the one extreme, scientism says that only hard, no-nonsense science produces all of our dependable, solid truth. At the opposite extreme, skepticism says that science produces no final, settled truth.

Yet, curiously, these opposite errors support exactly the same verdict on any worldview inquiry appealing to empirical and public evidence. On the one hand, scientism automatically and breezily dismisses any worldview arguments coming from philosophy, theology, or any other discipline in the humanities because such disciplines lack the validity and authority that science alone possesses. On the other hand, after skepticism has already judged all science to be awash in uncertainty and tentativeness, ambitious worldview inquiries are bound to receive this same verdict of impotence.

Returning to mainstream science, some scientists explore science's worldview import, other scientists exclude worldview issues in the name of methodological naturalism, and still other scientists have no interests or opinions on such matters whatsoever. This diversity of interests and temperaments hardly seems surprising.

Empirical method in the humanities

This whole book is about scientific method, but this one section is about a broader topic that may be termed *empirical method*, which subsumes scientific method as a special case. Empirical method concerns what can be known by means of empirical and public evidence, regardless of whether that evidence comes from the sciences or the humanities. Any persons interested in pushing empirical and public evidence to its limits must understand the structure and workings of empirical method, not merely scientific method.

The humanities are academic disciplines that study the human condition. They include the classics, languages, literature, history, law, philosophy, religion or theology, and the visual and performing arts. The humanities use a great variety of methods, including some use of empirical and public evidence.

The essence of scientific method is to appeal to empirical and public evidence to gain knowledge of great theoretical and practical value about the physical world. In greater detail than that single sentence can capture, this book's account of scientific method features the PEL model of full disclosure and the justification of truth claims based on that model, as summarized in Figures 5.1 and 5.3 – although this whole book is needed for a reasonably complete account of scientific method. But, clearly, empirical and public evidence also has roles in the humanities. Especially when empirical evidence is used in ambitious worldview inquiries, as contrasted with routine scientific or technological investigations, the combined perspectives of the sciences and the humanities yield the most reliable and beneficial results.

This section's extremely brief account of empirical method is relevant in this book on scientific method for at least three reasons. First, understanding how public evidence and standard reasoning support truth claims in multiple contexts across the sciences and the humanities gives students their best chance of deeply understanding rationality within science itself. Comparing and contrasting stimulates real comprehension. Second, the AAAS (1990) vision of science as a liberal art calls for a humanities-rich understanding of science, which is promoted greatly by grasping the empirical method that spans the sciences and the humanities. Third, the scientism that is decisively renounced by mainstream science, but still finds frequent expression especially at the popular level, is best discredited by conscious awareness of projects in the humanities that also appeal to empirical evidence.

All of the disciplines in the humanities contribute to a meaningful worldview. But among these many academic disciplines, natural theology is a prominent example of using empirical method to address worldview questions by means of public evidence.

The article on natural theology by MacDonald (1998) in the *Routledge Encyclopedia of Philosophy* characterizes this discipline. "Natural theology aims at establishing truths or acquiring knowledge about God (or divine matters generally) using only our natural cognitive resources." He further explained that "The phrase 'our natural cognitive resources' identifies both the methods and data for natural theology: it relies on standard techniques of reasoning and facts or truths in principle available to all human beings just in virtue of their possessing reason and sense perception." Natural theology considers arguments both for and against theism, with proponents of diverse perspectives sharing a common impartial methodology.

The implicit contrast is with revealed theology, which instead relies on a revelation or scripture taken as authoritative or inspired within a given faith community. However, a scripture may have some contents and aspects that are verifiable independently with public evidence, so the relationship between natural and revealed theology is one of partial overlap.

Natural theology may be completely unknown to many students and professionals in the sciences. But this unfamiliarity does not negate the existence of this vigorous academic discipline, nor negate natural theology's character as a discipline that relies on empirical and public evidence. Two resources on natural theology may be mentioned for those who are interested. *The Blackwell Companion to Natural Theology* provides a recent and scholarly overview of natural theology (Craig and Moreland 2009). Its chapters review the ontological, cosmological, and moral arguments and the arguments from evil, consciousness, reason, religious experience, and miracles. The ongoing Gifford lectures on natural theology – endowed by Lord Gifford more than a century ago in Scotland's four ancient universities – are frequently published in readily available books.

Gifford lectures by eminent scientists, theologians, philosophers, and other scholars engage an astonishing and intriguing diversity of arguments and evidence. These renowned lectures on natural theology have included scientists Simon Conway Morris, Richard Dawkins, Freeman Dyson, Sir John Eccles, Sir Arthur Eddington, Werner Heisenberg, Michael Polanyi, Martin Rees, and Carl Sagan; theologians Karl Barth, Rudolf Bultmann, Stanley Hauerwas, Jurgen Moltmann, Reinhold Niebuhr, Albert Schweitzer, and Paul Tillich; scientist-theologians Ian Barbour, Stanley Jaki, and Sir John Polkinghorne; philosophers

Marilyn Adams, Sir Alfred Ayer, John Dewey, Antony Flew, Étienne Gilson, Alasdair MacIntyre, Mary Midgley, Alvin Plantinga, Paul Ricoeur, Eleonore Stump, Richard Swinburne, and Alfred Whitehead; and scholars Noam Chomsky, Frederick Copleston, Jaroslav Pelikan, and Arnold Toynbee.

The literature on natural theology – both historical and contemporary – is largely from prestigious academic publishers, and it is simply enormous. Evaluation of natural theology's empirical evidence is outside the scope of this book and it requires considerable effort. By stark contrast, evaluation of natural theology's empirical method is within the scope of this book on scientific method for the three reasons indicated near the beginning of this section, and this evaluation is easy work. It requires merely one longish paragraph, as follows.

The PEL model, which applies to all disciplines and inquiries using empirical and public evidence to support truth claims, specifies three requirements. (1) *Appropriate Presuppositions.* MacDonald's definition of natural theology does not mention presuppositions explicitly, but the context makes two things abundantly clear. On the one hand, natural theology's arguments support conclusions either for or against theistic beliefs, so avoidance of circular reasoning necessarily prohibits natural theology's presuppositions from containing any worldview distinctives. On the other hand, "facts or truths in principle available to all human beings just in virtue of their possessing reason and sense perception" just is public and empirical evidence. Accordingly, like natural science, natural theology must also presuppose the existence and comprehensibility of the physical world. Hence, natural science and natural theology have identical presuppositions. (2) *Admissible and Relevant Evidence.* The admissibility of empirical evidence depends on a methodological consideration, namely, appropriate presuppositions, as already mentioned. And the relevance of empirical evidence depends on whether a given item or collection of admissible evidence bears differentially on the credibilities of the competing hypotheses. To count as relevant evidence in public discourse, the evidence must constitute facts established to everyone's satisfaction, and the interpretation of the evidence must also be settled, which involves agreement over how likely (at least approximately) the observed facts would be were each of the hypotheses true. That is, disputes concern which worldview hypothesis is true or likely, but not the facts, and not the interpretations of the facts. Relevance must be judged on a case-by-case basis and hence is a matter for detailed empirical investigation, rather than a methodological consideration to be resolved by a single decision yielding a comprehensive verdict. (3) *Standard and Impartial Logic.* The logic that natural theology uses "relies on standard techniques of reasoning." The implicit contrast is with special pleading that biases an argument toward the favored conclusion. Natural theology uses the same sorts of deductive and inductive logic as natural science. Logic is explored in the following three chapters, including Bayesian inference that is used extensively in natural theology.

Hypothesis tests using Bayesian methods treat all hypotheses symmetrically and impartially. As will be explained in following chapters, an exceedingly strong conclusion can emerge when the weight of the evidence grows exponentially with its amount. Some arguments in natural theology exemplify this particularly favorable situation. In review, reasons that count across worldviews satisfy three necessary and sufficient conditions: appropriate presuppositions, admissible and relevant evidence, and standard and impartial logic. Natural theology's methodology assures, once and for all, that natural theology has appropriate presuppositions, admissible evidence, and standard and impartial logic. That leaves only the relevance of the evidence for testing specified hypotheses to be judged on a case-by-case basis by means of careful empirical investigation.

Avoidance of circular reasoning is crucial for applications of empirical method in worldview inquiries. Unfortunately, circular reasoning can take much more subtle forms than its obvious archetype, "*X*; therefore *X*." For a first example of subtle circular reasoning, consider the question: Does evolution show that life emerged from random mutations and processes within a purposeless universe? Atheists or agnostics such as Richard Dawkins (1996, 2006) typically presume that random processes like gene mutations must be purposeless. But theist Francis Collins (2006:205, also see 80–82) believes in a sovereign God who inhabits eternity, so "God could be completely and intimately involved in the creation of all species, while from our perspective, limited as it is by the tyranny of time, this would appear a random and undirected process." Hence, there can be agreement about the facts of random mutations and yet disagreement about the interpretation of those facts as regards purposelessness. Until the interpretation of these facts has been settled in a manner that counts across worldviews, any assertion that randomness implies purposelessness constitutes subtle circular reasoning. Why? That implication of purposelessness depends crucially on a particular and supportive worldview, atheism, that is only one of the worldviews included in a conversation taking place in natural theology.

For a second and final example of subtle circular reasoning, consider the question: Can science explain everything? The claim that everything has a scientific, natural explanation has been a popular argument for atheism at least since medieval times. Thomas Aquinas (*c.* 1225–1274) expressed this objection to theism quite concisely: "it seems that we can fully account for everything we observe in the world while assuming that God does not exist" (Davies and Leftow 2006:24). But exactly what is this "everything" that science explains? Scientists in particular and people in general disagree about this "everything" that has happened in our world, largely because of worldview differences. For instance, an interesting exchange between Richard Dawkins, identified as a biologist and "an agnostic leaning toward atheism," and Simon Conway Morris, an evolutionary paleontologist and a Christian, was reported in *Scientific American* (Horgan 2005). Dawkins thought that neither the fine-tuning of the universe nor the origin of life requires an explanation involving God, whereas many theists judge otherwise. But Conway Morris "asserted that the resurrection and other miracles attributed to Christ were 'historically verifiable'," whereas atheists typically deny that such miracles really happened. Consequently, if an argument for either theism or atheism presupposes a particular and controversial account of this "everything" that has happened in our world and then claims that success in explaining "everything" supports this same worldview, then such an argument is merely a subtle instance of the argument form, "*X*; therefore *X*."

These two examples might prompt a suspicion that all arguments in natural theology, if inspected carefully enough, would reduce to circular reasoning. However, an inference from merely two examples, intentionally selected to illustrate potential problems, to a universal verdict on natural theology constitutes singularly bad inductive logic. The intent here is to stimulate careful assessment, not to justify breezy dismissal.

Historically, the weaker of science or theology often sought support from the stronger: "With the benefit of hindsight we can now see that over the course of the past 150 years a remarkable reversal has taken place. Whereas once the investigation of nature had derived status from its intimate connections with the more elevated disciplines of ethics and theology, increasingly during the twentieth century these latter disciplines have humbly sought associations with science in order to bask in its reflected glory" (Peter Harrison, in Dixon, Cantor, and Pumfrey 2010:28). Nevertheless, whatever legitimacy and success natural theology may have is *not* derived from its similarities with natural science, nor the reverse. "Reason interpreting experience uses many different methods, depending on the subject-matter and the point of view, but they all throw light on one another. Science, then, is not to be confused with other modes of thought, but neither is it to be entirely divorced from them" (Caldin 1949:135). Indeed, it is by understanding rational procedure in multiple instances, with each legitimated on its own merits, that one can best understand rationality within any of its applications.

Besides natural theology, other humanities also apply public and empirical evidence to worldview inquiries, including some arguments in philosophy. And because some religions or worldviews are based substantially on historical events, historical and archaeological evidence can have worldview import. On the other hand, literature, music, and art contribute greatly to cultures and worldviews, but not particularly by way of empirical evidence bearing on worldview hypotheses. In the special case of a scientific or historical inquiry that is especially rich in worldview import, at least as some scholars see it, philosophical and statistical analysis is often essential for a proper assessment of the bearing of the evidence on competing worldview hypotheses, including avoidance of subtle circular reasoning. The principal requirement for any worldview inquiry appealing to public and empirical evidence, whether it be pursued in natural theology or science or history, is that the action be in public evidence, not controversial presuppositions or biased logic. The very fact that ordinarily worldviews are highly comprehensive tends to implicate multiple possibilities for relevant evidence, so cumulative cases with multiple arguments are common. The inherent strength of a cumulative case, however, comes at the risk of diffuse and rambling argumentation with little action in any one spot. Consequently, a cumulative case is more engaging if at least one of its lines of argumentation is strong, even when considered singly.

In conclusion, mainstream science favors, and historical review exemplifies, science's contribution to a meaningful worldview. But empirical and public evidence from the humanities and sciences together is far more informative than from the sciences alone. The reward for the scientist who perceives scientific method to be an instance of empirical method more generally is the liberty to put empirical evidence to greater use.

Individual experience and worldviews

The preceding two sections concerned empirical and public evidence from the sciences and the humanities. But public evidence is not the sum total of influences on worldview convictions. People are also influenced by their individual experience, including experience that would not ordinarily count as empirical and public evidence.

For example, consider personal beliefs about whether miracles occur, which can influence worldview convictions substantially. To be clear, what is meant by miracles here is real, decidedly supernatural miracles – not the "miracle" of seeing one's own child born or the "miracle" of getting that dream job. Many persons believe in miracles, either from direct observation or from dependable reports from trusted family and friends, as well as from historical miracle reports in a scripture that is trusted and authoritative within a given religious tradition. And many other persons have encountered nothing whatsoever that seems beyond the ordinary workings of the physical world.

Because worldview convictions are so controversial within the scientific community (Easterbrook 1997; Larson and Witham 1999; Ecklund 2010), it is inappropriate for scientific organizations to take positions on which worldview is true. Furthermore, only scientific evidence is within the provenance and competence of scientific organizations, and yet many scholars, including many scientists, believe for good reasons that a wider survey than science alone can offer is required to reach the most reliable and robust conclusions about worldviews.

On the other hand, because mainstream science asserts that science contributes to a meaningful worldview, it is appropriate for individual scientists to argue that scientific evidence supports a particular worldview. When the worldview convictions of such scientists have also been influenced by the humanities, individual experience, or other significant factors, readers of their arguments will benefit from getting the whole story.

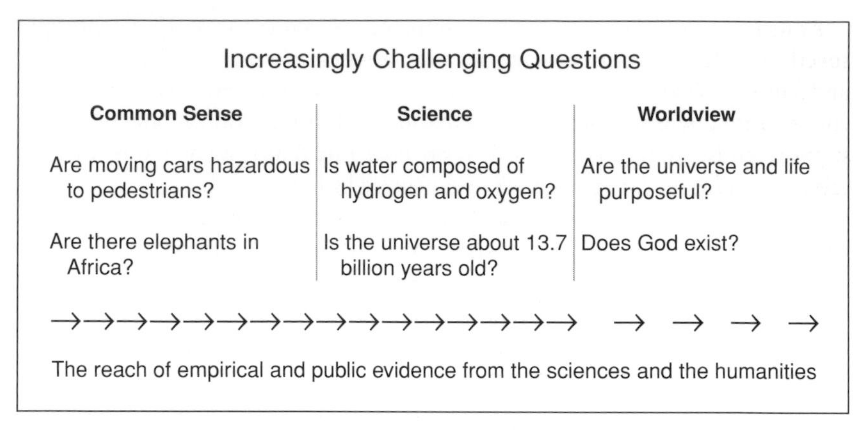

Figure 6.1 Increasingly challenging questions in the realms of common sense, science, and worldviews. Mainstream science presumes the competence of common sense to answer rudimentary questions, such as those listed in this figure, and affirms the competence of science to answer more difficult questions. But the scientific community lacks consensus on whether empirical and public evidence from the sciences and the humanities can answer challenging worldview questions. Hence, the reach of such evidence is depicted by continuous arrows extending through common sense and science, but by dashed arrows thereafter. All scientists follow along the continuous arrows, whereas only some scientists continue along the dashed arrows.

To understand the worldview diversity among individuals within the scientific community, a simple but helpful insight is that increasingly challenging questions arise as one progresses from common sense to science to worldview questions, as depicted in Figure 6.1. The underlying issue is the reach of empirical and public evidence from the sciences and the humanities. Such evidence could be of interest for various reasons. Some persons, whether a scientist or not, may think that empirical evidence is the only sort of evidence that really counts. Other persons, especially those with interests in the humanities, may have a broader conception of the sources of knowledge. In either case, a person may want to push empirical and public evidence to its limits, addressing questions as challenging as possible.

Progressing from left to right in this figure, the most rudimentary questions are in the realm of common sense at the left. One who gets that far with empirical and public evidence, having rejected radical skepticism, might well feel encouraged to take the next step: attempting more difficult questions in the realm of science. If that attempt fails, trying even more challenging worldview questions is bound to be futile. But if that attempt succeeds, one might well want to take the next step: attempting yet harder worldview questions, especially by engaging evidence from both the sciences and the humanities.

At its best, the conversation among individual scientists having diverse perspectives on the limits of empirical knowledge is significant, erudite, respectful, and fruitful. What an understanding of scientific method can contribute to that conversation, by drawing on the PEL model, is complete clarity that whatever support science may give to a specific worldview originates from admissible and relevant evidence, rather than from science's worldview-independent presuppositions and impartial logic. What an understanding of science's powers and limits can contribute is a perspective on the liberal art of science that appreciates the combined strength of the sciences and the humanities when tackling ambitious worldview questions, especially questions that methodological naturalism puts outside science's purview. And what a proper understanding of methodological naturalism can contribute is a stipulatory prohibition against invoking supernatural entities within natural science that (1) is not confused with asserting ontological naturalism, and (2) is not extended thoughtlessly to other disciplines outside natural science, such as natural theology, that have their own questions, evidence, and rules.

Logical roles and diagnoses

The basic components of scientific reasoning – identified by the PEL model as presuppositions, evidence, logic, and conclusions – represent four different logical roles. Different logical roles interact with worldviews in different ways. A statement's logical role is as important as its content.

The difference between "The universe is purposeless" and "The universe is purposeful" is obvious, marking out a vigorous debate. But equally different are "The universe is purposeless" in the logical role of a presupposition and this same "The universe is purposeless" in the role of a conclusion. As a worldview presupposition going beyond science's legitimate presuppositions, its function would be limited to self-congratulatory discourse among kindred spirits. But as a worldview conclusion from a sound argument with worldview-independent presuppositions and impressive evidence, its audience would be the larger world. Recognizing the importance of a statement's logical role, as well as its content, leads to the following several diagnoses.

If a worldview belief has logical roles as both a presupposition and a conclusion within a given discourse, then the diagnosis is circular reasoning in the service of empty dogmatism.

If an argument is unclear regarding whether its worldview belief has the logical role of a presupposition or a conclusion, then the diagnosis is amateurish discourse.

If an argument for a given worldview belief presumes or asserts that science exclusively is the only source of public and empirical evidence, then the diagnosis is the unmitigated scientism that is roundly repudiated as being outside mainstream science (AAAS 1989:26, 30, 133–135, 1990:24–25).

Finally, if a worldview belief emerges in the logical role of a conclusion from an argument also having appropriate presuppositions, admissible and relevant evidence that is public and empirical, and impartial logic, then the diagnosis is a legitimate argument meriting consideration.

Review of boundaries

Much could be said about the boundaries between science's powers and limits. The following five concise statements express the essence of these boundaries.

(1) The scientific community can build upon and move beyond common sense in providing much reliable and even certain knowledge about the physical world.

(2) Science cannot explain everything about the physical world because of fundamental and practical limits. Also, it cannot prove its own needed presuppositions.

(3) Science is worldview independent as regards its presuppositions and methods, but scientific evidence, or empirical and public evidence more generally, can have worldview import. Methodological considerations reveal this possibility and historical review demonstrates its actuality.

(4) It is appropriate for individual scientists to argue that scientific evidence supports a particular worldview, or else to claim that such arguments are illegitimate. But it is not appropriate for scientific organizations to advocate particular positions because worldview commitments are highly controversial within the scientific community and because the humanities also offer relevant evidence and arguments outside the competence of scientific organizations.

(5) Considerations that inform worldview choice include (1) empirical and public evidence from the sciences; (2) empirical and public evidence from the humanities, especially natural theology; and (3) the individual experience of a given person that is meaningful for that person, although it may not qualify as empirical and public evidence for the wider world. Accordingly, science has significant but limited competence for addressing worldview questions, including whether God exists and whether the universe is purposeful. The sciences without the humanities are lame, and public evidence without individual experience is dehumanizing.

Personal rewards from science

The intellectual, technological, and economic benefits from science are widely acknowledged by society. Likewise, the importance of science education for good citizenship in a scientific and technological age is widely appreciated. But another important value of science receives far too little attention: the personal rewards of science, that is, the beneficial effects of science on scientists' personal character and experiences of life.

As this chapter on science's powers and limits draws toward a close, these powers merit attention. Caldin (1949) explored this topic with rare wisdom and charm, so the following remarks draw much from him. Unfortunately, "the place of science in society is too often considered in the narrow setting of economic welfare alone, so that the potential contribution of science to the growth of the mind and will is under-estimated" (Caldin 1949:155). One reward from science is stimulation of rationality and wisdom:

> Now a knowledge of nature is part of wisdom, and we need it to live properly. . . . Science is, therefore, good "in itself," if by that we mean that it can contribute directly to personal virtue and wisdom; it is not just a means to welfare, but part of welfare itself. . . . Scientific life is a version of life lived according to right reason. . . . Consequently, the practice of science requires both personal integrity and respect for one's colleagues; tolerance for others' opinions and determination to improve one's own; and care not to overstate one's case nor to underrate that of others. . . . By studying science and becoming familiar with that form of rational activity, one is helped to understand rational procedure in general; it becomes easier to grasp the principles of all rational life through practice of one form of it, and so to adapt those principles to other studies and to life in general. Scientific work, in short, should be a school of rational life. (Caldin 1949:133–135)

Still another personal reward from science is cultivation of discipline, character, realism, and humility:

> It is not only the intellect that can be developed by scientific life, but the will as well. Science imposes a discipline that can leave a strong mark on the character as can its stimulation of the intellect. All who have been engaged in scientific research know the need for patience and buoyancy and good humour; science, like all intellectual work, demands (to quote von Hügel) "courage, patience, perseverance, candour, simplicity, self-oblivion, continuous generosity towards others, willing correction of even one's most cherished views." Again, like all learning, science demands a twofold attention, to hard facts and to the synthetic interpretation of them; and so it forbids a man to sink into himself and his selfish claims, and shifts the centre of interest from within himself to outside. But for scientists there is a special and peculiar discipline. Matter is perverse and it is difficult to make it behave as one wants; the technique of experimental investigation is a hard and chastening battle. Experimental findings, too, are often unexpected and compel radical revision of theories hitherto respectable. It is in this contact with "brute fact and iron law" that von Hügel found the basis of a modern and scientific asceticism, and in submission to this discipline that he found the detaching, de-subjectifying force that he believed so necessary to the good life. The constant friction and effort, the submission to the brute facts and iron laws of nature, can give rise to that humility and selflessness and detachment which ought to mark out the scientist. (Caldin 1949:135–136)

Summary

Understanding the boundary between science's powers and limits is a core component of scientific literacy. Some limitations are rather obvious, such as that science cannot explain everything about the world and that it cannot prove its own needed presuppositions.

Many scientists, philosophers, and other scholars have debated whether helping to inform large worldview issues, such as the purpose of life, is among science's powers or beyond its limits. However, the mainstream position represented by the AAAS is that contributing to a meaningful worldview is both a proper ambition of science and a historical reality of science. But the sciences are not alone in this endeavor. Many disciplines in the humanities also contribute to a meaningful worldview, including philosophy, theology, history, literature, and the visual and performing arts. In addition to public evidence from the sciences and the humanities, individual experience can also inform a person's worldview convictions, even though personal experience may not count as public evidence.

Empirical method uses empirical and public evidence from the sciences and the humanities to reach conclusions that can bear on worldviews. In assessing arguments for or against a given worldview, not only the content but also the logical role of statements matters. An argument merits consideration that presents its worldview in the logical role of a conclusion, emerging from appropriate presuppositions, empirical evidence, and impartial logic.

Among science's powers is a considerable ability to be of benefit to scientists' personal character and experiences of life. The essence of this benefit is the selflessness, detachment, and humility that result from deliberate and outward-looking attention given to the physical world.

Study questions

(1) The AAAS insists that understanding the boundary between science's powers and limits is a core requirement of scientific literacy. Have you received any instruction on these matters? If so, what was the message, did it make sense, and did it align with position papers on science from the AAAS and NRC? If not, what explanations might you suggest for its absence in your curriculum?

(2) Which components of science – presuppositions or logic or evidence – could potentially have worldview import? Explain all three of your verdicts. What diagnosis results if a worldview belief has logical roles of both a presupposition and a conclusion?

(3) The text argues that worldview convictions can be informed by three sources: the sciences, the humanities, and personal experiences. What is

a significant example of each? What does having three potential resources imply for science's role in forming worldview convictions?

(4) What is the distinction between scientific method and empirical method? How do they differ in their powers and limits, particularly in the range of hypotheses and evidence under consideration?

(5) Regardless of whether you are a student or a professional in the sciences or the humanities, what personal rewards by way of wisdom, discipline, and character have you gained from your experiences with science?

7

Deductive logic

The preceding five chapters are directed mainly at this book's purpose of cultivating a humanities-rich perspective on science. This is the first of five chapters directed mainly at this book's other purpose of increasing scientific productivity.

Logic is the science of correct reasoning and proof, distinguishing good reasoning from bad. Logic addresses the relationship between premises and conclusions, including the bearing of evidence on hypotheses.

In the context of logic, an "argument" is not a dispute but rather is a structured set of statements in which some statements, the premises, are offered to support or prove others, the conclusions. Many deductive systems, including arithmetic and geometry, are developed on a foundation of logic in the modern and unified vision of mathematics.

Of course, given the simple premises that "All men are mortal" and "Socrates is a man," one trusts scientists to reach the valid conclusion that "Socrates is mortal," even without formal study of logic. But given the more difficult problems that continually arise in science, the rate of logical blunders can increase substantially in the absence of elementary training in logic. Fortunately, most blunders involve a small number of common logical fallacies, so even a little training in logic can produce a remarkable improvement in reasoning skills.

The aim of this chapter differs from that of an ordinary text or course on logic. One short chapter cannot teach logic comprehensively. What it can do, however, is convey an insightful general impression of the nature and structure of deductive logic. Recall that the PEL model introduced in Chapter 5 identifies logic as one of the three essential inputs (along with presuppositions and evidence) required to support scientific conclusions. Consequently, the credibility of science depends on having a logic that is coherent and suitable for investigating the physical world.

This chapter distinguishes the two basic kinds of logic: deductive logic, explained in this chapter, and inductive logic, explored in Chapter 9. One branch of deduction, probability theory, is deferred to the next chapter. The history of logic is reviewed briefly, followed by basic accounts of propositional logic, predicate logic, and arithmetic. Common logical fallacies are analyzed to refine reasoning skills.

Deduction and induction

The distinction between deduction and induction can be explained in terms of three interrelated differences. Of these three differences, the one listed first is the fundamental difference, with the others being consequences or elaborations. Custom dictates distinct appellative terms for good deductive and inductive arguments. A deductive argument is valid if the truth of its premises guarantees the truth of its conclusions and is invalid otherwise. An inductive argument is strong if its premises support the truth of its conclusions to a considerable degree and is weak otherwise. The following deductive and inductive arguments, based on Salmon (1984:14), illustrate the three differences.

Valid Deductive Argument
Premise 1. Every mammal has a heart.
Premise 2. Every horse is a mammal.
Conclusion. Every horse has a heart.

Strong Inductive Argument
Premise 1. Every horse that has been observed has had a heart.
Conclusion. Every horse has a heart.

First, the conclusion of a deductive argument is already contained, usually implicitly, in its premises, whereas the conclusion of an inductive argument goes beyond the information present, even implicitly, in its premises. The technical terms for this difference are that deduction is nonampliative but induction is ampliative. For example, the conclusion of the foregoing deductive argument simply states explicitly, or reformulates, the information already given in its premises. All mammals have hearts according to the first premise, and that includes all horses according to the second premise, so the conclusion follows that every horse has a heart. On the other hand, the conclusion of the foregoing inductive argument contains more information than its premise. The premise refers to some group of horses that have been observed up to the present, whereas the conclusion refers to all horses, observed or not, and past or present or future.

Note that this difference, between ampliative and nonampliative arguments, concerns the relationship between an argument's premises and conclusions, specifically whether or not the conclusions contain more information than the premises. This difference does not pertain to the conclusions as such, considered in isolation from the premises – indeed, the foregoing two arguments have exactly the same conclusion.

Second, given the truth of all of its premises, the conclusion of a valid deductive argument is true with certainty, whereas even given the truth of all of its premises, the conclusion of an inductive argument is true with at most high probability. This greater certainty of deduction is a direct consequence of its being nonampliative: "The [deductive] conclusion must be true if the premises are true, *because* the conclusion says nothing that was not already stated by the premises" (Salmon 1984:15). The only way that the conclusion of a valid deductive argument can be false is for at least one of its premises to be false. On the other hand, the uncertainty of induction is a consequence of its being ampliative: "It is because the [inductive] conclusion says something not given in the premise that the conclusion might be false even though the premise is true. The additional content of the conclusion might be false, rendering the conclusion as a whole false" (Salmon 1984:15). For example, the foregoing inductive conclusion could be false if some other horse, not among those already observed and mentioned in this argument's premise, were being used for veterinary research and had a mechanical pump rather than a horse heart.

Deductive arguments are either valid or invalid on an all-or-nothing basis because validity does not admit of degrees. But inductive arguments admit of degrees of strength. One inductive argument might support its conclusion with a very high probability, whereas another might be rather weak.

The contrast between deduction's certainties and induction's probabilities can easily be overdrawn, however, as if to imply that induction is second-rate logic compared with deduction. Representing certain truth by a probability of 1 and certain falsehood by 0, an inductive conclusion can have any probability from 0 to 1, including values arbitrarily close to 1 representing certainty of truth (or 0 representing certainty of falsehood). Given abundant evidence, induction can deliver practical certainties, although it cannot deliver absolute certainties.

Third and finally, deduction typically reasons from the general to the specific, whereas induction reasons in the opposite direction, from specific cases to general conclusions. That distinction was prominent in Aristotle's view of scientific method (Losee 2001:5–8) and remains prominent in today's dictionary definitions. For instance, the *Oxford English Dictionary* defines "deduction" as "inference by reasoning from generals to particulars," and it defines "induction" as "The process of inferring a general law or principle from the observation of particular instances." Deduction reasons from a given model to expected data, whereas induction reasons from actual data to an inferred model, as depicted in Figure 7.1.

As encountered in typical scientific reasoning, the "generals" and "particulars" of deduction and induction have different natures and locations. The general models or theories exist in a scientist's mind, whereas the particular instances pertain to physical objects or events that have been observed. Often,

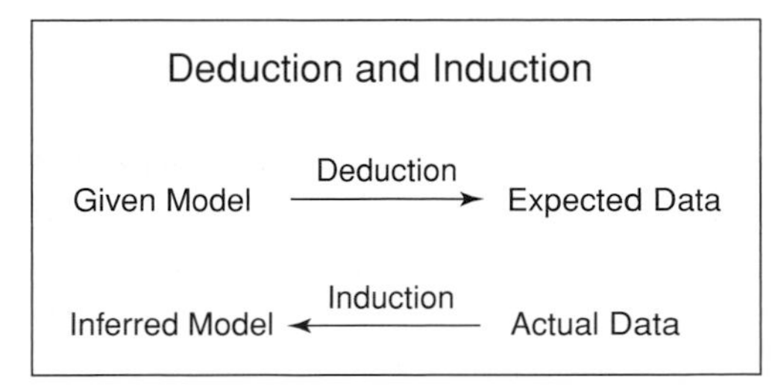

Figure 7.1 The opposite reasoning directions of deduction and induction. Deduction reasons from the mind to the world, whereas induction reasons from the world to the mind.

the observations or data comprise a limited sample, but the researchers are interested in the larger population from which the sample was drawn. For instance, a clinical trial may examine a representative sample of persons to reach conclusions pertaining to the whole population of persons suffering from a given disease.

Deduction is neither better than induction nor worse. Rather, they pursue answers to different kinds of questions, with deduction reasoning from a mental model to expected data, and induction reasoning from actual data to a mental model. Both are indispensable for science.

Historical perspective on deduction

Aristotle (384–322 BC) wrote extensively on logic. Although his works on some topics, including natural science, suffered much neglect until the early 1200s, his corpus on logic, the *Organon* or tool (of reasoning), fared better. Aristotle's logic built on ideas from Socrates and Plato. Epicurean, Stoic, and Pythagorean philosophers also developed logic and mathematics. Besides Greece, there were impressive ancient traditions in logic in Babylon, Egypt, India, and China. Largely because of Augustine's early influence, the Aristotelian tradition came to dominate logic in the West, so that tradition is emphasized here.

In his *Prior Analytics*, Aristotle taught that every belief comes through either deduction or induction. His syllogistic logic is the first deductive system, predating Euclid's geometry. Aristotle proposed an inductive–deductive model of scientific method that features alternation of deductive and inductive steps. This alternation, moving from mental model to physical world and back again, leads scientists to a mind–world correspondence – to truth. In this process, any discrepancy between model and world is to be resolved by adjusting the model to the world because the actual data in the inductive step have priority over the expected data in the deductive step. Assuming that the data are not faulty or excessively inaccurate, actual data contrary to a model's expectations imply that something is wrong with the model. Adjusting one's model to the world is the basis of scientific realism.

Euclid (*fl. c.* 300 BC) was the great master of geometry. Many truths of geometry were known before Euclid. For example, earlier Babylonians and Egyptians knew that the sum of the interior angles of a triangle equals 180 degrees. But that was known by empirical observation of numerous triangles, followed by inductive generalization. Their version of geometry was a practical art related to surveying, in line with the name "geometry," literally meaning "earth measure."

Euclid's *Elements of Geometry*, in one of the greatest paradigm shifts ever, instead demonstrated those geometrical truths by deduction from several axioms and rules. Euclid's geometry had five postulates concerning geometry, such as that a straight line can be extended in either direction, plus five axioms or "common notions" concerning correct thinking and mathematics in general, such as that the whole is greater than its parts. Euclid's combination of geometrical postulates and logical axioms represented a nascent recognition that logic underlies geometry. Countless theorems can be deduced from Euclid's postulates and axioms, including that the sum of the interior angles of a triangle equals 180 degrees.

Subsequently, non-Euclidean geometries were discovered by Thomas Reid (1710–1796), Nikolai Lobachevsky (1792–1856), János Bolyai (1802–1860), Bernhard Riemann (1826–1866), and others. This rendered Euclid's work *a* geometry rather than *the* geometry. In Reid's alternative geometry, the sum of the interior angles of a triangle equals more than 180 degrees.

Anicius Manlius Severinus Boethius (AD 480–524) translated, from Greek into Latin, many parts of Aristotle's logical works, Porphyry's *Introduction to Aristotle's Logic*, and parts of Euclid's *Elements*. His *On Arithmetic*, based on earlier work by Nicomachus of Gerasa, became the standard text on arithmetic for almost a millennium.

Peter Abelard (*c.* 1079–1142) wrote four books on logic. He and his students, John of Salisbury and Peter Lombard, greatly influenced medieval logic. The use of Arabic numerals was spread into Europe by Alexandre de Villedieu (*fl. c.* 1225), a French Franciscan, John of Halifax (or Sacrobosco, *c.* 1200–1256), an English schoolman, and Leonardo of Pisa (or Fibonacci, *c.* 1180–1250), an Italian mathematician. The modern mind can hardly imagine the tedium of multiplication or division using Roman numerals, or how few persons in medieval Europe could perform what we now regard as elementary calculations.

Albertus Magnus (*c.* 1200–1280) wrote 8,000 pages of commentary on Aristotle, including much logic. He also wrote a commentary on Euclid's *Elements*.

Robert Grosseteste (*c.* 1168–1253) founded the mathematical-scientific tradition at Oxford. He affirmed and refined Aristotle's inductive–deductive scientific method, which he termed the "Method of Resolution and Composition" for its inductive and deductive components, respectively. Also, his Method of Verification involved deriving the deductive consequences of a theory beyond the original facts on which the theory was based and then observing the actual outcome in a controlled experiment to check the theory's predictions. That method recognized the priority of data over theories, in accord with Aristotle. Grosseteste's Method of Falsification eliminated bad theories or explanations by showing that they imply things known to be false. To increase the chances of eliminating false theories, he recommended that conclusions reached by induction be submitted to the test of further observation or experimentation.

Putting all those methods together, the objective of Grosseteste's new science was to make theory bear on the world and the world bear on theory, thereby bringing theory into correspondence with the world. Grosseteste's scientific method sought to falsify and reject false theories, to confirm and accept true theories, and to discern which kinds of observational or experimental data would help the most in theory evaluation.

There is substantial similarity between Grosseteste's medieval science and modern science. "Modern science owes most of its success to the use of these inductive and experimental procedures, constituting what is often called 'the experimental method'. The . . . modern, systematic understanding of at least the qualitative aspects of this method was created by the philosophers of the West in the thirteenth century. It was they who transformed the Greek geometrical method into the experimental science of the modern world" (Crombie 1962:1). I concur with this assessment that a basically correct and complete scientific method emerged in the thirteenth century.

William of Ockham (*c.* 1285–1347) wrote a substantial logic text, the *Summa logicae*. The principle of parsimony is often called Ockham's razor because of his influential emphasis on this principle. Jean Buridan (*c.* 1295–1358) wrote the *Summulae de dialectica*, a then-modern revision and amplification of the earlier logic text by Peter of Spain (*fl.* first half of the thirteenth century), and two advanced texts, the *Consequentiae* and *Sophismata*.

René Descartes (1596–1650) was the founder of analytic geometry. Blaise Pascal (1623–1662) contributed to projective geometry, arithmetic, combinatorial analysis, probability, and the theory of indivisibles (a forerunner of integral calculus). He developed the first commercial calculating machine. Isaac Newton (1642–1727) and Gottfried Leibniz (1646–1716) invented calculus. Giuseppe Peano (1858–1932) devised axioms for arithmetic.

For millennia, the various branches of deduction – such as logic, arithmetic, and geometry – had been developed as separate and unrelated systems. Early great works aiming to unify logic and mathematics were the brilliant *Grundgesetze der Arithmetik* (*The Basic Laws of Arithmetic*) of Frege (1893) and the monumental *Principia Mathematica* of Whitehead and Russell (1910–13).

Table 7.1. Truth-table definitions for negation, conjunction, disjunction, implication, and equality

Assignments		Not	And	Or	Implies	Equals
A	B	$\sim B$	$A \wedge B$	$A \vee B$	$A \rightarrow B$	$A \equiv B$
T	T	F	T	T	T	T
T	F	T	F	T	F	F
F	T		F	T	T	F
F	F		F	F	T	T

Propositional logic

Propositional logic, also called statement calculus and truth-functional logic, is a rather elementary branch of deductive logic. Nevertheless, it is quite important because it pervades common-sense reasoning and scientific reasoning.

A simple proposition has a subject and a predicate, such as "This apple is red" or "Mary is coming." Propositional logic considers only declarative statements. Accordingly, every simple proposition has the property of having one or the other of two possible truth-values: true (T) and false (F). Note that the truth-value applies to the proposition as a whole, such as "This apple is red" being true for a red apple but false for a green apple. In propositional logic, as introduced in this section, there is no further analysis of the subject and predicate within a proposition. But, in predicate logic, to be explained in the next section, further analysis is undertaken. Hence, predicate logic is more complicated, subsuming propositional logic and adding new concepts and analysis.

Proposition constants represent specific simple propositions and are denoted here by uppercase letters like *A*, *B*, and *C* (except that T and F are reserved to represent the truth-values true and false). For example, "The barometer falls" can be symbolized by *B*, "It will rain" by *R*, and "It will snow" by *S*. Then, the compound sentence "If the barometer falls, then either it will rain or it will snow" can be expressed by "If *B*, then *R* or *S*."

The most common connectives or operators are "not," "and," "or," "implies," and "equals." They are also termed negation, conjunction, disjunction, implication, and equality. These five connectives are denoted here by these symbols: "$\sim$," "$\wedge$," "$\vee$," "$\rightarrow$," and "$\equiv$." The meanings of these connectives are specified by a truth table (Table 7.1).

"Not" is a unary operator applied to a single proposition. If B is true, then $\sim B$ is false; and if B is false, then $\sim B$ is true. That is, B and $\sim B$ have opposite truth-values. The other connectives are binary operators applied to two propositions. For example, "A and B," also written as "$A \wedge B$," is true when both A is true and B is true and is false otherwise. Simple propositions can be combined with connectives, such as $B \rightarrow (R \vee S)$ to symbolize the preceding compound proposition about a barometer. Parentheses are added as needed to avoid ambiguity. To simplify expressions, the conventions are adopted that negation has priority over other connectives and applies to the shortest possible sub-expression, and parentheses may be omitted whenever the order makes no difference.

Incidentally, two other logical operators, not already specified in Table 7.1, are important in computer design because they can be implemented with simple transistor circuits. Joint denial of A and B, expressed by "Neither A nor B" and symbolized by "$A \downarrow B$," equals the negation of "A or B" and hence is also named "Nor." An alternative denial of A and B, expressed by "Either not A or not B" and symbolized by "$A \mid B$," equals the negation of "A and B" and hence is also named "Nand." (To avoid potential confusion, note that this symbol "|" instead means "or" in several computer-programming languages.) Remarkably, all of the logical operators in Table 7.1 can be defined or replaced by joint denial alone, or by alternative denial alone. For instance, $\sim A$ is logically equivalent to $A \downarrow A$ or to $A \mid A$. Likewise, $A \wedge B$ is logically equivalent to $(A \downarrow A) \downarrow (B \downarrow B)$ or to $(A \mid B) \mid (A \mid B)$. Consequently, circuits using Nor and Nand operations are extremely useful in computers. Annually, the world produces more transistors than it produces grains of wheat or grains of rice.

Proposition variables stand for simple propositions and are denoted here by lowercase letters like p and q. Hence, the variable p could stand for the constant A or B or C. Proposition expressions are denoted here by script letters and are formed by one or more applications of two rules: (1) any proposition constant or variable is a proposition expression; and (2) if $\mathcal{A}$ and $\mathcal{B}$ are proposition expressions, then their negations are proposition expressions as well as their being combined by conjunction, disjunction, implication, and equality.

An argument is a structured, finite sequence of proposition expressions, with the last being the conclusion (ordinarily prefaced by the word "therefore" or the symbol "∴"), and the others the premises. The premises are intended to support or prove the conclusion. For example, *modus ponens* is a valid argument with two premises and one conclusion: A; A implies B; therefore B. Likewise, *modus tollens* is the valid argument: not B; A implies B; therefore not A. Incidentally, the full Latin names are *modus ponendo ponens* meaning "the way that affirms by affirming," and *modus tollendo tollens* meaning "the way that denies by denying." An argument is valid if under every assignment of truth-values to the proposition variables that makes all premises true, the conclusion is also true. Otherwise, the argument is invalid.

There are several methods for proving that an argument is valid or else invalid, as the case may be. Different methods all give the same verdict, but one

Formal Propositional Logic

(1) Language. The symbols used are as follows: ~, →, (,), p_1, p_2, p_3, and so on.
(2) Expressions. A well-formed formula (wff) is formed by one or more applications of two rules. (a) Each p_i is a wff. (b) If A and B are wffs, then $(\sim A)$ and $(A \rightarrow B)$ are wffs.
(3) Axioms. For any wffs A, B, and C, axioms are formed by the following three axiom schemes:
Axiom Scheme 1. $(A \rightarrow (B \rightarrow A))$
Axiom Scheme 2. $((A \rightarrow (B \rightarrow C)) \rightarrow ((A \rightarrow B) \rightarrow (A \rightarrow C)))$
Axiom Scheme 3. $(((\sim A) \rightarrow (\sim B)) \rightarrow (B \rightarrow A))$
(4) Rule. The rule, *modus ponens*, says from A and $(A \rightarrow B)$, infer B.
(5) Interpretation. The symbols "~" and "→" are the logical connectives negation and implication, which may have associated parentheses needed to specify the order of operations, and the symbols p_1, p_2, p_3, and so on, represent proposition variables having truth-values of either true or false. A proof is a sequence of wffs $A_1, \ldots, A_N, A$ such that each wff either is an axiom or follows from two previous members of the sequence by application of *modus ponens*. The final wff, A, is a theorem.

Figure 7.2 The elements of formal propositional logic. This logic is specified by its language, expressions, axioms, rule, and interpretation.

method may be easier to understand or use in a given instance than is another. The conceptually simplest method, directly reflecting the definitions of validity and invalidity, is to construct a truth table to determine whether or not each assignment of truth-values to the argument's proposition variables that makes all premises true also makes the conclusion true. Another method for proving validity is to deduce the argument as a theorem from the axioms and rules. But proof strategies are best left to any standard logic text.

Figure 7.2 presents a formal system for propositional logic, drawing on Hamilton (1978:28). Some liberty has been taken to simplify this presentation. The ordinary letters "*A*," "*B*," and "*C*" in this figure should actually be script letters to represent proposition expressions, not merely proposition constants as denoted by these ordinary letters elsewhere in this section.

Propositional logic is both sound and complete. Basically, this means that its rules are correct and that no additional rules are needed. Propositional logic is also decidable, meaning that any argument can be proven to be valid or else invalid. For example, consider the argument: $A \rightarrow B$; $B \rightarrow (C \vee D)$; $A \wedge \sim C$; $\therefore D$. Is it valid or invalid? Some time and effort are required to render the verdict, which turns out to be that this argument is valid. But before even starting to assess validity, it is already known and guaranteed in advance that the outcome is predetermined by the axioms of propositional logic and the answer is decidable.

Predicate logic

Predicate logic, also called first-order logic and predicate calculus, subsumes and surpasses prepositional logic. It adds two extensions. First, it distinguishes between a proposition's subject and predicate. In a conventional symbolism, the predicate is denoted by an uppercase letter and the subject is denoted by the following lowercase letter placed within parentheses. For instance, "This apple is red," with its predicate "is red" and subject "This apple," can be symbolized by $R(a)$. Second, predicate logic has the existential quantifier "some" denoted by "$\exists$" and the universal quantifier "all" denoted by "$\forall$." For example, $(\forall x)(A(x))$ means "All x are A" and $(\exists x)(A(x))$ means "Some x is A."

One kind or subset of predicate logic is syllogistic logic. A familiar example is the argument: All men are mortal; Socrates is a man; therefore, Socrates is mortal. Because of its deeper analysis distinguishing subjects and predicates and its inclusion of existential quantifiers such as "all," predicate logic can analyze this argument and declare this syllogism valid. But the simpler propositional logic cannot express or handle syllogisms.

A formal deductive system for predicate logic is about twice as complicated as the one shown in Figure 7.2 for propositional logic (Hamilton 1978:49–56, 71–72). By contrast, the range of theorems that predicate logic can prove is incomparably greater than the range for propositional logic. Accordingly, predicate logic supplies the powerful logic that lays the foundation upon which other branches of mathematics can be constructed, including arithmetic and probability.

Arithmetic

In the contemporary vision of deductive systems, numerous branches of mathematics, such as arithmetic, are all built on a foundation of predicate logic. To build a branch of mathematics on logic, two items must be added: an interpretation and some axioms. A formal language is abstract, and an interpretation attaches a particular meaning to some symbols of a formal language, such as arithmetic being about numbers. Additional axioms are needed because often a mathematical statement is true (or false) because of the mathematical meanings of its terms, rather than merely the logical arrangement of its terms. There are both logical truths and mathematical truths. Both require axioms.

Although syllogistic logic was axiomatized by Aristotle and geometry by Euclid more than two millennia ago, arithmetic was axiomatized only just over a century ago in 1889 by Giuseppe Peano. His axioms can be edited in various ways to make them somewhat more transparent or convenient. Figure 7.3 presents a formulation with nine axioms.

Peano Axioms for Arithmetic

1. 0 is a natural number.
2. For every natural number x, $x = x$.
3. For all natural numbers x and y, if $x = y$, then $y = x$.
4. For all natural numbers x, y, and z, if $x = y$ and $y = z$, then $x = z$.
5. For all a and b, if a is a natural number and $a = b$, then b is also a natural number.
6. For every natural number n, its successor $S(n)$ is a natural number.
7. For every natural number n, $S(n) = 0$ is false.
8. For all natural numbers m and n, if $S(m) = S(n)$, then $m = n$.
9. If K is a set such that (a) 0 is in K and (b) for every natural number n, if n is in K, then $S(n)$ is in K; then K contains every natural number.

Figure 7.3 The nine Peano axioms for arithmetic. The first axiom assumes that zero is a natural number, the next four describe the equality relation, the next three describe the successor function (where 1 is the successor of 0, 2 is the successor of 1, and so on), and the last axiom concerns the set of all natural numbers.

Arithmetic can be developed from the Peano axioms and the inherited predicate logic. For the most part, the meanings of the Peano axioms should be fairly obvious. For instance, axiom 2 says that every number equals itself, and axiom 8 says that if $m + 1 = n + 1$, then $m = n$.

To reiterate an important point about predicate logic from the preceding section in the present context of arithmetic, axioms fix in advance the outcome for subsequent theorems or calculations. For example, is $871 \times 592 = 515432$ correct? A little effort is required to check this calculation, but before even starting, the verdict has been predetermined by the arithmetic axioms. Actually, this calculation is incorrect, the proper value being 515632. Precisely because the rules of arithmetic are fixed before the game begins, arithmetic is meaningful and rational. If different persons could get different sums for $27 + 62$, then in such a world there would be no science, and no banks either.

Many persons may miss the wonder, but Albert Einstein asked "How is it possible that mathematics, a product of human thought that is independent of experience, fits so excellently the objects of physical reality?" (Frank 1957:85). Likewise, Potter (2000:17–18) expressed this wonder specifically as regards arithmetic, remarking that "it is not immediately clear why the properties of abstract objects [numbers] should be relevant to counting physical or mental ones. . . . One has only to reflect on it to realize that this link between experience, language, thought, and the world, which is at the very centre of what it is to be human, is truly remarkable."

Indeed, there is something wonderful about arithmetic's effectiveness. It may be noted, however, that the Peano axioms generate standard arithmetic, whereas there also exist equally internally coherent but different *non*standard arithmetics. For instance, in standard (Peano) arithmetic $2 + 2 = 4$. But, in the nonstandard ring arithmetic based on the circular and repeating arrangement of integers 0, 1, 2, 3, 0, 1, 2, 3, 0, and so on, the sum of interest becomes $2 + 2 = 0$. Likewise, in the ring arithmetic with just 0, 1, and 2 repeating, the sum becomes $2 + 2 = 1$. All three of these arithmetics are equally internally coherent, although they are also different from each other.

There are occasional practical uses for nonstandard arithmetics or geometries. For example, standard arithmetic says that $11 + 3 = 14$. But an ordinary clock is based on a circular arrangement of its integers from 1 to 12, so this ring arithmetic says that 3 hours after 11 o'clock, the time is 2 o'clock, or $11 + 3 = 2$. (Or, some clocks have instead the integers 1 to 24 written in a circle). For another example, ordinary surveying or earth-measure uses ordinary geometry. But airplane pilots traveling great distances use the non-Euclidean geometry that Reid invented (for studying optics in a roughly spherical eye) to follow the shortest great circle bearing on our spherical earth, thereby saving time and fuel.

But apart from these understandable exceptions, standard logic and arithmetic and geometry prevail in daily life. While properly appreciating the wonder of arithmetic, part of the reason that (standard) arithmetic fits with our experiences in the physical world is that the choice of standard over nonstandard arithmetic has been guided preemptively by our interests and needs as incarnate beings living in the physical world. That is, in choosing an arithmetic (or geometry or whatever), coherence is not our only criterion but also fit with our experiences of the world. Hence, in the mathematical world of coherent arithmetics, one can obtain $2 + 2 = 0$ or $2 + 2 = 1$ or $2 + 2 = 4$. However, in the physical world of actual objects and events, standard arithmetic is uniquely appropriate. Two apples plus two apples equals four apples.

Common fallacies

Ever since Aristotle's *Sophistical Refutations*, logicians have been providing helpful analyses and classifications of logical fallacies. Furthermore, science educators report that "all the standard logical fallacies, known since Aristotle's day, are routinely committed by science students" (Matthews 2000:331).

There are many fine books and resources on fallacies. But the book by Madsen Pirie, with its generous list of 79 fallacies, is outstanding because of its fun rhetoric in the guise of a naughty sophist. He explained: "This book is intended as a practical guide for those who wish to win arguments. It also teaches how to perpetrate fallacies with mischief at heart and malice aforethought.... I have given the reader recommendations on how and where the fallacy may be used to deceive with maximum effect.... In the hands of the wrong person this is more of a weapon than a book, and it was written with that wrong person in

Figure 7.4 The logical fallacy *argumentum ad lapidem*, argument to a stone. Samuel Johnson vigorously kicks a stone, attempting to refute the idea that the physical world does not exist, while an unimpressed George Berkeley observes. (This drawing by Carl R. Whittaker is reproduced with his kind permission.)

mind" (Pirie 2006:ix–x). This is the book that everyone needs as we set about the all-important business of getting our own way!

The study of fallacies best begins with its opposite, the study of right thinking. Knowing the genuine article makes its counterfeits more obvious. Recalling the PEL model in Chapter 5, the essence of scientific thinking is evidence that is *admissible* relative to the presuppositions and *relevant* relative to the hypotheses, as well as deductive and inductive *logic* to draw conclusions and weigh evidence. The three italicized words emphasize the principal opportunities for defects: inadmissible evidence, irrelevant evidence, and fallacious logic. The fourth and final category of fallacies reviewed in this section involves a personal rather than a procedural defect, failure of will to pursue the truth.

Inadmissible Evidence. The *argumentum ad lapidem* (argument to a stone) appeals to inadmissible evidence. This fallacy is named for a famous incident depicted in Figure 7.4. George Berkeley had argued that only minds and ideas exist, not physical objects and events, as mentioned in Chapter 5. When Dr. Samuel Johnson was told by James Boswell that this argument is impossible to refute, he vigorously kicked a stone, exclaiming "I refute it thus."

But as Pirie (2006:101–104) observed, Johnson was not so much refuting Berkeley's argument as ignoring it. Johnson was presuming a realist interpretation or ontology regarding the empirical evidence provided by sight or feel or sound or any other sense, which is precisely the matter in dispute given Berkeley's idealist ontology. As emphasized in Chapter 5, the existence and comprehensibility of the physical world are presuppositions of mainstream science, not conclusions of science (or philosophy either). To think otherwise is to commit the *argumentum ad lapidem* fallacy. Berkeley could accept that Johnson had an experience of kicking a stone and could even share that experience with him. But Berkeley would not infer from this experience the metaphysical theory that the stone has an independent physical existence. Presuppositions cut deeper than evidence.

Irrelevant Evidence. Several fallacies appeal to evidence that is admissible, given the common-sense presuppositions of mainstream science, but that evidence is irrelevant because it fails to bear on the credibilities of the various hypotheses under consideration. One such fallacy is the *argumentum ad hominem* (argument to the man), which attacks the person promoting the disliked idea rather than the idea itself. For instance, a theory could be attacked by saying its proponent is a teacher at a small community college.

Another fallacy is the *red herring*. This draws attention away from the original argument to some other matter that is irrelevant but provides an easier target for refutation.

An alluring fallacy for scientists is *unobtainable perfection*, or at least excessive perfection. This fallacy discredits a result by requiring greater accuracy or scope. For instance, if a paper under review compares methods *A* and *B*, a reviewer might say that it must also compare method *C* in order to be publishable. But simply to complain that more could be done is irrelevant because this is always the case. Rather, the relevant criteria are whether that paper adds to what was known before and whether it has some theoretical interest or practical value. Also, adding method *C* may be a good idea, but the editor might intervene and propose this as a suggestion or recommendation rather than a requirement.

Fallacious Logic. Most logical fallacies obtain their apparent plausibility from being subtle variations on other arguments that are valid. Logical fallacies are especially deceptive when their conclusions are already believed or desired.

Fallacies result from invalid variations on the valid argument *modus ponens*: *A*; *A* implies *B*; therefore *B*. The implication "*A* implies *B*" consists of the antecedent *A* and the consequent *B*. Hence, the valid argument *modus ponens* affirms the antecedent. Similarly, the valid argument *modus tollens* denies the consequent: not *B*; *A* implies *B*; therefore not *A*. But other variations are invalid. Affirming the consequent is the logical fallacy: *B*; *A* implies *B*; therefore *A*.

An example is: The plants are yellowish; if plants lack nitrogen, then they become yellowish; therefore the plants lack nitrogen. Likewise, denying the antecedent is also invalid: not *A*; *A* implies *B*; therefore not *B*. A common version of this fallacy is the argument from *missing evidence.* However, an observation of missing evidence *A* has no force in rejecting theory *B* unless it is supplemented with an additional argument showing that evidence *A* would be expected to exist, and perhaps even to be abundant, were theory *B* true. Furthermore, an honest evaluation of theory *B* would also consider whether some other kinds of evidence are relevant and available rather than just eagerly pursuing the easiest possible way to discredit *B*.

Syllogisms have 256 possible forms, of which only 24 are valid. An example of a valid syllogism is: Socrates is a man; all men are mortal; therefore Socrates is mortal. Because most forms are invalid, apart from some training in logic, syllogisms offer numerous opportunities for tricky fallacies.

A *false dilemma* mentions fewer alternatives than actually exist. In the false dilemma "*A* or else *B*; not *A*; therefore *B*," the logical form is valid, but the first premise "*A* or else *B*" is false because of additional possibilities such as *C*. For example, "Either apply nitrogen fertilizer or get yellowish plants" is a false dilemma for many reasons, including the possibilities that a particular soil already has adequate nitrogen without adding fertilizer, or that a virus causes yellowish plants despite adequate fertilizer. Of course, the opposite fallacy also occurs: the "optionitis" of believing that one has more options than reality (or feasibility) actually offers. Some dilemmas are real.

A variant on the false dilemma is the *straw-man argument.* The logical form is this same "*A* or else *B*; not *A*; therefore *B*," where *A* represents an opponent's position and *B* the favored position. However, the premise "not *A*" is supported by attacking the opponent's weakest evidence or a simplistic misrepresentation of the opponent's position. An honest refutation of the opponent's position must instead represent *A* accurately and tackle its strongest evidence and arguments.

Yet another variant on the false dilemma is the *argumentum ad ignorantiam* (argument from ignorance). This fallacy attempts to drive opponents to accept an argument unless they can find a better argument to the contrary. For example, an environmentalist might say "We cannot prove that this pesticide is safe, so we must assume that it is dangerous and outlaw its use." There may or may not be some other good arguments against this pesticide's safety, but an argument from ignorance is not a good reason. The implicit dilemma in an appeal to ignorance is "Give me a better argument, or else accept my argument." But the unmentioned third option is to admit current inability to construct a better argument while still either rejecting the offered argument or suspending judgment.

Failure of Will. Given the dishonorable nature of failure of will, this fourth and final category of fallacies is best discussed by adopting Pirie's guise as a naughty sophist. After all, adroit evasion of knowledge, while still giving every appearance of energetic pursuit of knowledge, requires considerable skill!

Three fallacies are useful to conceal failure of will: privileged cynicism, secret alliance, and personal exemption. Admittedly, these are imperfect means for putting a pretty face on failure of will. But these three fallacies work as well as can be expected, given the inherent challenges of this naughty business. Their principal merits are that these fallacies are unrivaled in their resistance to remediation, and sometimes they can even achieve self-deception!

An effective fallacy for implementing failure of will is *privileged cynicism*: When there is a spectrum of positive to negative opinions about something's merit, the most negative, skeptical, cynical opinion is privileged by being presented and perceived as automatically the view of the sophisticated elite – unlike the naïve and despised view of the ignorant commoners. For instance, the commoners (including most practicing scientists!) may think that much knowledge is readily attained and perfectly solid, whereas presumably the academic elite is steeped in postmodern rejection of knowledge claims, so privileged cynicism declares that *automatically* the latter group has the more sophisticated view. A skilled professor can wield this fallacy to encourage students in a cynical attitude that then becomes the students' passport into the alluring world of the cultural elite. Inside such a culture, cynicism equals sophistication.

This fallacy of privileged cynicism applies readily to science. Students can be lured easily into the mighty gratifying feeling that they, being superior to the gullible commoners, are getting the real, dirty story on what science is. The fallacy of privileged cynicism has great appeal to persons who already feel disappointed or disenfranchised in life for any reason.

A huge advantage of privileged cynicism is its ease. A lackluster high school student, let alone a bright college student, can learn five skeptical or cynical remarks in as many minutes. Furthermore, merely two or three pages suffice for a skilled writer to display a cynical view of science in all its glory, which seems to call for automatic assent from any reader wishing to be numbered among the sophisticates who are in the know. By contrast, a satisfactory account of actual scientific method takes work to write and work to read. Hence, the hard-won sophistication of a working scientist cannot possibly compete with the cynical version of "sophistication" in terms of being offered on the cheap.

A second fallacy for implementing failure of will is *secret alliance*. This wonderfully subtle fallacy involves fighting an intense battle not so much for its own intrinsic importance as for its strategic value in defending an ally in a larger war, while that ally receives so little explicit mention as to remain virtually a secret. Thereby, the real motivations for the battle are not obvious, perhaps even to many of the battle's most prominent combatants on both sides.

The main example in the realm of science is the notorious "science wars" reviewed in Chapter 4. The intensity of this intellectual war, augmented by melodramatic and inflammatory rhetoric, is astonishing and perhaps even mystifying. Why is it so intense? One might suspect that often the underlying motivations have not been expressed in an entirely forthright manner. Indeed, whether a person's intellectual verdict is that the prospects for human knowledge are dim or bright, and whether that person's emotional reaction to this verdict is sad or happy, are two separate matters. Rather than the usual giddy triumph over vanquished truth, why not express instead a crushing sadness over unrelenting ignorance? This love of ignorance and uncertainty demands some explanation!

Occasionally, there are revealing remarks that arouse suspicions of a secret alliance, although that alliance may be subtle enough to operate at an unconscious level. For example, philosopher Brown (1987:230) remarked, "I have offered here one detailed argument for the now familiar thesis that there is no fundamental methodological difference between philosophy and science.... [But] it has become progressively clearer that the sciences cannot provide certainty and have no a priori foundation.... [Admittedly,] earlier thinkers believed that both science and philosophy provide certain knowledge of necessary truths. We must conclude that neither do.... [The] human intellect... seems unable to grasp a final truth." So, chastened science has no truth, and now philosophy can enjoy the same!

For another example, science educator Meyling (1997) mentioned one of his high school students who began with a common-sense, realistic view of science, but in the end she accepted her teacher's "fallibilistic-pluralistic model of epistemology" of "existential uncertainty" and "the tentativeness of science." Meyling quoted her saying that "Truth is relative, we have to get used to that, there are only things that are more correct than others, but there is nothing that is absolutely correct.... When you think you know the truth, you force others to think and live that way.... This is a claim on absoluteness that cannot be justified – by no one and by no theory." Meyling commented that "I believe that this recognition is far more important than the knowledge about a whole set of scientific 'facts'," and he was particularly pleased that his student extended her new skeptical epistemology to the "ethical level." He mentioned a letter he had received, in which "Sir Karl R. Popper was very pleased with this quote." But Meyling's enthusiasm and Sir Karl's praise notwithstanding, some parents may feel that a science classroom is not a fitting place to encourage ethical relativism or skepticism in other persons' children.

The rather popular idea that science is the sole source and guardian of empirical evidence, and hence of all objective and public knowledge, is a mistake that can seemingly justify failure of will in other realms outside science. But this mistake cannot be supported by mainstream science, which maintains the exact opposite: that scientific thinking, with public evidence as its foremost feature, is also applicable beyond science itself in the humanities and everyday life. Nor can it be supported by insistence on methodological naturalism because this is a stipulatory convention within natural science that is inherently inappropriate in many other disciplines that also use empirical and public evidence. Nevertheless, this mistaken idea of scientism is easily motivated and long sustained by the most potent of fallacies, failure of will. Frankly, for those persons who heartily want empirical evidence to work for technological comforts *and* not to work for worldview inquiries, simplistic arguments – preferably expressed in a mere sentence or two – should provide welcome and adequate reassurance. On the other hand, for other persons who heartily want empirical evidence to work for technological comforts *and* scientific discoveries *and* worldview inquiries, energetic study of mainstream science and mainstream philosophy should prove fruitful. Getting the most knowledge and benefit from empirical and public evidence requires engaging both the sciences and the humanities, in alignment with the appealing AAAS (1990) vision of science as a liberal art participating in an exciting wider world.

A third and final fallacy for implementing failure of will is *personal exemption.* This fallacy involves mastering fallacies for the purpose of dismantling and evading other persons' arguments, while ignoring the responsibility of detecting and correcting one's own fallacies, as if one has a personal exemption from dealing with truth and reality. The following chapters on probability and statistics examine additional fallacies.

Summary

Logic is the science of correct reasoning and proof. It addresses the relationship between premises and conclusions, including the bearing of evidence on hypotheses. A deductive argument is valid if the truth of its premises entails the truth of its conclusions and is invalid otherwise. Formal deductive logic begins with a language, axioms, and rules and then derives numerous theorems.

As applied in science, deductive logic argues with certainty from an assumed model to particular expected data. By contrast, inductive logic argues with probability from particular actual data to an inferred general model. In its pursuit of realism and truth, scientific thinking alternates deduction, reasoning from mind to world, and induction, reasoning from world to mind.

The first deductive systems to be axiomatized were syllogisms by Aristotle and then geometry by Euclid. Medieval philosopher-scientists advanced deductive logic considerably. Arithmetic was finally axiomatized only just over a century ago by Peano. The modern vision of deduction, which unites all of its branches into a single unified system built on a base of predicate logic, began with stunningly brilliant work by Frege and by Whitehead and Russell.

The formal system for propositional logic presented here has three axioms and one rule. The axioms for predicate logic are about twice as complicated, but the resulting range of theorems that predicate logic can prove is incomparably greater than the range for propositional logic. Peano arithmetic is presented with nine axioms. Probability is another branch of deductive logic, but that topic is deferred to the next chapter.

Fallacies have received much interest since Aristotle. Fallacies are best understood and categorized after first recalling the key resources of scientific thinking: admissible and relevant evidence, and deductive and inductive logic. Accordingly, three major categories of fallacies are inadmissible evidence, irrelevant evidence, and fallacious logic. The fourth and final category of fallacies reviewed in this chapter involves a personal rather than a procedural defect: failure of will to pursue the truth.

Study questions

(1) What are the three interrelated differences between deductive and inductive arguments? Is deduction superior to induction, or are they complementary in scientific thinking?

(2) What are the truth-table definitions for the logical operators Nor and Nand? Why are these operators so extremely useful in computer circuits?

(3) What two main sorts of considerations inform axiom choice for any standard version of a deductive system, such as standard logic or standard arithmetic? What are some applications for nonstandard arithmetic and non-Euclidean geometry?

(4) What is the fallacious *argumentum ad lapidem*, the argument to a stone? Can you contrive an alluring example? How does this fallacy relate to science's presuppositions?

(5) Failure of will to pursue the truth can be implemented by various means, including privileged cynicism, secret alliance, and personal exemption. Give an example of each. Might failure of will be a contributing factor in attacks on science's rationality? Explain your answer.

8

Probability

Suppose the Smiths tell you that they have two children and show you the family photograph in Figure 8.1. One child is plainly a girl, but the other is obscured by being behind a dog so that its gender is not apparent. What is the probability that the other child is also a girl? This probability question might seem quite simple. But the fact that this problem appeared in the pages of *Scientific American* is a hint that it might actually be tricky (Stewart 1996). Later in this chapter, this problem will be solved, but for now, just remember your initial answer for comparison with the correct solution.

Probability is the branch of deductive logic that deals with uncertainty. Logic occupies three chapters in this book on scientific method, with this chapter being the middle one. The previous chapter concerned other branches of deductive logic: propositional logic, predicate logic, and arithmetic. The next chapter concerns inductive logic, also called statistics. Recall that the PEL model identifies three inputs needed to reach any scientific conclusions: presuppositions, evidence, and logic. Accordingly, science needs functional deductive and inductive logic.

Another reason why the study of probability is important is that errors in probability reasoning are among the most common and detrimental of all fallacies. Probability errors prompt physicians to administer suboptimal treatments. Probability errors prompt juries to render wrong verdicts. And probability errors also cost scientists plenty. Correct probability reasoning is important because scientific research and daily life alike are full of unavoidable practical decisions that must be made on the basis of imperfect information and uncertain inferences. And yet, probability reasoning is rather difficult because it involves precise distinctions and complex relationships that are often subtle and sometimes counter-intuitive. Consequently, some basic training in probability theory can confer substantial benefits.

Figure 8.1 Children in a family. The Smith family has two children. One is a girl, but the other is obscured by the family dog. Reasoning with conditional probability can calculate the probability that the hidden child is also a girl. (This drawing by Susan Bonners is reproduced with her kind permission.)

Probability concepts

There are two primary concepts of probability, one pertaining to events and one to beliefs. An *objective* or *physical* probability expresses the propensity of an event to occur. For example, upon flipping a fair coin, the probability of heads is 0.5 (because there are two possible events or outcomes, namely, heads or tails, and they are equally likely). A *subjective* or *personal* or *epistemic* probability is the degree of belief in a proposition warranted by the evidence. For example, given today's weather forecast, a given person may judge that the belief "It will rain today" has a 90% probability of being true. Of course, personal and physical probabilities are often interrelated, particularly because personal beliefs are often about physical events. Most events and propositions with low probabilities are not or will not be actualized or true, whereas most with high probabilities are or will be actualized or true.

The concept of probability occurs in a variety of different contexts. A single, unified probability theory needs to work in all of probability's diverse applications. Consider the following eight common-sense usages of the concept of probability:

(1) A fair coin has a probability of 0.5 of heads, and likewise 0.5 of tails; so the probability of tossing two heads in a row is 0.25.
(2) There is a 10% probability of rain tomorrow.
(3) There is a 10% probability of rain tomorrow given the weather forecast.
(4) Fortunately, there is only a 5% probability that her tumor is malignant, but this will not be known for certain until the surgery is done next week.
(5) Smith has a greater probability of winning the election than does Jones.
(6) I believe that there is a 75% probability that she will want to go out for dinner tonight.
(7) I left my umbrella at home today because the forecast called for only a 1% probability of rain.
(8) Among 100 patients in a clinical trial given drug *A*, 83 recovered, whereas among 100 other patients given drug *B*, only 11 recovered; so new patients will have a higher probability of recovery if treated with drug *A*.

All eight examples use the same word, "probability." To a first approximation, the meaning of probability is the same in all of these examples. Furthermore, in informal discourse, other words could be used with essentially the same meaning, such as "chance" or "likelihood." The common-sense meanings of these examples should be entirely clear to everyone. Nevertheless, these eight examples express a variety of distinguishable concepts.

Example 1 is essentially a definition or theoretical description of what is meant by a "fair" coin, that it has equal chances of landing heads or tails, whether or not any actual coin is exactly fair. It then states the deductive implication regarding tossing two heads in succession. Example 2 is solidly empirical, obviously purporting to convey information about the physical world, namely, about the probability of rain. Example 2 also differs from Example 1 in that the first example's event (a coin toss) is a repeatable event, both in theory and in practice, whereas the second example's event (rain in a particular place and day) is a singular, nonrepeatable event. Incidentally, Example 1 expresses its probabilities as numbers within the range of 0 to 1, whereas Example 2 multiplies such values by 100 to yield a percentage, but this cosmetic difference is not particularly significant.

Example 3 is based on Example 2 but adds an explicit statement about the evidence in support of its assertion. Hence, the unconditional probability in Example 2 is a function of only one thing, the event of rain; whereas the conditional probability in Example 3 is a function of two things, the event of rain given the evidence of the weather forecast. Example 4 expresses a 5% probability of malignancy. In fact, however, the tumor is either benign or malignant – it is *not* 5% malignant and 95% benign. Hence, this probability value of 5% must refer to the present state of knowledge, as contrasted with the actual status of the tumor. This interpretation is reinforced by the expectation that further knowledge after surgery next week will modify this probability, hopefully to a zero probability of malignancy rather than to the dreaded 100%. Hence, the concept of probability must be capable of handling the addition of new evidence to an existing body of old evidence.

Example 5 says that Smith has a better chance of winning the election than does Jones, but it does not specify whether Smith's chance is great (say, more than 0.5) or small. The problem is that no information is offered about the presence or absence of other candidates. Also note that this example uses no numbers to express its probabilities but rather merely expresses a comparative relationship of one probability being "greater" than another. Example 6 is the first that explicitly recognizes the existence and role of the person expressing a probability judgment, "*I* believe that. . . ." Furthermore, this example is subjective – evidently other persons might come up with different estimates. For instance, someone with no knowledge whatsoever of this woman's plans might pick a probability of 50% to represent maximal uncertainty or ignorance. Hence, the concept of probability needs to be able to model ignorance as well as knowledge.

Example 7 shows probability taking a role not only in personal inferences and beliefs but also in personal decisions and actions. It weighs the personal cost or bother of carrying an umbrella against the potential benefit of not getting soaked. Finally, Example 8 uses data on two recovery rates to derive a conclusion about a probability judgment. Its logical progression is therefore the reverse of that in the first example. Example 1 is representative of deductive thinking that begins with a model or theory (about a fair coin) and then derives conclusions regarding expected observations (of two heads in succession). By contrast, Example 8 is representative of inductive thinking that progresses in the reverse direction. It begins with specific actual observations (regarding 83 and 11 recoveries) and then supports a general theory (about two drugs' relative merits). Hence, the concept of probability is used in both deductive and inductive settings.

In review, a satisfactory theory of probability must encompass events and beliefs, theoretical and physical entities, repeatable and singular events, numerical and comparative expressions, unconditional and conditional probabilities, old and new evidence, knowledge and ignorance, inferences and decisions, and deductive and inductive contexts. Probability concepts must be sophisticated enough to handle complex scientific problems and yet be sensible enough to express simple common-sense applications.

Four requirements

Probability theory progresses by selecting axioms and then deriving theorems. But what do we want this theory to do? What are its intended applications?

What are the requirements for a satisfactory theory? These requirements are best addressed near the outset of this chapter, even before axiom choice in the next section. Four requirements are specified here, largely following the exceptionally wise and practical book on probability by Sir Harold Jeffreys (1983).

(1) General. An adequate theory of probability must provide a general method suitable for all of its intended applications, including the eight examples of probability concepts in the preceding section. As will be explained in the next chapter, the inductive reasoning in statistics requires no additional axioms beyond those for the deductive reasoning in probability – although decision theory does require one additional axiom. Hence, this chapter's choice of axioms is intended to cover both probability and statistics.

(2) Coherent. The theory of probability and statistics must be coherent or self-consistent. It must not be possible to derive contradictory conclusions from the axioms and any given dataset. Furthermore, all axioms must be stated explicitly, and all subsequent theorems must follow from them. The number of axioms must be small in order to minimize the number of apparently arbitrary choices and to give the foundations great simplicity and clarity.

(3) Empirical. Probability conclusions must be dominated by empirical evidence, not by any presuppositions. Accordingly, probability axioms must be applicable to the physical world but not say specific things about it. For example, consider the scientific finding that the probability is 89.28% that a radioactive decay of a potassium-40 atom emits a positron. Probability theory can be used in this context because it contains no presuppositions about this particular probability, thereby leaving conclusions free to be determined by the evidence. Rather, the only legitimate presupposition of science (including probability theory), which is necessary to render empirical evidence admissible, is that the physical world is real and comprehensible, as was explained in Chapter 5.

(4) Practical. Probability theory must be practical, applicable to real experiences and experiments within reach of human endowments and capacities. It must not require impossible experiments or calculations. The theory must provide for occasional revisions of erroneous scientific inferences because some mistakes are inevitable. What is required is not perfection but rather recoverability in the light of better analysis or more data. Furthermore, all theories of deduction, including probability, have been shown to have some limitations and imperfections, given sufficient statistical and philosophical inspection – so this must be accepted as a permanent and irremediable situation. But it is better to have a probability theory that can do only 99.999% of what all scientists and philosophers could want and yet is tidy and robust, rather than going after that last 0.001% with a dauntingly erudite and disgustingly complex theory that would still be short of perfection.

Kolmogorov Axioms for Probability

1. The probability of event E is a non-negative real number $P(E) \geq 0$.
2. The probability of the conjunction of all possibilities Ω is $P(\Omega) = 1$.
3. For mutually exclusive events $E_1, E_2, \ldots$, the probability of this conjunction of events equals the sum of the individual probabilities: $P(E_1 \cup E_2 \cup \ldots) = P(E_1) + P(E_2) + \ldots$.

Figure 8.2 The three Kolmogorov axioms for probability. They were formulated by Andrey Kolmogorov in 1933.

Probability axioms

Some notation is needed to express probability axioms. The probability of event or belief X is denoted by $P(X)$. The negation of X is denoted by $\sim X$. The members of a set are listed in brackets, such as $\{1, 2\}$ being the set composed of the integers 1 and 2. The union of X and Y is designated by $X \cup Y$, the set containing everything that belongs to X or Y or both. The intersection of X and Y is designated by $X \cap Y$, the set containing everything belonging to both X and Y. For example, if $X = \{1, 2\}$ and $Y = \{2, 3, 4\}$, then the union is $X \cup Y = \{1, 2, 3, 4\}$ and the intersection is $X \cap Y = \{2\}$. By definition, two sets are mutually exclusive if they have no members in common, such as $\{2\}$ and $\{3, 4\}$. Also, by definition, sets are jointly exhaustive if there are no other possibilities. The universal set of all possible outcomes is denoted by Ω.

Figure 8.2 gives the Kolmogorov (1933) axioms for probability. Remarkably, from these three simple axioms, as well as the inherited axioms of predicate logic and arithmetic, all probability theorems can be derived.

For instance, if the probability of X is 0.7, what is the probability of $\sim X$? None of these three axioms can provide an answer for this question. However, the required theorem can be derived readily from these axioms in order to calculate the answer. The union $X \cup \sim X$ is a universal set of all possible outcomes Ω, so axiom 2 yields $P(X \cup \sim X) = P(\Omega) = 1$. Because X and $\sim X$ are mutually exclusive, axiom 3 yields $P(X \cup \sim X) = P(X) + P(\sim X)$. Combining these two results gives $P(X) + P(\sim X) = 1$, and finally rearranging terms yields $P(\sim X) = 1 - P(X)$. Hence, if the probability of rain equals 0.7, then the probability of no rain equals 0.3.

A conditional probability is the probability of X given Y, or X conditional on Y, and is denoted by $P(X \mid Y)$ where the vertical bar "|" means "given." It can be defined in terms of unconditional probabilities: $P(X \mid Y) = P(X \cap Y) / P(Y)$, provided that $P(Y)$ does not equal 0. As a simple example, assume that a class has six girls with blue eyes and four with brown eyes, and five boys with blue eyes and eight with brown eyes. The conditional probability that a student has blue eyes, given that the student is a girl, is $P(\text{blue} \mid \text{girl}) = 6 / (6 + 4)$, or 0.6.

The conditional probability that a student has blue eyes, given that the student is a boy, is P(blue | boy) = 5 / (5 + 8), or about 0.38. For comparison, the unconditional probability that a student has blue eyes is P(blue) = (6 + 5) / (6 + 4 + 5 + 8), or about 0.48.

Conditional probabilities can be surprisingly tricky. Returning to the opening question about the Smith family photograph shown in Figure 8.1: If the Smiths have two children, one of whom is a girl but the other is obscured by the family dog, what is the probability that the other is a girl? Because the obscured child is a specific child, the correct answer is simply 1/2. The same answer applies to any other specific child, such as the youngest child or the oldest child.

However, a different question can have a different answer. Consider instead the question: If the Millers have two children, one of whom is a girl, what is the probability that the other is a girl? Let B and G denote boy and girl, and make the simplifying assumptions that P(B) = P(G) = 1/2 and that gender is an independent factor (although, in fact, boys are slightly more numerous than girls and gender is not independent in the rare case of identical twins – but the topic here is probability theory rather than reproductive biology). Then there are four equally probable gender sequences for the Miller children with the letters arranged in order of birth: BB, BG, GB, and GG. Because we know that the Millers have at least one girl, the sequence BB is eliminated. That leaves three equally likely cases, and in just one of those cases (GG) is the other child also a girl. Hence, the required conditional probability that the other child is a girl is actually 1/3. Therefore, far from being *equally* likely that the other child is a boy or a girl, actually it is *twice* as likely that the other child is a boy.

At this point, you might recall your initial answer to the probability problem about the Smith family photograph, which was posed at the start of this chapter, in order to check whether you had gotten it right. But, even if your answer was correct, was it more than a lucky guess? Was the reasoning behind your answer adequately precise so that you could distinguish and solve both of these probability problems? The crucial difference between these two problems is that the Smith family has two specific children, the visible child and the obscured child, and only the obscured child might be a boy; whereas the Miller family has two unspecific children, so either child might be a boy.

The Kolmogorov axioms are not uniquely suitable for probability because other axiom sets can be chosen that are equivalent and exchangeable, supporting the same probability theorems. For instance, Salmon (1967:59–60) used four axioms expressed with conditional probabilities, the first being that $P(X \mid Y)$ is a single number $0 \leq P(X \mid Y) \leq 1$. These axioms are equally suitable. But some choices may render the proof of a given theorem somewhat harder or easier.

Even as there are nonstandard logics, nonstandard arithmetics, and non-Euclidean geometries, so also there are nonstandard probability theories resulting from unusual axioms. Burks (1977:99–164) gave examples. The standard and nonstandard probability theories are equally internally coherent or consistent, although they contradict each other. A probability theory based on the Kolmogorov axioms or on any equivalent and exchangeable set of axioms suits the four requirements specified in the previous section, but a nonstandard theory does not.

Probability axioms serve to enforce coherence among a set of probability assignments and to derive certain probabilities from others. But they do little to provide probability assignments. Instead, that job is done by probability rules. The main one is the so-called straight rule of induction. It says that "If *n As* have been examined and *m* have been found to be *Bs*, then the probability that the next *A* examined will be a *B* is *m* / *n*" (Earman 2000:22). For instance, if 200 university students are surveyed and 146 report that they got a flu shot, then the probability for another student having gotten this shot is 146/200 or 73%. Of course, one would trust this estimate or prediction more if the survey had a random sample of the students – rather than all athletes, or all women, or all graduate students – because special subgroups might introduce a bias.

Besides enforcing coherence, probability axioms also make probabilities meaningful. Given the Kolmogorov or equivalent axioms, probabilities are scaled in the range 0 to 1, so $P(X) = 0$ means X is impossible, $P(X) = 1$ means X is certain, and $P(X) = 0.5$ means that X and $\sim X$ are equally likely. But without probability axioms, $P(X) = 0.3$ or $P(X) = 817.7$ or whatever would be utterly meaningless, communicating nothing about the probability of X. Likewise, given the probability axioms and theorems, $P(X) = 0.7$ has a clear implication for $P(\sim X)$; whereas, without a coherent probability theory, there would be no implication whatsoever. Incidentally, the same sentiments apply to arithmetic. That four apples plus three apples equals seven apples is meaningful given the coherent and meaningful context provided by standard arithmetic axioms and theorems; but, without that context, an isolated arithmetic assertion about seven apples would lack meaning utterly. This larger context may be informal and implicit in common sense or may be formal and explicit in probability theory; but, in either case, coherence is essential for meaning in any kind of deductive or inductive reasoning.

Bayes's theorem

For millennia, there had been interest in quantifying probabilities for gambling and other applications, but exact mathematical formulations developed relatively recently. Early contributors were Pierre de Fermat (1601–1665), Blaise Pascal (1623–1662), Christiaan Huygens (1629–1695), Jakob Bernoulli (1654–1705), Abraham de Moivre (1667–1754), and Daniel Bernoulli (1700–1782). They explored probability in its deductive setting, reasoning from a given model to expected observations.

The first person to explore probability in its inductive setting, reasoning in the opposite direction from actual observations to the model, was Thomas Bayes (1702–1761), whose seminal paper was published posthumously in Bayes (1763) by his friend Richard Price. Soon thereafter, Pierre-Simon Laplace (1749–1827) independently discovered Bayes's theorem in 1774 and further developed this inductive reasoning, which was called "inverse probability." But then, in the 1920s, an alternative approach called *frequentist statistics* was developed, but that story is better told in the next chapter. The seminal paper by Bayes (1763) is readily available on the Internet and has also been reproduced by Barnard (1958) and Swinburne (2002:117–149).

A simple form of Bayes's theorem is:

$$P(A|B) = [P(B|A) \times P(A)]/P(B). \quad (8.1)$$

Each term has a conventional name. $P(A \mid B)$ is the conditional probability of A given B, also called the posterior probability of A. $P(B \mid A)$ is the conditional probability of B given A, also called the likelihood. $P(A)$ is the unconditional probability of A, also called the prior probability of A. $P(B)$ is the unconditional or prior probability of B.

Bayes's theorem can be derived easily from the definition of conditional probability. Recall the definition from the previous section that $P(A \mid B) = P(A \cap B) / P(B)$, so likewise $P(B \mid A) = P(A \cap B) / P(A)$. Rearranging and combining these two equations yields $P(A \mid B) \times P(B) = P(A \cap B) = P(B \mid A) \times P(A)$. Finally, dividing the left and right sides of this equation by $P(B)$, provided that $P(B)$ does not equal 0, yields the simple form of Bayes's theorem stated previously.

The salient feature of Bayes's theorem is that it relates the reverse conditional probabilities $P(A \mid B)$ and $P(B \mid A)$. These two quantities always have different meanings and usually have different numerical values, sometimes wildly different.

In an important application, let H denote a hypothesis from some theory or model and E denote some evidence or data. Then, a quantity of the form $P(H \mid E)$ represents inductive reasoning from given evidence to an inferred hypothesis, whereas the reverse conditional probability $P(E \mid H)$ represents deductive reasoning from a given hypothesis to expected evidence. Recall that these opposite reasoning directions of deduction and induction were depicted in Figure 7.1.

Bayes's theorem is used by statisticians for many purposes, including estimating quantities and testing hypotheses. A convenient form for testing two competing hypotheses H_1 and H_2 in light of evidence E is the ratio form:

$$\frac{P(H_1|E)}{P(H_2|E)} = \frac{P(E|H_1)}{P(E|H_2)} \times \frac{P(H_1)}{P(H_2)}. \quad (8.2)$$

This equation may be read as the posterior ratio equals the likelihood ratio times the prior ratio. For example, if initial considerations give hypotheses H_1 and H_2 a prior ratio of 1/5 favoring H_2 and a new experiment gives them a likelihood ratio of 200/1 favoring H_1, then the posterior ratio of 40/1 reverses the initial preference to instead favor H_1.

This middle term, the likelihood ratio $P(E \mid H_1) / P(E \mid H_2)$, is also called the *Bayes factor*. It constitutes an alternative to posterior probabilities or ratios for reporting the results from a Bayesian analysis of some evidence (Kass and Raftery 1995). A large Bayes factor would favor H_1, a small factor would favor H_2, and a factor near 1 would be rather uninformative.

Another form of Bayes's theorem for two hypotheses is convenient for solving some probability problems:

$$P(H_1|E) = \frac{P(E|H_1) \times P(H_1)}{[P(E|H_1) \times P(H_1)] + [P(E|H_2) \times P(H_2)]}. \tag{8.3}$$

Note that in order to solve for $P(H_1 \mid E)$ on the left, four quantities must be known, as specified on the right. However, in the special case that H_1 and H_2 are mutually exclusive and jointly exhaustive, either $P(H_1)$ or $P(H_2)$ suffices for determining both of them because of the trivial relationship that $P(H_1) + P(H_2) = 1$. Hence, to solve for $P(H_1 \mid E)$, the required probabilities are $P(E \mid H_1)$, $P(E \mid H_2)$, and either $P(H_1)$ or $P(H_2)$.

A nice little application of Bayes's theorem from statistician Sir Ronald A. Fisher (1973:18–20) concerns black and brown mice. The gene for black fur (*B*) is dominant over the gene for brown (*b*), so the homozygous *BB* and heterozygous *Bb* are black, whereas the homozygous *bb* is brown. Now suppose that a female, known to be heterozygous *Bb* (because her parents were *BB* and *bb*), is mated with a heterozygous male. From basic Mendelian genetics for a diploid organism such as mice, the expectation for the offspring from this mating between two black heterozygous parents is one black homozygous *BB* to two black heterozygous *Bb* to 1 brown homozygous *bb*. Interest now focuses on one of her black daughters. The competing hypotheses are H_{BB} that this black daughter is homozygous and H_{Bb} that she is heterozygous. From Mendelian genetics, the prior probabilities are $P(H_{BB}) = 1/3$ and $P(H_{Bb}) = 2/3$, so the prior ratio is $P(H_{BB}) / P(H_{Bb}) = 1/2$. Further suppose that this black daughter is mated with a brown male (*bb*) and she has a litter with seven offspring, all black. The likelihoods resulting from this experimental evidence are $P(\text{litter} \mid H_{BB}) = 1$ whereas $P(\text{litter} \mid H_{Bb}) = 1 / 2^7 = 1/128$, so the likelihood ratio or Bayes factor is $P(\text{litter} \mid H_{BB}) / P(\text{litter} \mid H_{Bb}) = 128$ in favor of H_{BB}. Finally, multiplying the prior ratio from the background information by the Bayes factor from the experimental data gives the posterior ratio $P(H_{BB} \mid \text{litter}) / P(H_{Bb} \mid \text{litter}) = 128/2 = 64$ that favors H_{BB}. The posterior probabilities are $P(H_{BB} \mid \text{litter}) = 64 / (1 + 64) = 64/65$ and $P(H_{Bb} \mid \text{litter}) = 1 / (1 + 64) = 1/65$. This example of Bayesian inference is exceptionally tidy because both the prior information from the pedigree and the experimental evidence from the litter are objective and public.

Furthermore, given the reasonably strong evidence from this single litter, any further evidence from additional litters would probably confirm the initial verdict. Indeed, the probability that this initial conclusion H_{BB} is actually false is 1/65, or only about 1.5%. But, if this experiment was repeated and that mouse produced several more litters, an exceedingly strong conclusion would result because the weight of the evidence grows exponentially with its amount. For instance, five additional and similar litters would give a posterior ratio (or a Bayes factor) of more than 10^{12} in favor of H_{BB}, which would render H_{BB} practically certain. On the other hand, if the mouse in question were heterozygous, then the investigation would be easier because a single brown offspring would prove the hypothesis H_{Bb} with certainty without any calculations being needed. This definitive conclusion that H_{Bb} is true would be expected to emerge rather quickly because at least one brown mouse among N offspring is expected with probability $1 - 2^{-N}$.

The advantage of posterior probabilities is that they address most directly the foremost question of scientists and scholars about competing hypotheses, namely, which hypothesis is probably true given the evidence. But the advantage of the Bayes factor is that if the prior probabilities are highly controversial or rather inscrutable, then any persons can compute their own posterior probabilities from the reported Bayes factor and their own personal prior probabilities. Also, in the special though fairly frequent case that the Bayes factor is huge – especially 10^{10} or more – the Bayes factor delivers an exceedingly strong verdict on its own without the additional labor of calculating and defending particular values for the prior probabilities.

In typical applications, the evidence E is public and settled, whereas the background information in the prior is personal and controversial (unlike the tidy mouse example with clear information determining the prior probabilities). For instance, the shared evidence E could be from a published clinical trial concerning two medications X and Y, whereas the prior information of individual physicians could be their own experiences of success or failure from giving patients those medications, which might vary considerably from physician to physician. The public evidence E needs to be reasonably strong in order to have greater influence than one's personal prior information (which might be contrary to the clinical trial) and thereby to convince most persons, or at least to interest them.

From its start in 1763 to the present day, Bayes's theorem has been applied extensively in wonderfully diverse contexts across the sciences and the humanities. In his introduction to Bayes's paper, Richard Price touted Bayes's contribution as "a sure foundation for all our reasonings concerning past facts, and what is likely to be hereafter," and said that "it is also a problem that has never before been solved" (Bayes 1763). Bayes's own examples concerned the positions of balls on a table and the proportions of blanks and prizes in a lottery. But Price's introduction also mentioned the potential application in philosophical or theological arguments "from final causes for the existence of the Deity," for which this new inverse probability is "more directly applicable" than the previous sorts of probability reasoning. Bayes's paper was published in the *Philosophical Transactions* of the Royal Society, the oldest scientific society in the world, and both the Rev. Bayes and the Rev. Price were fellows of that prestigious society. The wide range of interests and applications that they foresaw for probability in this new context of inductive reasoning were characteristic of the philosopher-scientists of their times. Indeed, the second Charter of 1663 of the Royal Society (which replaced the first Charter of 1662 with greater privileges by which the Society has since been, and continues to be, governed) stated the purpose of "further promoting by the authority of experiments the sciences of natural things and of useful arts, to the glory of God the Creator, and the advantage of the human race." McGrayne (2011) has written a brilliant and entertaining account of how Bayes's theorem has contributed to many important developments, including cracking the German's Enigma code during World War II, discovering how genes are controlled and regulated, and implementing spam filters for email. At present, Bayes's theorem is used extensively in science and technology, as well as in philosophy and the social sciences. Countless consumer goods are what they are today in part because of Bayes's theorem helping scientists to optimize their inferences and decisions during the research and development that has improved those products. Likewise, Bayes's theorem has added considerable clarity to many important conversations in the humanities.

Probability distributions

A dozen probability distributions are used frequently and dozens more occasionally. A few of the most important ones are mentioned here. A probability distribution specifies the probability y over the range of the variable x. The height of a probability curve is adjusted such that the area under the curve is 1, in keeping with the second Kolmogorov axiom that $P(\Omega) = 1$.

The uniform distribution is the simplest one. Over the range $0 \leq x \leq 1$, its probability is $y = 1$ inside this range, and $y = 0$ elsewhere. More generally, over the range $a \leq x \leq b$, the probability is $y = 1 / (b - a)$ inside this range, and $y = 0$ elsewhere. For instance, over the range $-1 \leq x \leq 1$, the probability is $y =$ 0.5 inside this range, and $y = 0$ elsewhere.

The normal or Gaussian distribution is the most prominent probability distribution. It has the equation:

$$f(x) = \frac{1}{\sqrt{2\pi\sigma^2}} e^{-\frac{(x-\mu)^2}{2\sigma^2}} \tag{8.4}$$

where μ is the mean, which locates the peak of this familiar bell-shaped curve, and σ^2 is the variance, which measures the width or spread of the distribution. The square root of the variance, σ, is called the standard deviation. The standard normal distribution has $\mu = 0$ and $\sigma^2 = 1$. For the standard normal distribution, from plus to minus 1 standard deviation accounts for about 68.27% of the area under the curve, 2 standard deviations for 95.45%, and 3 for 99.73%.

The central-limit theorem states that under mild conditions, the sum of a large number of random variables is distributed approximately normally. For instance, an easy method for generating random variables with a nearly standard normal distribution is to sum 12 random variables from the uniform distribution with range $0 \leq x \leq 1$ and then subtract 6. Observational errors are often caused by the cumulative effects of a number of uncontrolled factors, giving these errors an approximately normal distribution. For N replicate observations with errors having a standard deviation of σ, the standard error of the mean of those N replicates equals $\sigma / N^{0.5}$. The normal distribution has the advantage of being very tractable mathematically.

The log-normal distribution applies to a random variable whose logarithm is normally distributed. Whereas the normal distribution arises from summing a number of random variables, the log-normal distribution arises from multiplying a number of random variables, all of which are positive. An amazing number of things in the physical, biological, and social sciences approximate a log-normal distribution. Examples include the diameter of ice crystals in ice cream or oil drops in mayonnaise, the abundance of species (bacteria, plants, or animals), latency periods for many human diseases, city sizes, and household incomes.

The binomial distribution describes the number of successes in a sequence of N independent trials, each of which yields success with probability p and failure $(1 - p)$. A simple example is tosses of a fair coin, with heads and tails equally probable. As the number of coin tosses becomes large, the binomial distribution approximates the normal distribution.

Another distribution is the Poisson distribution, which expresses the probability of a number of events occurring within a fixed period of time if these events occur with a known average rate and independently of the time since the last event. For instance, it applies to decays of atoms in a sample of a radioactive material.

Many common probability distributions, including the normal, binomial, and Poisson distributions, belong to an important class, the exponential family. Accordingly, mathematical results proven for the exponential family have a wide range of applications. However, for this book on scientific method, this brief section on probability distributions must suffice. Any further study of probability distributions is better left to texts on probability and statistics.

Permutations and combinations

Many probability problems, especially in the context of games and gambling, involve a number of possible outcomes that are equally probable. Such problems can be solved by counting numbers of permutations or combinations.

For R events or experiments such that the first event has N_1 possible outcomes, and for each of those the second event has N_2 possible outcomes, and so on up to the Rth with N_R outcomes, there is a total of $N_1 \times N_2 \times \cdots \times N_R$ possible outcomes. For example, how many different license plates could be made with three digits followed by three letters? The solution is

$$10 \times 10 \times 10 \times 26 \times 26 \times 26 = 17{,}576{,}000. \tag{8.5}$$

A permutation is a distinct ordered arrangement of items. For example, for the set of letters *A* and *B* and *C*, all possible permutations are *ABC*, *ACB*, *BAC*, *BCA*, *CAB*, and *CBA* – which number 6, because the first choice has 3 options, the second 2, and the third 1, for a total of $3 \times 2 \times 1 = 6$ permutations. The general rule is that for N entities, there are $N \times (N - 1) \times (N - 2) \times \cdots \times 3 \times 2 \times 1$ permutations. This number is called "N factorial" and is denoted by $N!$ For example, $1! = 1$, $3! = 6$, and $5! = 120$. Also, by definition $0! = 1$. As a simple probability problem, presuming that these three letters are drawn at random, what is the probability of a drawing starting with the letter *A*? It is 1/3 because there are two permutations satisfying this condition (*ABC* and *ACB*) and there are six permutations all told.

Sometimes the entities are not all unique, as in the set *A*, *A*, *B*, and *C*, which has two letters *A* that are alike. For N objects, of which N_1 are alike, N_2 are alike, and so on up to N_R alike, there are $N! \,/\, (N_1! \times N_2! \times \cdots \times N_R!)$ permutations. Hence, this set of letters has $4! \,/\, (2! \times 1! \times 1!) = 24/2 = 12$ permutations. Presuming that these four letters are drawn at random, what is the probability of a drawing starting with both *A*s? It is 1/6 because there are 2 permutations satisfying this condition (*AABC* and *AACB*) and there are 12 permutations all told.

A combination is a particular number for each of several different entities or outcomes for which the order does not matter. For instance, consider selecting three items from the five items *A*, *B*, *C*, *D*, and *E*. There are five ways to select the first item, four for the second, and three for the third, so there are $5 \times 4 \times 3 = 60$ permutations that distinguish different orderings. But every group of three items, such as *A* and *B* and *C*, gets counted $3! = 6$ times, as was explained earlier. So the number of combinations, which do not distinguish different orderings, of three items selected from five is 60/6 or 10 and all of them are equally probable. The general rule is that N objects taken R at a time have $N! \,/\, ((N - R)! \times R!)$ possible combinations. Presuming that three letters are drawn from these five at random, what is the probability of drawing a combination that includes the letters *A* and *B*? It is 3/10 because there are 3 combinations satisfying this condition (*ABC*, *ABD*, and *ABE*) and there are 10 combinations all told.

Probability fallacies

There is a substantial literature on probability fallacies in medicine, law, science, and other fields. An especially common one is the *base rate fallacy*, also called the *false positive paradox*, which results from neglecting the base rate or prior probabilities. But because that neglect automatically leads to confusion between $P(H|E)$ and $P(E|H)$, an equally suitable name would be the *reversed conditionals fallacy*. Stirzacker (1994:25) proposed an example that I have expressed concisely as follows:

A simple medical problem involves three facts and one question. The facts are: (1) A rare disease occurs by chance in 1 in every 100,000 persons. (2) If a person has the disease, a fairly reliable blood test correctly diagnoses the disease with probability 0.95. (3) If a person does not have the disease, the test gives a false diagnosis of disease with probability 0.005. If the blood test says that a person has the disease, what is the probability that this is a correct diagnosis?

Most people, including many physicians, answer that the probability of disease is about 95%. However, from plugging the numbers into Bayes's theorem, surprisingly the correct answer is $(0.95 \times 0.00001) / [(0.95 \times 0.00001) + (0.005 \times 0.99999)]$, or only about 0.2%. This is *drastically* different from 95%, and it strongly supports the exact opposite conclusion! As Stirzaker (1994:26) remarked, "Despite appearing to be a pretty good test, for a disease as rare as this the test is almost useless." For every real instance of the disease detected by that test, there would be more than 500 false positives, so the results could hardly be taken seriously. At best, such a test might offer economical screening before administering another more expensive, definitive test.

What went wrong to give that incorrect answer? Let H_W and H_S be the hypotheses that the person is actually well or sick, and let E be the evidence of a blood test indicating disease. What is given is that $P(H_S) = 0.00001$ and hence $P(H_W) = 0.99999$, that $P(E \mid H_S) = 0.95$, and that $P(E \mid H_W) = 0.005$; and what is required is $P(H_S \mid E)$. The incorrect answer results from ignoring all but one fact, that $P(E \mid H_S) = 0.95$, and assuming erroneously that the reverse conditional probability $P(H_S \mid E)$ has this same value. But ignoring the base rate is a fallacy. Indeed, all three facts given in this problem are needed to obtain the correct solution. That most people in general, and many doctors in particular, make this common blunder in probability reasoning is alarming and potentially dangerous.

Two additional probability fallacies are problematic in the context of law, as emphasized in a seminal paper by Thompson and Schumann (1987): the *Prosecutor's Fallacy* and the *Defense Attorney's Fallacy*. This context is tremendously important because, increasingly, criminal cases have involved scientific evidence and statistical arguments. They gave an example that I have expressed concisely as follows:

A simple legal example of probability reasoning involves two facts and one question. The facts are: (1) various other kinds of information give a prior probability of 0.1 that the suspect is guilty of committing a murder; (2) a sample of the murderer's blood found at the scene of the crime matches the suspect's rare blood type found in only one person in 100. The question is: How much weight should be given to this evidence from the laboratory blood test?

Thompson and Schumann presented this probability problem to 73 college undergraduates and asked them to evaluate the following two arguments. The first is an example of the Prosecutor's Fallacy and the second of the Defense Attorney's Fallacy. I have edited their texts slightly.

The prosecution argued that the blood test is highly relevant. The suspect has the same blood type as the attacker. This blood type is found in only 1% of the population, so there is only a 1% chance that the blood found at the scene came from someone other than the suspect. Since there is only a 1% chance that someone else committed the crime, there is a 99% chance that the suspect is guilty.

The defense argued that the evidence about blood types has very little relevance. Admittedly only 1% of the population has the rare blood type. But the city where the crime occurred has a population of 200,000 so this blood type would be found in about 2,000 persons. Therefore, the evidence merely shows that the suspect is one out of 2,000 persons in the city who might have committed the crime. A one-in-2,000 chance or 0.05% probability has little relevance for proving that this suspect is guilty.

What do you think of these two arguments? Of the 73 college students, about 29% thought the prosecution's argument is correct and 69% thought the defense's argument is correct (including a few students who judged both arguments correct). In fact, both of these arguments are fallacious.

Let H_G and H_I denote the hypotheses that the suspect is guilty or innocent and E denote the evidence of the matching blood type. We are given that $\mathrm{P}(H_G) = 0.1$, so $\mathrm{P}(H_I) = 0.9$, and also that $\mathrm{P}(E \mid H_I) = 0.01$. The prosecution's argument assumes that $\mathrm{P}(H_G \mid E) = 1 - \mathrm{P}(E \mid H_I)$. But this is a fallacy because these reversed conditional probabilities are not properly related (in addition to the problem that the prior information is ignored).

On the other hand, the defense's argument adds a new fact, that the population of the city is 200,000. But that fact is irrelevant. This number does not appear when this problem is framed properly by Bayes's theorem. That it is irrelevant can be seen easily by attempting to insert such a number into the previous problem, which is analogous, about a blood test for a rare disease. Whether the patient lives in a large city of 10,000,000 persons or a small town of 350 persons is obviously irrelevant as regards this patient's diagnosis.

What is the probability of guilt given the match, $\mathrm{P}(H_G \mid E)$, according to Bayes's theorem? Again, we know $\mathrm{P}(H_G)$, $\mathrm{P}(H_I)$, and $\mathrm{P}(E \mid H_I)$. However, we are not supplied any value for $\mathrm{P}(E \mid H_G)$, so not enough information has been given to solve this problem. Nevertheless, for the sake of argument, if we assume that $\mathrm{P}(E \mid H_G)$ is quite close to 1, then by Bayes's theorem the answer is

$(0.01 \times 0.9) / [(0.01 \times 0.9) + (1 \times 0.1)] = 0.009/0.109$, or about 8% probability that the suspect is guilty, given the prior information and the blood test. Note that this value is in between the 99% probability of the prosecution and the 0.05% probability of the defense. Also note that the probability of matching blood types $P(E \mid H_I)$ is required by Bayes's theorem in order to calculate the reverse conditional probability $P(H_I \mid E)$, so the defense's argument that the former has little relevance is clearly fallacious.

Thompson and Schumann concluded on a somber note. "The use of mathematical evidence is likely to increase dramatically in the future . . . and legal professionals will increasingly face difficult choices about how to deal with it. Because their choices will turn, in part, on assumptions about the way people respond to mathematical evidence, now is an opportune time for social scientists to begin exploring this issue. Our hope is that social scientists . . . will be able to answer the key underlying behavioral questions so that lawyers and judges may base decisions about mathematical evidence on empirical data rather than unguided intuitions."

Whether an application of probability theory is in medicine or law or science or whatever, such precisely are the choices: reliable inferences based on empirical evidence, or else unreliable inferences based on unguided intuitions. Even a basic education in probability theory, such as this brief chapter provides or at least begins, can reduce probability fallacies.

Summary

Probability is the propensity for an event to occur or for a proposition to be true. A convenient scaling for probabilities ranges from 0 to 1, with 0 representing certain falsity, 1 representing certain truth, and intermediate values representing uncertainty.

To do business with physical reality, probability theory must meet four basic requirements. It must be general, suitable for all of its intended applications, including deductive probability reasoning and inductive statistics reasoning. It must be coherent. It must not make specific assertions about the physical world in advance of observation and experimentation so that probability conclusions can be dominated by empirical evidence, not by any inappropriate presuppositions. And it must be practical, applicable to real experiences and experiments within reach of human endowments and capacities.

Probability theory is built on predicate logic and arithmetic, requiring the three Kolmogorov probability axioms (or an equivalent set of axioms). Probability axioms serve to enforce coherence among a set of probability assignments and to make probabilities meaningful. The definition of conditional probability is $P(X \mid Y) = P(X \cap Y) / P(Y)$, provided that $P(Y)$ does not equal 0. In addition to axioms, probability theory also needs rules to assign probabilities, the principal one being the straight rule of induction. And, for certain problems, counting numbers of permutations and combinations of equally probable events can provide probability assignments.

Bayes's theorem is easily derived from the definition of conditional probability. The salient feature of Bayes's theorem is that it relates the reverse conditional probabilities $P(A \mid B)$ and $P(B \mid A)$. A simple form of Bayes's theorem is: $P(A \mid B) = [P(B \mid A) (P(A)] / P(B)$. Two additional forms are given that are useful for solving various problems. For hypotheses H_1 and H_2 and evidence E, the most common report from a Bayesian hypothesis test is the posterior probabilities $P(H_1 \mid E)$ and $P(H_2 \mid E)$. But an alternative report that is sometimes used is the Bayes factor $P(E \mid H_1) / P(E \mid H_2)$. Since its publication in 1763, Bayes's theorem has had countless applications across the sciences and the humanities.

Several common probability distributions are defined, including the uniform, normal, and log-normal distributions. The normal distribution is particularly important because the central-limit theorem states that under mild conditions, the sum of a large number of random variables is distributed approximately normally. Observational errors often follow an approximately normal distribution.

Probability fallacies are rampant in medicine, law, science, and other fields. An especially common one is the base rate fallacy that results from neglecting the base rate or prior probabilities, which then leads to confusion between the reverse conditional probabilities $P(H \mid E)$ and $P(E \mid H)$, where H denotes some hypothesis and E denotes some evidence. Additional probability fallacies that occur in the context of law are called the Prosecutor's Fallacy and the Defense Attorney's Fallacy. A basic understanding of probability theory helps one make reliable inferences based on empirical evidence rather than unreliable inferences based on unguided intuitions.

Study questions

(1) The Bakers have three children, of whom two are boys. What is the probability that their other child is a girl?

(2) What four requirements underlie a choice of probability axioms? List the three Kolmogorov axioms. What roles do these axioms serve?

(3) State Bayes's theorem in the ratio form. Give names to the three ratios, as well as a basic explanation of what each term means. What are the differences between the reverse conditional probabilities $P(H \mid D)$ and $P(D \mid H)$ as regards both meaning and numerical value?

(4) What does the central-limit theorem state about the normal distribution? Why do experimental errors often approximate the normal distribution?

(5) A rare disease affects 170 in every 100,000 persons. If a person has this disease, a diagnostic test detects it with probability 0.85; whereas if a person does not have this disease, a false positive occurs with probability 0.003. If the test indicates that a person has this disease, what is the probability that this is a correct diagnosis?

9

Inductive logic and statistics

The logic that is so essential for scientific reasoning, being the "L" portion of the PEL model, is of two basic kinds: deductive and inductive. Chapter 7 reviewed deductive logic, and Chapter 8 probability, which is a branch of deductive logic. This chapter reviews inductive logic, with "statistics" being essentially the term meaning applied inductive logic.

A considerable complication is that statisticians have two competing paradigms for induction: Bayesian and frequentist statistics. At stake are scientific concerns, seeking efficient extraction of information from data to answer important questions, and philosophical concerns, involving rational foundations and coherent reasoning.

This chapter cannot possibly do what entire books on statistics do – present a comprehensive treatment. But it can provide a prolegomenon to clarify the most basic and pivotal issues, which are precisely the aspects of statistics that scientists generally comprehend the least. The main objectives are to depict and contrast the Bayesian and frequentist paradigms and to explain why inductive logic or statistics often functions well despite imperfect data, imperfect models, and imperfect scientists. Extremely important research in agriculture, medicine, engineering, and other fields imposes great responsibilities on statistical practice.

Historical perspective on induction

This section gives a brief history of induction from Aristotle to John Stuart Mill, with more recent developments deferred to later sections. Aristotle (384–322 BC) had a broad conception of induction. Primarily, induction is reasoning from particular instances to general conclusions. That is ampliative reasoning from observed to unobserved, from part to whole, from sample to population. Aristotle cautioned against hasty generalizations and noted that a single counter example suffices to nullify a universal generalization. He carefully distinguished induction from deduction, analogy, and isolated examples.

One of Aristotle's most influential contributions to the philosophy of science was his model of scientific logic or reasoning, the inductive–deductive method. Scientific inquiry alternates inductive and deductive steps. From observations, induction provides general principles, and with those principles serving as premises, deduction predicts or explains observed phenomena. Overall, there is an advance from knowledge of facts to knowledge of an explanation for the facts.

Epicurus (341–271 BC) discussed the fundamental role of induction in forming concepts and learning language in his doctrine of "anticipation." From repeated sense perceptions, a general idea or image is formed that combines the salient, common features of the objects, such as the concept of a horse derived from numerous observations of horses. Once stored in memory, this concept or anticipation acts as an organizing principle or convention for discriminating which perceptions or objects are horses and for stating truths about horses.

Robert Grosseteste (*c.* 1168–1253) affirmed and refined Aristotle's inductive–deductive method, which he termed the Method of Resolution and Composition for its inductive and deductive components, respectively. But he added to Aristotle's methods of induction. His purposes were to verify true theories and to falsify false theories. Causal laws were suspected when certain phenomena were frequently correlated, but natural science sought robust knowledge of real causes, not accidental correlations. "Grosseteste's contribution was to emphasize the importance of *falsification* in the search for true causes and to develop the method of verification and falsification into a systematic method of experimental procedure" (Crombie 1962:84). His approach used deduction to falsify proposed but defective inductions. As mentioned in the earlier chapter on deduction, Grosseteste's Method of Verification deduced consequences of a theory beyond its original application and then checked those predictions experimentally. His Method of Falsification eliminated bad theories by deducing implications known to be false.

Grosseteste clearly understood that his optimistic view of induction required two metaphysical presuppositions about the nature of physical reality: the uniformity of nature and the principle of parsimony or simplicity. Without those presuppositions, there is no defensible method of induction in particular or method of science in general.

In essence, at Oxford in 1230, Grosseteste's new scientific method – with its experiments, Method of Resolution and Composition, Method of Verification, Method of Falsification, emphasis on logic and parsimony, and commonsense presuppositions – was the paradigm for the design and analysis of scientific experiments. Science's goal was to provide humans with truth about the physical world, and induction was a critical component of scientific method.

Roger Bacon (*c.* 1214–1294) promulgated three prerogatives of experimental science, as mentioned in Chapter 3. Of those, the first two concerned induction. His first prerogative was that inductive conclusions should be submitted to further testing. That was much like his predecessor Grosseteste's Method of Verification. His second prerogative was that experiments could increase the amount and variety of data used by inductive inferences, thus helping scientists to discriminate between competing hypotheses.

John Duns Scotus (*c.* 1265–1308), at Paris, reflected Oxford's confidence about inductive logic. He admired Grosseteste's commentaries on Aristotle's *Posterior Analytics* and *Physics* but disagreed on some points. Duns Scotus admitted that, ordinarily, induction could not reach evident and certain knowledge through complete enumeration, and yet he was quite optimistic that "probable knowledge could be reached by induction from a sample and, moreover, that the number of instances observed of particular events being correlated increased the probability of the connexion between them being a truly universal and causal one. . . . He realized that it was often impossible to get beyond mere empirical generalizations, but he held that a well-established empirical generalization could be held with certainty because of the principle of the uniformity of nature, which he regarded as a self-evident assumption of inductive science" (Crombie 1962:168–169).

Building on an earlier proposal by Grosseteste, Duns Scotus offered an inductive procedure called the Method of Agreement. "The procedure is to list the various circumstances that are present each time the effect occurs, and to look for some one circumstance that is present in every instance" (Losee 2001:29–30). For example, if circumstances *ABCD*, *ACE*, *ABEF*, and *ADF* all gave rise to the same effect *x*, then one could conclude that *A* could be the cause of *x*, although Duns Scotus cautiously refrained from the stronger claim that *A* must be the cause of *x*. The Method of Agreement could promote scientific advances by generating plausible hypotheses that merited further research to reach a more nearly definitive conclusion.

Henry of Ghent (*c.* 1217–1293), in contrast to Duns Scotus, believed that real knowledge had to be about logically necessary things, not the contingent things of which the physical world is composed. Had his view prevailed, science in general and induction in particular would now be held in low philosophical esteem.

William of Ockham (*c.* 1285–1347) further developed inductive logic along lines begun earlier by Grosseteste and Duns Scotus. He added another inductive procedure, the Method of Difference. "Ockham's method is to compare two instances – one instance in which the effect is present, and a second instance in which the effect is not present" (Losee 2001:30–31). For example, if circumstances *ABC* gave effect *x*, but circumstances *AB* did not, then one could conclude that *C* could be the cause of *x*. But Ockham was cautious in such claims, especially because he realized the difficulty in proving that two cases differed in only one respect. As a helpful, although partial, solution, he recommended comparing a large number of cases to reduce the possibility that an unrecognized factor could be responsible for the observed effect *x*.

Nicholas of Autrecourt (*c.* 1300–1350) had the most skeptical view of induction among medieval thinkers, prefiguring the severe challenge that would come several centuries later from David Hume. "He insisted that it cannot be established that a correlation which has been observed to hold must continue to hold in the future" (Losee 2001:37). Indeed, if the uniformity of nature is questioned in earnest, then induction is in big trouble. Recall that Grosseteste had recognized that induction depended on the uniformity of nature.

Sir Francis Bacon (1561–1626) so emphasized induction that his conception of scientific method is often known as Baconian induction. He criticized Aristotelian induction on three counts: haphazard data collection without systematic experimentation; hasty generalizations, often later proved false; and simplistic enumerations, with inadequate attention to negative instances.

Bacon discussed two inductive methods. The old and defective procedure was the "anticipation of nature," with "anticipation" reflecting its Epicurean usage, which led to hasty and frivolous inductions. The new and correct procedure was the "interpretation of nature." Inductions or theories that were acceptable interpretations "must encompass more particulars than those which they were originally designed to explain and, secondly, some of these new particulars should be verified," that is, "theories must be larger and wider than the facts from which they are drawn" (Urbach 1987:28). Good inductive theories would have predictive success.

René Descartes (1596–1650) deemed Bacon's view untenable, so he attempted to invert Bacon's scientific method: "But whereas Bacon sought to discover general laws by progressive inductive ascent from less general relations, Descartes sought to begin at the apex and work as far downwards as possible by a deductive procedure" (Losee 2001:64). Of course, that inverted strategy shifted the burden to establishing science's first principles, which had its own challenges.

Sir Isaac Newton (1642–1727) developed an influential view of scientific method that was directed against Descartes's attempt to derive physical laws from metaphysical principles. Rather, Newton insisted on careful observation and induction, saying that "although the arguing from Experiments and Observations by Induction be no Demonstration of general Conclusions, yet it is the best way of arguing which the Nature of Things admits of" (Losee 2001:73). Newton affirmed Aristotle's inductive–deductive method, which Newton termed the "Method of Analysis and Synthesis" for its deductive and inductive components, respectively. "By insisting that scientific procedure should include both an inductive stage and a deductive stage, Newton affirmed a position that had been defended by Grosseteste and Roger Bacon in the thirteenth century, as well as by Galileo and Francis Bacon at the beginning of the seventeenth century" (Losee 2001:73).

In Newton's scientific method, induction was extremely prominent, being no less than one of his four rules of scientific reasoning: "In experimental philosophy we are to look upon propositions collected by general induction from phænomena as accurately or very nearly true, notwithstanding any contrary hypotheses that may be imagined, till such time as other phænomena occur, by which they may either be made more accurate, or liable to exceptions" (Williams and Steffens 1978:286).

John Stuart Mill (1806–1873) wrote a monumental *System of Logic* that covered deductive and inductive logic, with a subtitle proclaiming a connected view of the principles of evidence and the methods of scientific investigation. Like Francis Bacon, Mill recommended a stepwise inductive ascent from detailed observations to general theories. He had four (or five) inductive methods for discovering scientific theories or laws that were essentially the same as those of Grosseteste, Duns Scotus, and Ockham. Despite his enthusiasm for induction, Mill recognized that his methods could not work well in cases of multiple causes working together to produce a given effect. Mill wanted not merely to discover scientific laws but also to justify and prove them, while carefully distinguishing real causal connections from merely accidental sequences. But his justification of induction has not satisfied subsequent philosophers of science.

More recent developments in inductive logic will be discussed later in this chapter. During the twentieth century, induction picked up a common synonym: statistics. Statistics *is* inductive logic. The historically recent advent of statistical methods, digital computers, and enormous databases has stimulated and facilitated astonishing advances in induction.

Bayesian inference

For a simple example of Bayesian inference about which hypothesis is true, envision joining an introductory statistics class as they perform an experiment. The professor shows the class an ordinary fair coin, an opaque urn, and some marbles identical except for color, being either blue or white. Two volunteers, students Juan and Beth, are appointed as experimentalists. Juan receives his instructions and executes the following: He flips the coin without showing it to anyone else. If the coin toss gives heads, he is to place in the urn one white marble and three blue marbles. But if the coin toss gives tails, he is to place in the urn three white marbles and one blue marble. Juan knows the urn's contents, but the remainder of the class, including Beth and the professor, know only that exactly one of two hypotheses is true: either H_B, that the urn contains one white marble and three blue marbles, or else H_W, that it contains three white marbles and one blue marble.

The class is to determine which hypothesis, H_B or H_W, is probably true, by means of the following experiment: Beth is to mix the marbles, draw one marble, show its color to the class, and then replace it in the urn. That procedure is to be repeated as necessary. The stopping rule is to stop when either hypothesis reaches or exceeds a probability of 0.999. In other words, there is to be at most

Marble Experiment: Problem

Setup
Flip a fair coin.
If heads, place in an urn 1 white and 3 blue marbles.
If tails, place in an urn 3 white and 1 blue marbles.

Hypotheses
H_B: 1 white and 3 blue marbles (WBBB).
H_W: 3 white and 1 blue marbles (WWWB).

Purpose
To determine which hypothesis, H_B or H_W, is probably true.

Experiment
Mix the marbles, draw a marble, observe its color, and replace it, repeating this procedure as necessary.

Stopping Rule
Stop when a hypothesis reaches a posterior probability of 0.999.

Figure 9.1 A marble experiment's setup, hypotheses, and purpose.

only 1 chance in 1,000 that the conclusion will be false. This marble problem is summarized in Figure 9.1.

The ratio form of Bayes's rule is convenient. Here it is recalled, with the earlier generic hypothesis labels "1" and "2" replaced by more informative labels, namely, "*B*" meaning mostly blue marbles (one white and three blue) and "*W*" meaning mostly white marbles (three white and one blue).

$$\frac{\mathrm{P}(H_B|E)}{\mathrm{P}(H_W|E)} = \frac{\mathrm{P}(E|H_B)}{\mathrm{P}(E|H_W)} \times \frac{\mathrm{P}(H_B)}{\mathrm{P}(H_W)} \tag{9.1}$$

Table 9.1 gives the data from an actual experiment with blue and white marbles and analyzes the data using this equation. From the coin toss, the prior odds for H_B:H_W are 1:1, so the prior probability $\mathrm{P}(H_B) = 0.5$, and this is also the posterior probability $\mathrm{P}(H_B \mid E) = 0.5$ before the experiment has generated any evidence.

The likelihood odds $\mathrm{P}(E \mid H_B)$:$\mathrm{P}(E \mid H_W)$ arising from each possible empirical outcome of drawing a blue or a white marble are as follows. Recalling that H_B has three of four marbles blue, but H_W has only one of four marbles blue, a blue draw is three times as probable given H_B as it is given H_W. Because $\mathrm{P}(\text{blue} \mid H_B) = 3/4 = 0.75$ and $\mathrm{P}(\text{blue} \mid H_W) = 1/4 = 0.25$, a blue draw contributes likelihood odds of 0.75:0.25 or 3:1 for H_B:H_W, favoring H_B. By similar reasoning, a white

Table 9.1 Bayesian analysis for an actual marble experiment, assuming prior odds for H_B:H_W of 1:1. The experiment concludes upon reaching a posterior probability of 0.999.

Draw	Outcome	Posterior H_B:H_W	Posterior $P(H_B \mid E)$
	(Prior)	1:1	0.500000
1	White	1:3	0.250000
2	Blue	1:1	0.500000
3	White	1:3	0.250000
4	Blue	1:1	0.500000
5	Blue	3:1	0.750000
6	Blue	9:1	0.900000
7	Blue	27:1	0.964286
8	Blue	81:1	0.987805
9	Blue	243:1	0.995902
10	White	81:1	0.987805
11	Blue	243:1	0.995902
12	White	81:1	0.987805
13	Blue	243:1	0.995902
14	Blue	729:1	0.998630
15	Blue	2187:1	0.999543

draw contributes likelihood odds of 1:3 against H_B. Furthermore, because each draw is an independent event after remixing the marbles, individual trials combine multiplicatively in an overall experiment. For example, two blue draws will generate likelihood odds in favor of H_B of 3:1 times 3:1, which equals 9:1. Thus, in a sequential experiment, each blue draw will increase the posterior odds for H_B:H_W by 3:1, whereas each white draw will decrease it by 1:3.

Applying this analysis to the data in Table 9.1, note that the first draw is a white marble, contributing likelihood odds of 1:3 against H_B. Multiplying those likelihood odds of 1:3 by the previous odds (the prior) of 1:1 gives posterior odds of 1:3, decreasing the posterior probability to $P(H_B \mid E) = 0.25$, where the evidence at this point reflects one draw. In this sequential experiment, the posterior results after the first draw become the prior results at the start of the second draw. The second draw happens to be blue, contributing likelihood odds of 3:1 favoring H_B, thereby bringing the posterior probability $P(H_B \mid E)$ back to the initial value of 0.5.

Moving on to the sixth draw, the previous odds are 3:1, and the current blue draw contributes likelihood odds of 3:1, resulting in posterior odds of 9:1 favoring H_B and hence a posterior probability of $P(H_B \mid E) = 0.9$. Finally, after 15 draws, the posterior probability happens to exceed the stopping rule's preselected value of 0.999, so the experiment stops, and hypothesis H_B is accepted with more than 99.9% probability of truth. Incidentally, in this particular instance of an actual marble experiment, the conclusion was indeed correct because the urn actually did contain three blue marbles and one white marble, as could have been demonstrated easily by some different experiment, such as drawing out all four marbles at once.

Table 9.1 illustrates an important feature of data analysis: results become more conclusive as an experiment becomes larger. During the first six draws, H_B has two wins, two losses, and two ties, so the results are quite inconclusive, and the better-supported hypothesis never reaches a probability beyond 0.9. Indeed, at only one draw and again at three draws, this experiment gives mild support to the false hypothesis! But draws 5 to 15 all give the win to H_B, which is actually true, finally with a probability greater than 0.999.

This particular experiment reached its verdict after 15 draws, but how long would such experiments be on average? A simple approximation, regardless whether H_B or H_W is true, is that on average each four draws give three draws that support the true hypothesis and one draw that supports the false hypothesis. Hence, on average, for four draws, two draws cancel out and two support the true hypothesis. Let M denote the margin of blue draws over white draws. Then, the posterior odds H_B:H_W equal 3^M:1, which exceed 999:1 or 99.9% confidence favoring H_B when $M = 7$, or exceed 1:999 favoring H_W when $M = -7$. Because half the data cancel and half count, the length L required for a margin of ± 7 averages about $2 \times 7 = 14$ draws. Hence, the particular experiment in Table 9.1, having 15 draws, is about average.

For M equal to 2, 3, 4, or 5, a more exact calculation gives the average length L as 3.2, 5.6, 7.8, or 9.9 draws, but thereafter the approximation that $L \approx 2M$ is quite accurate. For instance, if only 1 chance of error in 1,000,000 were to be tolerated, that would require a margin of 13 because $3^{13} = 1{,}594{,}323$ and hence an average length of about 26 draws. Because the weight of this experimental evidence grows exponentially with its amount, an exceedingly high probability of truth is readily attainable.

Furthermore, this exponential increase in the weight of the evidence confers robustness to this Bayesian analysis were this experiment to encounter various problems and complications that can plague real-world experiments. Problems can be disastrous but not necessarily so because weighty evidence can surmount considerable difficulties. Four substantial but surmountable problems are described here: controversial background information, messy data, wrong hypotheses, and different statistical methods.

Controversial Background Information. The foremost objection to Bayesian inference has been that frequently the background information that determines the prior probabilities is inadequate or even controversial. This perceived deficiency prompted the development of an alternative statistical paradigm, frequentist statistics, which will be described in the next section.

Recalling the mouse experiment in the previous chapter, which is analogous to the marble experiment in this chapter, Fisher judged that "the method of Bayes could properly be applied" because the pedigree information for these mice supplied "cogent knowledge" of the prior probabilities (Fisher 1973:8). On the other hand, "if knowledge of the origin of the mouse tested were lacking, no experimenter would feel he had warrant for arguing as if he knew that of which in fact he was ignorant, and for lack of adequate data" to determine the prior probabilities "Bayes' method of reasoning would be inapplicable" (Fisher 1973:20). Fisher's tale of the black and brown mice was a moral tale that waxed sermonic in its conclusion that "It is evidently easier for the practitioner of natural science to recognize the difference between knowing and not knowing than this seems to be for the more abstract mathematician," that is, for the Bayesian statistician (Fisher 1973:20).

For the present marble experiment, the prior probabilities $P(H_B)$ and $P(H_W)$ are known precisely because of the setup information about a coin toss. But what happens if no background information is given, so the prior probabilities are unknown and potentially controversial?

A particularly unfavorable case results from assigning a small prior probability to what is actually the true hypothesis, such as $P(H_B) = 0.1$ when H_B is true. Prior odds for $H_B{:}H_W$ of 1:9 require an additional likelihood odds of 9:1 to move the odds back to the 1:1 starting point of the original setup, which entails an average of about four draws. Hence, this unfavorable prior increases the original average length of the experiment from 14 to $14 + 4 = 18$ draws. Likewise, were the prior odds extremely challenging, such as $P(H_B) = 0.001$ when H_B is true, the experimental effort increases to about 28 draws. Consequently, prior probabilities that are unfavorable to the truth result in more work, but the truth is still attainable.

A Bayesian statistician has essentially two alternatives for dealing with inadequate prior information. One alternative is to supply a noninformative prior, namely, $P(H_B) = P(H_W) = 0.5$, and also show what range of prior probabilities still leaves the conclusion unaltered, given the data at hand. If the data are strong, the conclusion may be robust despite a vague prior. The other alternative is to report the Bayes factor $P(E \mid H_B) \,/\, P(E \mid H_W)$ instead of the posterior probabilities because this avoids prior probabilities altogether. Either way, in the favorable case that the weight of the evidence grows exponentially with its amount, exceptionally strong evidence can be attainable that provides for a reliable and convincing conclusion.

Messy Data. In the original experimental procedure, the student, Beth, faithfully showed the class the marble resulting from each draw, ensuring quality data. But what happens if instead a weary and fickle Beth works alone and observes the drawn marble's color accurately only half of the time, whereas she reports blue or white at random the other half of the time? Can these messy data still decide between H_B and H_W with confidence?

The original margin between blue and white draws of seven draws allows the probability of a false conclusion to climb to 0.027 with the messy data. To maintain the specified 0.001 probability of error or 0.999 probability of truth, now the required margin increases to 14 and the average length of the experiment increases to about 56 draws. Hence, in this case, increased data quantity can compensate for decreased data quality. Of course, more pathological cases would be disastrous, such as unrecognized problems causing serious bias in the data. Frequently, scientific experiments are rather messy but not downright pathological, so the remedy of more data works.

Wrong Hypotheses. Certainly, one of the deepest problems that a scientific inquiry can possibly encounter is that the truth is not even among the hypotheses under consideration. For instance, consider this marble experiment with its setup specifying the hypotheses H_B with one white and three blue marbles, or else H_W with three white and one blue marbles. But what happens if by mistake or by mischief the experimentalist, Juan, puts two white and two blue marbles in the urn? Now the true hypothesis, H_E denoting equal numbers of both colors, is not even under consideration.

When H_E is true but only H_B and H_W are considered, on average, the experiment will require 49 draws until a margin of 7 draws declares H_B or else H_W true. But such a long experiment is suspicious, given that a length around 14 draws is expected. The length of the experiment has a wide variability around its average of 49 draws, with 70 or more draws occurring 22% of the time, which is extremely suspicious. But, on the other hand, only 20 or fewer draws occur 23% of the time, which would not be alarming. However, if the experiment were repeated several times, most likely the results would be weird: some unbelievably long experiments, contradictory conclusions favoring H_B about half of the time and H_W the other half, and frequencies for both blue and white draws near 0.5 for the pooled data. An unsuspected problem may escape detection after just one run but probably not after three or four runs, and almost certainly not after 10 or 20 runs.

The data are likely, at least eventually, to embarrass a faulty paradigm and thereby precipitate a paradigm shift. Even rather severe mistakes can be remediable. Scientific discovery is like a hike in the woods: you can go the wrong way for a while and yet still arrive at your destination at the end of the day.

Different Statistical Methods. Sometimes various scientists working on a given project adopt or prefer different statistical methods for various reasons, including debates between advocates of the Bayesian and frequentist approaches to statistics. How do statistical debates affect science? Can scientists get the same answers even if they apply different statistical methods to the data?

The short answer is that small experiments generating few data can leave scientists from different statistical schools with different conclusions about which hypothesis is most likely to be true. Rather frequently, scientists have only rather limited data, so the choice of a first-rate, efficient statistical procedure is important. However, as more data become available, the influence of differences in statistical methods diminishes. Eventually, everyone will come to the same conclusion, even though they differ in terms of the particular calculations used and the exact confidence attributed to the unanimous conclusion.

Many additional challenges could be encountered beyond these four problems. For instance, closer hypotheses would be harder to discriminate between than H_B and H_W having widely separated probabilities of 0.75 and 0.25 for drawing a blue marble. The new hypotheses H_1 with three white and five blue marbles and H_2 with five white and three blue marbles give closer probabilities of 0.625 and 0.375. Now the average length of the experiments is about 56 draws to maintain the 0.999 probability of a true conclusion. Hence, data quantity can compensate for yet another potential challenge.

In conclusion, numerous problems can be overcome by the simple expedient of collecting more data, assuming that this option is not too expensive or difficult. This favorable outcome is especially likely when the weight of the evidence increases exponentially with the amount of the evidence.

Frequentist inference

Historically, the Bayesian paradigm preceded the frequentist paradigm by about a century and a half, so the latter was formulated in reaction to perceived problems with its predecessor. Principally, the frequentist paradigm sought to eliminate the Bayesian prior because it burdened scientists with the search for additional information that often was unavailable, diffuse, inaccurate, or controversial. Frequentists such as Sir Ronald A. Fisher, Jerzy Neyman, and Egon S. Pearson wanted to give scientists a paradigm with greater objectivity.

Frequentist statistics designates one hypothesis among those under consideration as the null hypothesis. Ordinarily, the null hypothesis is that there is no effect of the various treatments, whereas one or more alternative hypotheses express various possible treatment effects. A null hypothesis is either true or false, and a statistical test either accepts or rejects the null hypothesis, so there are four possibilities. A Type I error event is to reject a true null hypothesis, whereas a Type II error event is to accept a false null hypothesis.

The basic idea of frequentist hypothesis testing is that a statistical procedure with few Types I and II errors provides reliable learning from experiments. Type I errors can be avoided altogether merely by accepting every null hypothesis regardless of what the data show, and Type II errors can be avoided by rejecting every null. Hence, there is an inherent trade-off between Types I and II errors, so some compromise must be struck.

The ideal way to establish this compromise is to evaluate the cost or penalty for Type I errors and the cost for Type II errors and then balance those errors so as to minimize the overall expected cost of errors of both kinds. In routine

	True	False	
Accept	91	2	93
Reject	4	3	7
	95	5	100

Figure 9.2 Hypothetical example of error events and rates. A null hypothesis is either true or false, and a test, involving experimental data and a statistical inference, is used to accept or reject the null hypothesis, so there are four possible outcomes, with counts as shown. Also shown are row totals, column totals, and the grand total of 100 tests. To reject a true null hypothesis is a false-positive error event, whereas to accept a false null hypothesis is a false-negative error event.

practice, however, scientists tend to set the Type I error rate at some convenient level and not to be aware of the accompanying Type II error rate, let alone the implied overall or average cost of errors.

Figure 9.2 provides a concrete example. Such numbers could represent the results when a diagnostic test accepts or rejects a null hypothesis of no disease and, subsequently, a definitive test determines for sure whether the null is true or false.

Understand that an error *event* and an error *rate* are two different things. To reject a true null hypothesis is a Type I error event, and there are four such events. To accept a false null hypothesis is a Type II error event, and there are two such events. The Type I error rate α is P(reject | true) $= 4/95 \approx 0.0421$ and the Type II error rate β is P(accept | false) $= 2/5 = 0.4$. Note that the Type I error rate is $4/95$, not $4/7$ and not $4/100$.

Another important quantity for frequentists is the p-value, defined as the probability of getting an outcome at least as extreme as the actual observed outcome under the assumption that the null hypothesis is true. To calculate the p-value, one envisions repeating the experiment an infinite number of times and finds the probability of getting an outcome as extreme as or more extreme than the actual experimental outcome under the assumption that the null hypothesis is true. The smaller the p-value, the more strongly a frequentist test rejects the null hypothesis. It has become the convention in the scientific community to call rejection at a p-value of 0.05 a "significant" result and rejection at the 0.01 level a "highly significant" result.

To illustrate the calculation of a p-value, Table 9.2 analyzes the marble experiment from a frequentist perspective that was previously analyzed in Table 9.1 from a Bayesian perspective. Let the null hypothesis be H_W, that the urn contains three white marbles and one blue marble, and let the alternative hypothesis be

Table 9.2 Frequentist analysis for an actual marble experiment, assuming that the null hypothesis H_W is true and the experiment stops at 15 draws. At an experimental outcome of 11 blue draws (and 4 white draws), which is marked by an asterisk and is the same outcome as in Table 9.1, the conclusion is to reject H_W at the highly significant p-value of 0.000115.

Blue Draws	Probability	p-value
0	0.01336346101016	1.00000000000000
1	0.06681730505079	0.98663653898984
2	0.15590704511851	0.91981923393905
3	0.22519906517118	0.76391218882054
4	0.22519906517118	0.53871312364936
5	0.16514598112553	0.31351405847818
6	0.09174776729196	0.14836807735264
7	0.03932047169656	0.05662031006068
8	0.01310682389885	0.01729983836412
9	0.00339806545526	0.00419301446527
10	0.00067961309105	0.00079494901001
11	0.00010297168046	0.00011533591896 *
12	0.00001144129783	0.00001236423850
13	0.00000088009983	0.00000092294067
14	0.00000004190952	0.00000004284084
15	0.00000000093132	0.00000000093132

H_B, that it contains one white and three blue marbles. (In this case, neither hypothesis corresponds to the idea of no treatment effect, so H_W has been chosen arbitrarily to be the null hypothesis, but the story would be the same had H_B been designated the null hypothesis instead.)

Table 9.2 has three columns of numbers. The first column lists, for an experiment with 15 draws, the 16 possible outcomes, namely, 0 to 15 blue draws (and, correspondingly, 15 to 0 white draws). This analysis takes H_W as the null hypothesis, and under this assumption that the urn contains three white marbles and one blue marble, the probability of a blue draw is 0.25. So experiments with 15 draws will average $15 \times 0.25 = 3.75$ blue draws. Accordingly, were this experiment repeated many times, outcomes of about 3 or 4 blue draws would be expected to be rather frequent, whereas 14 or 15 blue draws would be quite rare. To upgrade this obvious intuition with an exact calculation using the probability theory explained in the preceding chapter, an outcome of b blue draws and w white draws from a total of $n = b + w$ draws can occur with $n! / (b! \times w!)$ permutations, and the probability of each such outcome is

$0.25^b \times 0.75^w$. For example, the probability of 5 blue and 10 white draws is $[15! / (5! \times 10!)] \times 0.25^5 \times 0.75^{10} \approx 0.165146$.

These probabilities, for all possible outcomes from b values of 0 to 15, are listed in the second column of Table 9.2. Finally, the third column is the p-value, obtained for b blue draws by summing the probabilities for all outcomes with b or more blue draws. For example, the p-value for 0 blue draws is 1, because it is the sum of all 16 of these probabilities, whereas the p-value for 14 blue draws is the sum of the last two probabilities. For the particular marble experiment considered here, the actual outcome was 11 blue draws, and an asterisk draws attention to the corresponding p-value of 0.000115. The conclusion, based on this extremely small p-value, is to reject H_W as a highly significant result.

Unlike Bayesian analysis, which requires specification of prior probabilities in order to do the calculations, the frequentist analysis requires no such input, and thereby it seems admirably objective. So even if we know nothing about the process whereby the urn receives either the one white and three blue marbles or the reverse, we can still carry on unhindered with this wonderfully objective analysis! Or, so it seems.

Most persons who have read this section thus far probably have not sensed anything ambiguous or misleading in this frequentist analysis. It all seems so sensible. Besides, this statistical paradigm has dominated in scientific research for the previous several decades, so it hardly seems suspect. Nevertheless, there are some serious difficulties.

One problem is that, frequently, the error rate of primary concern to scientists is something other than the Type I or Type II error rates. The False Discovery Rate (FDR) is defined as the probability of the null hypothesis being true given that it is rejected, P(true | reject), which equals $4/7 \approx 0.5714$ for the example in Figure 9.2. It has the meaning here of the probability that a diagnosis of disease is actually false. Unfortunately, scientists often use the familiar Type I error rate P(reject | true) when their applications actually concern the FDR, which is the reverse conditional probability P(true | reject), which always has a different meaning and usually has a different value.

A worrisome feature of p-values is the strange influence accorded to the rule specifying when an experiment stops, which must be specified because every experiment must stop. The implicit stopping rule needed to make Table 9.2 comparable with Table 9.1 is that the experiment stops at 15 draws. But other stopping rules could result in exactly these same data, such as stop at 11 blue draws or stop at 4 white draws. However, these three rules differ in the imaginary outcomes that would result as the frequentist envisions numerous repetitions of the experiment. Consequently, exactly the same data can generate different p-values by assuming different stopping rules. For instance, Berger and Berry (1988) cited a disturbing example in which frequentist analyses of a single experiment gave p-values of 0.021, 0.049, 0.085, and any other value up to 1 just by assuming different stopping rules. So a p-value depends on the experimental data *and* the stopping rule. Consequently, it depends on the actual experiment that did occur *and* an infinite number of other imaginary experiments that did not occur. Different stopping rules are generating different stories about just what those other imaginary experiments are, thereby changing p-values. But such reasoning seems bizarre and problematic, opening the door to unlimited subjectivity, quite in contradiction to the frequentists' grand quest for objectivity.

Another problem with p-values is that they usually overestimate, but can also underestimate, the strength of the evidence because they are strongly affected by the sample size. Raftery (1995) explained that Fisher's choice of Type I error rates α of 0.05 and 0.01 for significant and highly significant results were developed in the context of agricultural experiments with typical sample sizes in the range of 30 to 200, but these choices are misleading for sample sizes well outside this range. Contemporary experiments in the physical, biological, and social sciences often have sample sizes exceeding 10,000, for which the conventional $\alpha = 0.05$ will declare nearly all tests significant. For instance, from Raftery's Table 9, $\alpha = 0.053$ for a "significant" result with 50 samples corresponds to $\alpha =$ 0.0007 with 100,000 samples, which is drastically different by a factor of almost 100. Unfortunately, many scientists are unaware of the adjustments in α that need to be made for sample size. Because of these problems with p-values, their use is declining, particularly in medical journals.

Bayesian methods have a tremendous advantage of computational ease over frequentist methods for models fitting thousands of parameters, which are becoming increasingly common in contemporary science. Also, some theorems (called complete class theorems) prove that even if one's objective is to optimize frequentist criteria, Bayesian procedures are often ideal for that (Robert 2007).

For introductory exposition of frequentist statistics, see Cox and Hinkley (1980); for Bayesian statistics, see Gelman et al. (2004) or Hoff (2009). For more technical presentations of Bayesian statistics, see the seminal text by Berger (1985) and the more recent text by Robert (2007).

Bayesian decision

The distinction between inference and decision is that inference problems pursue true beliefs, whereas decision problems pursue good actions. Clearly, inference and decision problems are interconnected because beliefs inform decisions and influence actions. Accordingly, decision problems incorporate inference sub-problems.

Many decisions are too simple or unimportant to warrant formal analysis, but some decisions are difficult and important. Formal decision analysis provides a logical framework that makes an individual's reasoning explicit, divides a complex problem into manageable components, eliminates inconsistencies in a person's reasoning, clarifies the options, facilitates clear communication with others also involved in a decision, and promotes orderly and creative problem-solving. Sometimes life requires easy and quick decisions but at other times it demands difficult and careful decisions. Accordingly, formal decision methods are supplements to, not replacements for, informal methods. On the one hand, even modest study of formal decision theory can illuminate and refine ordinary informal decisions. On the other hand, simple common-sense decision procedures provide the only possible ultimate source and rational defense for a formal theory's foundations and axioms.

The basic structure of a decision problem is as follows. Decision theory partitions the components or causes of a situation into two fundamentally different groups on the basis of whether or not we have the power to control a given component or cause. What we can control is termed the "action" or choice. Obviously, to have a choice, there must exist at least two possible actions at our disposal. What we cannot control is termed the "state" or, to use a longer phrase, the state of nature. Each state-and-action combination is termed an "outcome," and each outcome is assigned a "utility" or "consequence" that assesses the value or benefit or goodness of that outcome, allowing negative values for loss or badness, and assigning zero for indifference. These possible consequences can be written in a consequences matrix, a two-way table with columns labeled with states and rows labeled with actions. There is also information on the probabilities of the states occurring, resulting from an inference sub-problem with its prior probabilities and likelihood information. If the state of nature were known or could be predicted with certainty, determining the best decision would be considerably easier; having only probabilistic information about the present or future state causes some complexity, uncertainty, and risk. Finally, the information on consequences and probabilities of states is combined in a decision criterion that assigns values to each choice and indicates the best action.

Figure 9.3 presents a simple example of a farmer's cropping decision. There are three possible states of nature, which are outside the farmer's control: good, fair, or bad weather. There are three possible actions among which the farmer can choose: plant crop A, plant crop B, or lease the land.

Beginning at the lower left portion of Figure 9.3, we know something about the probabilities of the weather states. We possess old and new data on the weather, summarized in the priors and likelihoods. For example, the old data could be long-run frequencies based on extensive historical climate records, indicating prior probabilities of 0.30, 0.50, and 0.20 for good, fair, and bad weather. The new data could be a recent long-range weather forecast that happens to favor good weather, giving likelihoods of 0.60, 0.30, and 0.10 for good, fair, and bad weather. Bayesian inference then combines the priors and likelihoods to derive the posterior probabilities of the weather states, as shown near the middle of the figure. Multiplying each prior by its corresponding

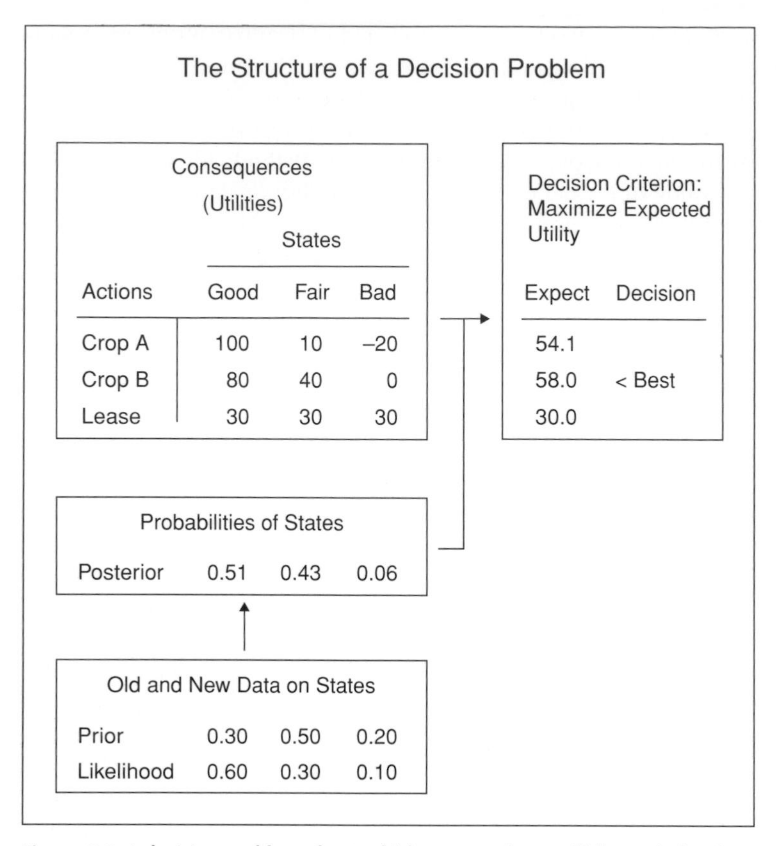

Figure 9.3 A decision problem about which crop to plant, which concludes that crop B is the best choice.

likelihood gives values of 0.18, 0.15, and 0.02, for a total of 0.35, and division of those three values by their total yields the posterior probabilities, namely, approximately 0.51, 0.43, and 0.06 for good, fair, and bad weather. So far, this is a standard inference problem. But a decision problem is more complicated, with two additional components, as explained next.

The upper left portion of Figure 9.3 shows the matrix of consequences or utilities. The outcome for any given growing season is specified by its particular state-and-action combination. The three possible states are good, fair, and bad weather, and the three possible actions are to plant crop A, plant crop B, or lease the land, for a total of $3 \times 3 = 9$ possible outcomes. The consequences matrix shows the utility or value of each possible outcome, using a positive number for a utility or gain, a negative number for a loss, or a zero for indifference. For example, in a given year, the outcome might be fair weather for crop B, which has a utility of 40, where this number represents profit in dollars per acre or whatever.

Finally, the upper right portion of Figure 9.3 specifies a decision criterion, which is to maximize the expected utility. The expected utility is the average or predicted utility, calculated for each possible action by multiplying the utility for each state by its corresponding probability and summing over the states. For example, the expected utility for crop A is $(100 \times 0.51) + (10 \times 0.43) + (-20 \times 0.06) \approx 54.1$. Likewise, the expected utility for crop B is 58.0 and that for leasing is 30.0. The largest of these three values is 58.0, indicating that planting crop B is the best decision to maximize the expected utility.

This example illustrates a frequent feature of decision problems: different penalties for different errors can cause the best decision to differ from the best inference. Bayesian inference gives the greatest posterior probability of 0.51 to good weather, and good weather favors the choice of crop A. But Bayesian decision instead chooses crop B, with its largest expected utility of 58.0, primarily because fair weather is rather likely and would involve a tremendous reduction in crop A's utility.

Both probability and statistics require only the three Kolmogorov probability axioms (and the inherited predicate logic and arithmetic axioms), but decision theory requires the addition of one more axiom, such as the axiom of desirability of Jeffrey (1983:80–81). In essence, it says that the utility or desirability of an action equals the average of the utilities for its various outcomes weighted by their probabilities, as was done in Figure 9.3.

Because of different attitudes toward risk, decision criteria other than maximized expected utility may be appropriate and preferable. For example, one might prefer to minimize the worst possible utility, which in this case would favor leasing the land (because the worst possible utility from leasing would be 30, whereas crop A could be as bad as –20, and crop B as bad as 0). Sometimes the response to the expected utility is nonlinear, such as a strong response to utilities below some minimum needed for survival, but a mild response to differences among utilities that merely distinguish various levels of luxury. Furthermore, decisions can be evaluated in terms of not only their average but also their variability around that average, with large variability implying much uncertainty and risk. Sometimes a relatively minor compromise in the average can gain a substantial reduction in the variability, which is the basis for the insurance industry.

Decisions may have several criteria to be optimized simultaneously, probably with some complicated trade-offs and compromises. For example, a farmer might want to optimize income, as in Figure 9.3, but also want to rotate crops to avoid an epidemic buildup of pest populations and want to diversify crops to stagger the workload during busy seasons. Those other constraints might result in a decision, say, to plant 60% crop B and 40% crop A, which would reduce the expected utility slightly to $(0.6 \times 58.0) + (0.4 \times 54.1) \approx 56.4$.

Although decision problems are more complex than inference problems, in practice, they often are easier than inference problems because the necessity to take some action can allow even small probability differences to force sensible decisions. For example, other things being equal, even a slightly higher probability that a particular medicine is effective or a particular airplane is safe will suffice to generate strong preferences. So odds of merely 60:40 can force practical decisions. Because most probability reasoning is motivated by the practical need to make good decisions, not merely by theoretical interests, even rather weak data and small probability differences can still significantly inform and influence decisions.

Induction lost and regained

This chapter's account of inductive logic has been, on the whole, rather confident and cheerful. However, a tremendous philosophical battle has raged over induction from ancient Greek skeptics to the present, with David Hume's critique being especially well known. Without doubt, inductive logic has suffered more numerous and drastic criticisms than all of the other components of scientific reasoning combined. Dozens of books, mostly by philosophers, have been written on the so-called problem of induction.

Unfortunately, the verdict of history seems to be that "the salient feature of attempts to solve Hume's problem is that they have all failed" (Friedman 1990:28). Broad's oft-quoted aphorism says that induction is "the glory of science and the scandal of philosophy" (Broad 1952:143), and Whitehead (1925:25) called induction "the despair of philosophy." Howson (2000:14–15, 2) concluded that "Hume's argument is one of the most robust, if not the most robust, in the history of philosophy," and it simply is "actually correct."

Hume's critique of induction appeared in his anonymous, three-volume *A Treatise of Human Nature*, which was a commercial failure and drew heavy criticism from his fellow Scottish philosophers Thomas Reid and James Beattie. Subsequently, his admirably brief *An Enquiry Concerning Human Understanding* reformulated his critique, and that punchy book was a great success. Because Hume's advertisement in the latter work dismisses the former as a juvenile work, the discussion here follows the usual custom of examining just the *Enquiry*.

Hume's argument in Chapters 4 and 5 of his *Enquiry* has three key premises followed by the conclusion: (1) Any verdict on the legitimacy of induction must result from deductive or inductive arguments, because those are the only kinds of reasoning. (2) A verdict on induction cannot be reached deductively. No inference from the observed to the unobserved is deductive, specifically because nothing in deductive logic can ensure that the course of nature will not change. (3) A verdict cannot be reached inductively. Any appeal to the past successes of inductive logic, such as that bread has continued to be nutritious and that the sun has continued to rise day after day, is but worthless circular reasoning when applied to induction's future fortunes. Therefore, because deduction and induction are the only options, and because neither can reach a verdict on induction, the conclusion follows that there is no rational justification for induction. Incidentally, whereas the second premise, that of no deductive link from the past to the future, had been well known since antiquity, the third premise, that of no (legitimate and noncircular) inductive link from the past to the future, was Hume's original and shocking innovation.

Induction suffered a second serious blow in the mid 1950s, two centuries after Hume, when Goodman (1955) propounded his "new riddle" of induction. "The new riddle of induction has become a well-known topic in contemporary analytic philosophy. . . . There are now something like twenty different approaches to the problem, or kinds of solutions, in the literature. . . . None of them has become the majority opinion, received answer, or textbook solution to the problem" (Douglas Stalker, in Stalker 1994:2).

Briefly, Goodman's argument ran as follows. Consider emeralds examined before time *t*, and suppose that all of them have been green (where *t* might be, say, tomorrow). The most simple and foundational inductive procedure, called the *straight rule of induction*, says that if a certain property has been found for a given proportion of many observed objects, then the same proportion applies to all similar unobserved objects as well as to individual unobserved objects. For example, if numerous rolls of a die have given an outcome of 2 with a frequency of nearly 1/6, then inductive logic leads us to the conclusion that the frequency of that outcome in all other rolls will also be 1/6, and likewise that the probability of any particular future roll giving that outcome will be 1/6. Similarly, those observations before time *t* of many emeralds that are all green support the inductive conclusion that all emeralds are green, as well as the prediction that if an emerald is examined after time *t*, it too will be green.

Then Goodman introduced a new property, "grue," with the definition that an object is grue if it is examined before time *t* and is green, or if it is not examined before time *t* and is blue. Admittedly, this is a rather contrived property, and the philosophical discussion of grue is quite technical and rather perplexing. But the main point is that Goodman showed that only some properties are appropriate (projectable) for applications of the straight rule of induction, but others are not. So how can one decide in a nonarbitrary manner which properties are projectable? Apart from clear criteria to discern when the straight rule is applicable, there is a danger that it will be used when inappropriate, thereby "proving" too much, even including contradictory conclusions.

All too predictably, Hume had complained that all received systems of philosophy were defective and impotent for justifying even the simple straight rule of induction. Goodman's complaint, however, was the exact opposite. His concern was not that induction proves too little but rather that it proves too much: a method that can prove anything proves nothing. Understand that

Goodman, like his predecessor Hume, was not intending to wean us from common sense, such as causing us to worry that all of our emeralds would turn from green to blue tomorrow. Rather, he was deploying the new riddle to wake us to the challenge of producing a philosophically respectable account of induction.

Finally, and perhaps most important, the great generality of those old and new problems of induction must be appreciated. Hume and Goodman expressed their arguments in terms of time: past and future, or before and after time *t*. But thoughtful commentators have discerned their broader scope. Gustason (1994:205) assimilated Hume's argument to a choice among various standard and nonstandard inductive logics. Accordingly, the resulting scope encompasses any and all inductive arguments, including those concerning exclusively past outcomes.

Howson (2000:30–32) followed Goodman in interpreting Goodman's argument as a demonstration that substantial prior knowledge about the world enters into our (generally sensible) choices about when to apply induction and how much data to require. Couvalis (1997:48) has cleverly said it all with a singularly apt example: "Having seen a large number of platypuses in zoos and none outside zoos, we do not infer that all platypuses live in zoos. However, having seen a small number of platypuses laying eggs, we might infer that all platypuses lay eggs." Similarly, Howson (2000:6, 197) observed that scientists are disposed to draw a sweeping generalization about the electrical conductivity of copper from measuring current flow in a few samples. But, obviously, many other scientific generalizations require enormous sample sizes.

Responding first to Hume's critique of induction, the role of common sense is critical. Hume said that we need not fear that doubts about induction "should ever undermine the reasonings of common life" because "Nature will always maintain her rights, and prevail in the end over any abstract reasoning whatsoever," and "Custom . . . is the great guide of human life" (Beauchamp 1999:120, 122). Hume's conclusion is not that induction is shaky but rather that induction is grounded in custom or habit or instinct, which we share with animals, rather than in philosophical reasoning. But Hume's argument depends on a controversial assumption that common sense is located outside philosophy rather than being an integral part and foundation of philosophy.

Indeed, when philosophy's roots in common sense are not honored, a characteristic pathology ensues: instead of natural philosophy happily installing science's presuppositions once, at the outset, by faith in a trifling trinket of common-sense knowledge, a death struggle with skepticism gets repeated over and over again for each component of scientific method, including induction. The proper task, "to explain induction," swells to the impossible task, "to defeat skepticism and explain induction." If Hume's philosophy cannot speak in induction's favor, that is because it is a truncated version of philosophy that has exiled animal habit rather than having accommodated our incarnate human nature as an integral component of philosophy's common-sense starting points, as Reid had recommended.

Plainly, all of the action in Hume's attack on induction derives ultimately from the concern that the course of nature might change, but that is simply the entrance of skepticism. His own examples include such drastic matters as whether or not the sun will continue to rise daily and bread will continue to be nutritious. Such matters are nothing less than philosophy's ancient death fight with skepticism! They are nothing less than the end of the world! In the apocalypse proposed by those examples, not only does induction hang in the balance but also planetary orbits and biological life. As Himsworth (1986:87–88) observed in his critique of Hume, if the course of nature did change, we would not be here to complain! So as long as we are here or we are talking about induction, deep worries about induction are unwarranted. Consequently, seeing that apocalypse as "the problem of induction" rather than "the end of the world" is like naming a play for an incidental character. The rhetoric trades in obsessive attention to one detail.

Turning next to Goodman's new riddle of induction, it shows that although the straight rule of induction is itself quite simple, judging whether or not to apply it to a given property for a given sample is rather complicated. These judgments, as in the example of platypuses, draw on general knowledge of the world and common sense. Such broad and diffuse knowledge resists tidy philosophical analysis.

Summary

Induction reasons from actual data to an inferred model, whereas deduction reasons from a given model to expected data. Both are important for science, composing the logic or "L" portion of the PEL model. Probability is the deductive science of uncertainty, whereas statistics is the inductive science of uncertainty.

Aristotle, medieval philosopher-scientists, and modern scholars have developed various inductive methods. But not until the publication in 1763 of Bayes's theorem was the problem finally solved of relating conditional probabilities of the form $P(E \mid H)$, the probability of evidence E given hypothesis H, found in deduction with reverse conditional probabilities of the form $P(H \mid E)$ required in induction. Bayes's theorem was illustrated with a simple example regarding blue and white marbles drawn from an urn. Inductive conclusions can be robust despite considerable difficulties with controversial background information, messy data, wrong hypotheses, and different statistical methods. Particularly when the weight of the evidence grows exponentially with the amount of the evidence, increased data quantity can often compensate for decreased data quality. However, the need to specify prior probabilities, which can be unknown or even controversial, prompted the development of an alternative paradigm intended to be more applicable and objective.

Frequentist inference, which is a competitor to Bayesian inference, was illustrated with the same marble experiment. Frequentist methods seek to minimize Type I errors, rejecting a true null hypothesis, and Type II errors, accepting a false null hypothesis, but this is challenging because of the inevitable trade-off between these two kinds of errors. Statistical significance is assessed by *p*-values that express the probability of getting an outcome as extreme, or more extreme, than the actual experimental outcome under the assumption that the null hypothesis is true. But sometimes error rates other than the Type I and Type II error rates are more relevant, particularly the False Discovery Rate. And *p*-values have been criticized because their strange dependence on stopping rules and imaginary outcomes undermines the presumed pursuit of objectivity and because their actual significance depends strongly on the number of samples.

Bayesian decision theory was illustrated with a simple example of a farmer's cropping decision. Whereas inference problems pursue true beliefs, decision problems pursue good actions. Decision theory requires one more axiom beyond those already needed for probability and statistics, an axiom of desirability saying in essence that the utility or desirability of an action equals the average of the utilities for its various outcomes weighted by their probabilities.

Inductive logic has received far more philosophical criticism than all of the other components of scientific method combined. David Hume argued that philosophy cannot justify any inductive procedures, including the simple straight rule of induction. More recently, Goodman's new riddle of induction showed the exact opposite, that the straight rule of induction can be used to prove anything – which is equally problematic. But given common-sense presuppositions, induction can be defended and implemented effectively.

Study questions

(1) Recall that the vertical bar in a conditional probability is read as "given," so $P(A \mid B)$ means the probability of *A* given *B*. Let *H* denote a hypothesis and *E* denote some evidence. How do $P(H \mid E)$ and $P(E \mid H)$ differ in meaning and in numerical value? How are they related by Bayes's theorem? How do they pertain to scientists' main research questions?

(2) List several kinds of problems that sometimes plague scientific experiments. How can inductive logic or statistics reach reliable and robust conclusions despite such problems?

(3) Define and compare Type I, Type II, and False Discovery Rate (FDR) error rates. Suppose that your research has two steps: an inexpensive initial screening for numerous promising candidates, followed by a very expensive

final test for promising candidates. Which kind of error rate would be most relevant for the initial screening and why?

(4) Does either the Bayesian or frequentist paradigm have a legitimate claim overall to greater objectivity and, if so, for exactly what reasons? What is the relative importance of statistical paradigm and evidential strength in achieving objectivity?

(5) Describe Hume and Goodman's riddles of induction. What are your own responses to these riddles? Do they undermine induction or not? What role do common-sense presuppositions play in a defense of induction?

10

Parsimony and efficiency

The principle of parsimony recommends that from among theories fitting the data equally well, scientists choose the simplest theory. It has four common names, also being called the principle of simplicity, the principle of economy, and Ockham's razor (with Ockham sometimes latinized as Occam).

This book's account of science's evidence, which is the "E" portion of the PEL model, takes the form primarily of this chapter's detailed analysis of parsimony. Mere data or observations become evidence when they are brought to bear on hypotheses or theories. This impact of data on theory is guided by several criteria, including the fit of the data with the theory and the parsimony of the theory. Most aspects of evidence are rather obvious to scientists and most evidence is gathered by means of specialized techniques used only within a given discipline. Accordingly, most of what needs to be said about scientific evidence is in the domain of specialized disciplines rather than general principles. However, the one great exception is parsimony, which is not obvious to many scientists, and yet considerations of parsimony pervade all of the sciences, so it is among science's general principles.

Parsimony is not an unusually difficult topic, compared with the ordinary topics routinely studied by scientists. Also, because parsimony pervades all of science, it is easy to find interesting examples and productive applications. Nevertheless, the implementation of parsimony has always faced serious obstacles. In the first place, many scientists seem inclined to think that only a few words, such as "Prefer simpler models," can exhaust the subject. Such complacency does not motivate further study and new insight. Also, the literature on parsimony is scattered in philosophy, statistics, and science, but few scientists read widely in those areas. Yet, each of those disciplines provides distinctive elements that must be combined to achieve a full picture.

In some areas of science and technology, such as in signal processing, the principle of parsimony has already been well understood to great advantage. But, in most areas, a superficial understanding of parsimony has been a serious deficiency of scientific method, costing scientists billions of dollars annually in wasted resources. Frequently, a parsimonious model that costs a few seconds of computer time can provide insight and increase accuracy as much as would the collection of more data that would cost thousands or millions of dollars. If more scientists really understood parsimony, science and technology would gain considerable momentum.

Historical perspective on parsimony

Parsimony has been discussed with two distinct but related meanings. On the one hand, parsimony has been considered a feature of nature, that nature chooses the simplest course. On the other hand, parsimony has been deemed a feature of good theories, that the simplest theory that fits the facts is best. These are ontological and epistemological conceptions, respectively, concerning nature itself and humans' theories about nature.

The venerable law of parsimony, the *lex parsimoniae*, has a long history. Aristotle (384–322 BC) discussed parsimony in his *Posterior Analytics*: "We may assume the superiority *ceteris paribus* [other things being equal] of the demonstration which derives from fewer postulates or hypotheses" (McKeon 1941:150). He used parsimony as an ontological principle in rejecting Plato's Forms. Plato (*c.* 427–347 BC) believed that both the perfect Form of a dog and individual dogs existed, but Aristotle held the more parsimonious view that only individual dogs existed. Hence, even something as elemental as the tendency in Western thought to regard individual physical objects as being thoroughly real derives from an appeal to parsimony. Likewise, in his influential commentary on Aristotle's *Metaphysics*, Averroes (Ibn Rushd, 1126–1198) regarded parsimony as a real feature of nature.

Robert Grosseteste (*c.* 1168–1253), who greatly advanced the use of experimental methods in science, also emphasized parsimony, as here in commenting on Aristotle: "That is better and more valuable which requires fewer, other circumstances being equal, just as that demonstration is better, other circumstances being equal, which necessitates the answering of a smaller number of questions for a perfect demonstration or requires a smaller number of suppositions and premisses from which the demonstration proceeds" (Crombie 1962:86). Grosseteste held parsimony not merely as a criterion of good explanations or theories but more fundamentally as a real, objective principle of nature. Thomas Aquinas (*c.* 1225–1274) also espoused a rather ontological version of parsimony, writing that "If a thing can be done adequately by means of one, it is superfluous to do it by means of several; for we observe that nature does not employ two instruments where one suffices" (Hoffmann, Minkin, and Carpenter 1996).

William of Ockham (*c.* 1285–1347) probably is the medieval scholar best known to modern scientists, through the familiar principle of parsimony, often called Ockham's razor. "It is quite often stated by Ockham in the form: 'Plurality is not to be posited without necessity' (*Pluralitas non est ponenda sine necessitate*), and also, though seldom: 'What can be explained by the assumption of fewer things is vainly explained by the assumption of more things' (*Frustra fit per plura quod potest fieri per pauciora*)" (Boehner 1957:xxi).

Just what does this principle mean? "What Ockham demands in his maxim is that everyone who makes a statement must have a sufficient reason for its truth, 'sufficient reason' being defined as either the observation of a fact, or an immediate logical insight, or divine revelation, or a deduction from these" (Boehner 1957:xxi). However, Ockham's principle of sufficient reason tends to reach modern scientists in a somewhat thinner version of parsimony, merely saying something like: "one should not complicate explanations when simple ones will suffice" (Hoffmann et al. 1996). Ockham insisted that parsimony was an epistemological principle for choosing the best theory, in contrast to his predecessor Robert Grosseteste and his teacher John Duns Scotus, who had interpreted parsimony as also an ontological principle for expecting nature to be simple. In Ockham's view, "This principle of 'sufficient reason' is epistemological or methodological, certainly not an ontological axiom" (Boehner 1957:xxi).

Nicolaus Copernicus (1473–1543) inherited the geocentric cosmology of Aristotle and Ptolemy. It fit the data within observational accuracy, accorded with the common-sense feeling that the earth was unmoving, and enjoyed the authority of Aristotle. However, its one major flaw was lack of parsimony, with its complicated cycles and epicycles for each planet. Consequently, Copernicus offered a new theory: that the earth revolved on its axis daily and journeyed around the sun annually. His main argument featured parsimony: the heliocentric model was simpler, involving fewer epicycles, and the various motions were interlinked in a harmonious system. "I found at length by much and long observation, that if the motions of the other planets were added to the rotation of the earth, and calculated as for the revolution of that planet, not only the phenomena of the others followed from this, but that it so bound together both the order and magnitudes of all the planets and the spheres and the heaven itself that in no single part could one thing be altered without confusion among the other parts and in all the Universe. Hence for this reason . . . I have followed this system" (Dampier 1961:110).

Isaac Newton (1642–1727) further anchored parsimony's importance with the four rules of reasoning in his monumental and influential *Philosophiae Naturalis Principia Mathematica* (Cajori 1947:398–400). Parsimony was the first rule, expressed in a vigorously ontological version concerning nature that echoed words of Aristotle and Duns Scotus: "We are to admit no more causes of natural things than such as are both true and sufficient to explain their appearances." Newton explained: "To this purpose the philosophers say that Nature does nothing in vain, and more is in vain when less will serve; for

Nature is pleased with simplicity, and affects not the pomp of superfluous causes." Again, parsimony, in a distinctively epistemological version concerning theories about causes, was the second of Newton's rules, corollary to the first: "Therefore to the same natural effects we must, as far as possible, assign the same causes," such as for "respiration in a man and in a beast" and "the reflection of light in the earth, and in the planets." Even his third and fourth rules about experiments and induction rested on the presupposition that nature "is wont to be simple."

Henri Poincaré (1854–1912) related parsimony to generalization: "Let us first observe that any generalization implies, to a certain extent, belief in the unity and simplicity of nature. Today, ideas have changed and, nevertheless, those who do not believe that natural laws have to be simple, are obliged to behave as if it was so. They could not avoid this necessity without rendering impossible all generalization, and consequently all science" (A. Sevin, in Hoffmann et al. 1996). Yet Poincaré also appreciated the subtlety of simplicity, bringing counterpoint with his view that "simplicity is a vague notion" and "everyone calls simple what he finds easy to understand, according to his habits" (A. Sevin, in Hoffmann et al. 1996).

More recently, Albert Einstein (1879–1955) employed parsimony in his discovery of general relativity: "Perhaps the scientist who most clearly understood the necessity for an assumption about the simplicity of [scientific] laws was Albert Einstein. In an informal conversation he once told me about his thoughts in arriving at The General Theory of Relativity. He said that after years of research, he arrived at a particular equation which, on the one hand, explained all known facts and, on the other hand, was considerably simpler than any other equation that explained all these facts. When he reached this point he said to himself that God would not have passed up the opportunity to make nature this simple" (Kemeny 1959:63). Likewise, Einstein spoke of "the grand aim of all science, which is to cover the greatest possible number of empirical facts by logical deductions from the smallest possible number of hypotheses or axioms" (Nash 1963:173). He also remarked that "Everything should be made as simple as possible, but not simpler" (Hoffmann et al. 1996).

Historically, philosophers and scientists have been the scholars who have written about parsimony. More recently, statisticians have also explored this subject, offering two new and important results. First, simple theories tend to make reliable predictions. Second, Bayesian analysis automatically gives simpler theories higher prior probabilities of being true, thereby favoring simpler theories (Jefferys and Berger 1992).

In line with these statisticians, the philosopher Richard Swinburne also saw simplicity as evidence of truth and of reliable predictions: "I seek . . . to show that – other things being equal – the simplest hypothesis proposed as an explanation of phenomena is more likely to be the true one than is any other available hypothesis, that its predictions are more likely to be true than are those of any other available hypothesis, and that it is an ultimate a priori epistemic principle that simplicity is evidence of truth" (Swinburne 1997:1).

The 1950s was a great decade for parsimony. Statistician Charles Stein published a seminal paper in 1955 that began the literature explaining how parsimonious models can gain accuracy and efficiency. But typical applications require millions or billions of arithmetic steps. For such calculations, reasonably affordable and available digital computers needed transistors, rather than the clumsy vacuum tubes used previously. Physicists John Bardeen, Walter Brattain, and William Shockley co-invented the transistor in 1947 and were awarded the 1956 Nobel Prize in physics. Transistors became increasingly available during the 1950s. Computer programmer John Backus with several associates invented FORTRAN, the first high-level programming language, in 1957. This language made it much easier for statisticians and scientists to use computers. These three resources – new statistical theory, fast transistor circuits, and convenient programming languages – allowed breakthroughs that went far beyond the earlier insights on parsimony from brilliant philosophers such as Aristotle and William of Ockham.

During the subsequent decades, there have been astonishing advances in statistical theory and computing power. Consequently, there are tremendous opportunities to put parsimony to work for gaining accuracy and efficiency, improving predictions and decisions, increasing repeatability, favoring truth, and accelerating progress. Parsimony has had an intriguing history but, more important, it will have an exciting future.

Preview of basic principles

This chapter's primary means for exploring parsimony are the following three examples of parsimony at work in science. But simplicity is a complicated topic! Accordingly, this section first previews five basic principles.

Signal and Noise. Data are imperfect, mixtures of real signal and spurious noise. These terms, "signal" and "noise," originated in the context of radio communication, where a receiver picks up the signal from a transmitter plus noise from various natural and human sources. But, in statistics, these terms are used more generally to refer to treatment or causal effects and random errors. Hence, the data equal the signal plus the noise.

There is a fundamental difference between signal and noise in that the signal ordinarily is relatively simple, caused mostly by only a few major treatment differences or causal factors, whereas the noise typically is extremely complex, caused by numerous small uncontrolled factors. Because the signal is parsimonious, it can be captured or fitted readily by an appropriate parsimonious model, but complex noise inevitably requires a complex model.

Model Families. Parsimony is important throughout science, particularly because generalization requires an appeal to parsimony (at least implicitly), as Poincaré emphasized. But parsimony is especially applicable for the common situation in which scientists are considering a model family for analyzing a given dataset. A model family is a sequence of models of the same mathematical form that include more and more parameters. Three such families are used in this chapter.

First, a familiar family is the polynomial family. Let x be the independent variable; y the dependent variable; and a, b, c, and d be constants. Then, the members of the polynomial family are the constant model, $y = a$; the linear model, $y = a + bx$; the quadratic model, $y = a + bx + cx^2$; the cubic model, $y = a + bx + cx^2 + dx^3$; and so on. The data structure that the polynomial model addresses is N paired observations of x and y. Given N pairs, the polynomial family has N members with 1 to N terms, the highest-order term being a constant times x^{N-1}. The highest member is called the *full or saturated model.* The full model automatically fits the data perfectly, whereas in general the lower-order models fit approximately. For instance, with 7 data points, the highest powers of x are $0, 1, 2, \ldots, 6$ in the 7 increasingly complex (decreasingly parsimonious) members of the polynomial family.

Second, principal components analysis (PCA) is a common analysis in numerous applications in science and technology, and it also involves a model family. The data structure that PCA addresses is a two-way data table with R rows and C columns. For example, plant breeders often measure yields for G genotypes tested in E environments. PCA provides a suitable model family for such data. A data table with R rows and C columns may be conceptualized geometrically as R points in a C-dimensional space, with each point's coordinates specified by the C values in its row (or the reverse, with C points in an R-dimensional space). The first principal component is the least-squares line through this high-dimensional cloud of points, meaning that perpendicular projections of these points onto that line maximize the sum of squared distances along that line (and simultaneously minimize the sum of squared distances off that line). The first two principal components specify the least-squares plane in this cloud of points, so they are often graphed to show the structure of the data because they provide the best two-dimensional view of this high-dimensional cloud. Likewise, the first three principal components provide the best three-dimensional approximation to the original cloud, and so on for increasingly complex members of the PCA model family. A common variant of PCA, called doubly-centered PCA, first subtracts the average for each row from each datum in that row, and the same for columns. In agricultural applications, its most common name is the Additive Main effects and Multiplicative Interaction (AMMI) model. For G genotypes and E environments, the highest or full member of the AMMI family has the lesser of $G - 1$ or $E - 1$ principal components. The lowest member, denoted by AMMI-0, has no principal components but rather only the additive effects, namely, the grand mean, the genotype deviations from the grand mean, and the environment deviations from the grand mean (Gauch 1992:85–96). For instance, the members of the AMMI family for a 7 × 10 data matrix are the seven models AMMI-0, AMMI-1, AMMI-2, and so on, up to AMMI-6, which is the full or saturated model that is also denoted by AMMI-F.

Third and finally, multiple linear regression is an extremely popular statistical method that also involves a model family. The data structure that it addresses is a number M of observations or samples for which a dependent variable $\boldsymbol{Y}$ is to be predicted or estimated on the basis of N predictor variables measured for each observation, $\boldsymbol{X}_1, \boldsymbol{X}_2, \ldots, \boldsymbol{X}_N$, where these variables are bolded to indicate that they are vectors of length M. For instance, the data may comprise measurements at several farms of wheat yield, rainfall, soil nitrogen, average August temperature, and altitude, and the objective is to predict wheat yield at each farm from these other four measurements. Multiple linear regression constitutes a model family because there are many choices about which predictors to include and which to exclude in order to get the most accurate predictions. Given N predictors, there are 2^N members of the multiple regression model family because each predictor can be in or out. Ordinarily, a rather parsimonious choice will be best. For instance, the best predictor of wheat yield might use only rainfall and soil nitrogen, while discarding the other two variables.

Statistical Criteria. Given a model family applied to noisy data, some statistical criterion must be specified to determine the best choice. An important goal is predictive accuracy, but this must be implemented by a specific procedure or equation. Statisticians have devised two basic kinds of strategies.

One strategy involves data resampling techniques, such as cross-validation and the bootstrap. A portion of the data is selected at random, typically about 10% to 25%, and is set aside temporarily to serve as validation data while the model family is fitted to the remaining data. The member with the smallest mean square prediction error for the validation data is selected. Typically, this procedure is repeated many times with different randomizations and the results are averaged for greater accuracy. Often, the final results are based on the selected member of the model family being applied to the recombined, entire dataset.

The other strategy for model choice is to use the Bayesian Information Criterion (BIC), also called the Schwarz Bayesian Criterion (SBC) invented by Schwarz (1978), which approximates the logarithm of the Bayes factor discussed in Chapter 9. The similar Akaike Information Criterion (AIC), or one of the less common alternatives, can also be used. The equations for BIC or AIC contain two terms, one rewarding model fit to the data and another penalizing model complexity (or rewarding model parsimony). Thereby, such criteria strike a balance between fit and parsimony intended to optimize predictive accuracy.

The statistical literature on model choice is extensive and technical (McQuarrie and Tsai 1998; Joo, Wells, and Casella 2010). Data resampling techniques are

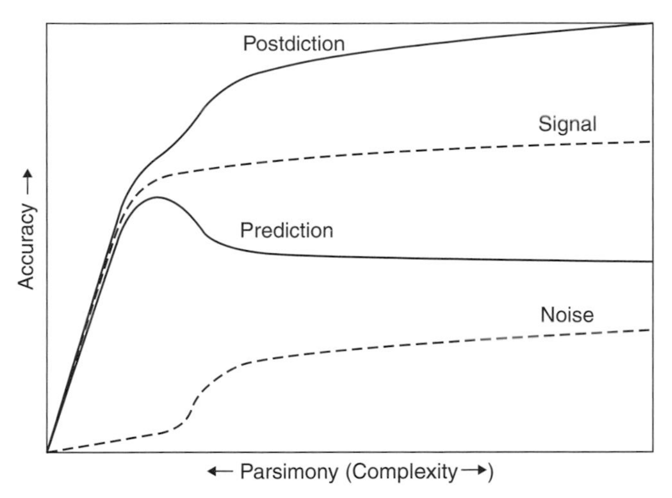

Figure 10.1 Predictive and postdictive accuracies of models differing in terms of parsimony. The abscissa represents more parsimonious models to the left (and more complex models to the right), and the ordinate shows model accuracy. Imperfect data are mixtures of real signal and spurious noise. Signal is recovered quickly at first as models become more complex, but thereafter signal is recovered slowly. By contrast, noise is recovered slowly at first while much signal is being recovered, then for a brief time noise is recovered more quickly, but thereafter slowly. Postdictive accuracy increases as the signal plus the noise, so it always increases for more and more complex models. But predictive accuracy increases as the signal minus the noise, so it rises to a maximum for some relatively parsimonious model, and thereafter declines.

rather popular. When several criteria for model selection are compared, often BIC wins. For example, Piepho and Gauch (2001) compared 14 model-selection criteria for simulated genetics data for which the true model was known by construction, and BIC performed best. However, for extremely large models, AIC typically outperforms BIC.

Ockham's Hill. Given noisy data analyzed by a model family that is evaluated by a statistical criterion, Figure 10.1 shows what happens. The abscissa depicts a sequence of increasingly parsimonious models moving toward the left (or increasingly complex models moving toward the right). For example, this could be a polynomial family, with its most simple model (the constant model) at the extreme left; then the more complex linear, quadratic, cubic, and higher models progressing toward the right; and finally its most complex model (the full model) at the extreme right. The full model has as many parameters as the data, and its estimates automatically equal the data exactly (such as a quadratic equation with its three parameters automatically going through three data points exactly). The ordinate shows model accuracy or goodness of fit.

Consider first the dashed lines for signal and noise. Because only a few main causal factors determine most of the signal, the relatively simple signal is recovered quickly in early-model parameters and then slowly thereafter. But the response for noise is more involved, with recovery initially slow, then briefly rapid, and then again slow. The initial focus on signal suppresses recovery of noise at first. But after most of the signal has been captured, the focus then shifts to the noise and chance correlations in the noise can be exploited briefly by statistical analyses to accelerate the recovery of noise. Then, after that opportunity has been largely exhausted, noise is recovered slowly.

The data are usually a limited sample from a larger population of interest, such as several hundred persons in a clinical trial who are afflicted by a disease that strikes millions. This distinction between a sample and a population leads to the further distinction between the goal of accurately fitting just the sample data, termed *postdiction*, and the goal of accurately fitting the entire population from which the sample was drawn, termed *prediction*. Nearly always, a scientist's objective is prediction rather than merely postdiction. This distinction between prediction and postdiction is shown by the solid lines in Figure 10.1. It has subtler and greater implications than many scientists realize.

On the one hand, the goal in postdiction is to model or fit the sample data, with no serious concern about a larger population or about the distinction between signal and noise. Recovery of signal and recovery of noise are rewarded alike. Accordingly, the line for postdictive accuracy is depicted as the signal line plus the noise line. The full model at the extreme right automatically recovers all of the signal and noise.

On the other hand, the goal in prediction is to model the entire population of interest. Recovery of signal is rewarded, whereas recovery of noise is penalized because noise is idiosyncratic and has no predictive value. Accordingly, the line for predictive accuracy is depicted as the signal line minus the noise line. Note that the lines for postdiction and prediction are different because of noise. Were the noise negligible, these two lines would be the same.

Quite importantly, these two lines have different shapes, reaching their peaks of maximum accuracy at different places. Postdictive accuracy is automatically maximized by the most complex, full model at the extreme right. But predictive accuracy is maximized for some relatively parsimonious model closer to the left, rather than at the extreme right where the full model equals the data. This means that parsimonious models can be more predictively accurate than their data! That is the principal message of this chapter. The shape of this response for predictive accuracy was given the apt name "Ockham's hill" by MacKay (1992), in honor of William of Ockham.

Designed Experiments. Scientific experiments generally have two designs: a treatment design and an experimental design (Gauch 2006). The treatment design specifies the deliberately controlled factors of scientific interest, such as different environments or different genotypes in an agricultural trial. The experimental design specifies how the treatments are allocated to the experimental units, which usually involves randomization and replication to reduce bias and increase accuracy.

For example, a yield trial could test G genotypes in E environments using R replications, for a total of GER observations. The two-way factorial of G genotypes by E environments constitutes the treatment design, whereas the R replications are involved in the experimental design. Ordinarily, the replications are organized in some specific scheme, such as subdividing the field used for an agricultural trial into a number of subunits or blocks that are smaller and more compact than the field as a whole and, hence, hopefully are rather uniform.

Blocks can be complete, including all treatments; or incomplete, including only some. Importantly, statistical analysis of an incomplete block design pursues two purposes, reducing the estimated errors to increase statistical significance, and adjusting treatment estimates closer to their true values. But complete blocks are less helpful, pursuing only the first of those two purposes.

However, for the present chapter on parsimony, the important message is simply that experiments have *two* designs, the treatment design and the experimental design, and *both* can provide opportunities to gain accuracy. Most scientists are abundantly aware of the value of replication to gain accuracy, although many would not realize that incomplete blocks are more aggressive than complete blocks by virtue of adjusting estimates closer to their true values. But precious few scientists in many disciplines, including agriculture and medicine, are aware of the *other* opportunity to gain accuracy by parsimonious modeling of the treatment design for many common designs. That other opportunity is what this chapter is about. Neglect of this other opportunity is regrettable because its potential for accuracy gain is often several times greater than the potential from replicating and blocking. It is ironic how often scientists implement the smaller of these opportunities to gain accuracy, when the larger opportunity is available but neglected. Of course, best practices require exploiting both opportunities.

Curve fitting

The first of this chapter's three examples of parsimony is curve fitting using the polynomial model family. The salient features of this example are that the true model is already known exactly, and the noise is also known exactly. Knowing both the signal and noise exactly allows for an unusually penetrating analysis, elucidating principles that subsequently can be recognized in more complex and realistic settings. Obviously, to get an example with signal and noise known exactly requires that we place ourselves in a very unusual position. Such an example must be constructed by us, not offered to us by nature. Accordingly, it must come from mathematics, not from the empirical sciences.

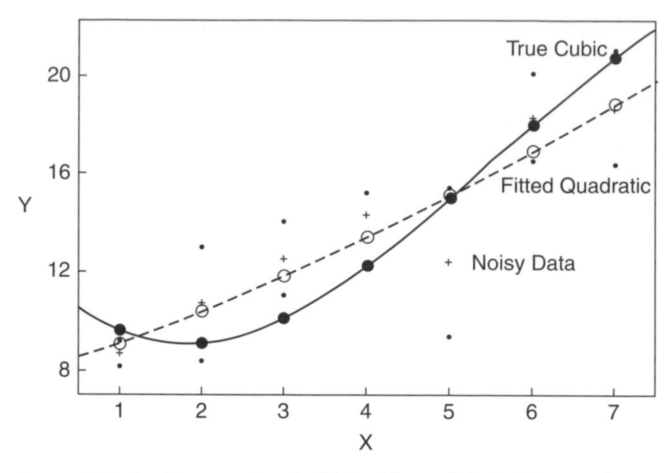

Figure 10.2 A cubic equation (solid line) is modeled with a quadratic equation (dashed line). Values of the true cubic equation are shown at seven levels of *X* (●). Noisy data are generated at each level for two replicates (.) and their average (+), with these averages having an S/N ratio of 5. Also shown are values for the quadratic equation fitted to the noisy data (○). Note that at every level except the sixth, the fitted quadratic's values are closer to the true values than are the data, the averages over replications. Remarkably, this parsimonious model is more accurate than its noisy data. It achieves a substantial statistical efficiency of 2.10, meaning that on average the quadratic model based on only two replications is slightly more accurate than averages based on twice as much data, four replications. So modeling helps predictive accuracy as much as would collecting twice as much data, but modeling is far more cost-effective when the data are expensive. (Adapted from Gauch 1993 and reproduced with kind permission from *American Scientist*.)

Figure 10.2 shows a cubic equation, $y = 12.00 - 3.50x + 1.17x^2 - 0.07x^3$, and its values at seven levels, $x = 1, 2, \ldots, 7$. By construction, this cubic equation is the true model or signal, known exactly. To mimic imperfect experimental data, random noise is added that is also known exactly. This noise has a normal distribution adjusted to have a variance of 0.2 times that of the cubic equation's data, which constitutes the signal. By definition, the signal-to-noise (S/N) ratio is the ratio of these variances, namely, $1/0.2 = 5$ in this instance. Frequently, experiments are replicated, which is represented here by showing these noisy data as averages of two replicates (that have twice as much variance as do their averages). Finally, this figure also shows the least-squares quadratic equation fitted to these noisy data, $y = 7.95 + 1.13x + 0.06x^2$.

Note that at every level except the sixth, the fitted quadratic's values are closer to the true values than are the data, the averages over replications. Some persons may find this outcome surprising but, indeed, this model is more accurate than

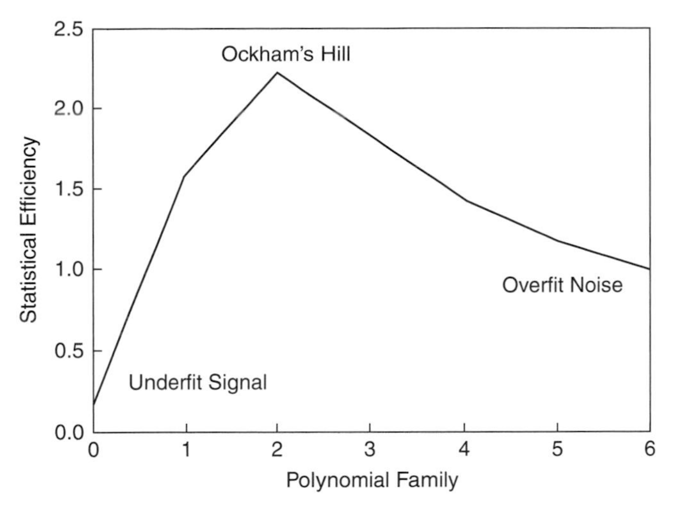

Figure 10.3 Ockham's hill for the noisy cubic data having an S/N ratio of 5, using the polynomial family encompassing the constant model up to the to sixth-order model. The quadratic model is at the peak of Ockham's hill, achieving the greatest statistical efficiency of 2.22. To the left of the peak, excessively simple models underfit real signal; to the right, excessively complex models overfit spurious noise.

its data, even though this fitted quadratic model is not the true cubic model! The sum of squares (SS) of differences between the data and true values is 24.67, and the SS of differences between the quadratic model and true values is 11.75. By definition, the statistical efficiency is the ratio of these values, 24.67/11.75 $\approx$ 2.10. A statistical efficiency of 1 means that a model has the same predictive accuracy as the data, whereas a statistical efficiency above or below 1 implies better or worse accuracy. Because the full model's estimates equal the actual data, its statistical efficiency is automatically exactly 1. Also, a statistical efficiency of 2 or 3 means that a model achieves the same accuracy as would the full model's estimates (namely, averages over replications) based on twice or thrice as many replications. Because this experiment has two replications, the quadratic model is as accurate as would be averages based on $2 \times 2.10 = 4.20$ replications. So, modeling increases accuracy as much as would collecting twice as much data!

The case shown in Figure 10.2 invites three generalizations. Instead of measuring performance with just one set of random noise values, what would be the performance averaged over numerous repetitions with different noise? Instead of presenting results for just the quadratic model, what would be the results for the entire polynomial family? And what would happen at various S/N levels?

Figure 10.3 shows statistical efficiencies for the entire polynomial family for noisy cubic data with an S/N ratio of 5. There are seven data points, so the polynomial family encompasses the constant model (marked 0 on the abscissa), linear model (1), quadratic model (2), and so on, up to the sixth-order model (6). The statistical efficiency of the quadratic model for the single case analyzed in Figure 10.2 was 2.10, but this figure shows that the average over numerous repetitions with different noise is slightly different, 2.22. Figure 10.3 shows the typical response, Ockham's hill, which was previewed earlier in the line for prediction in Figure 10.1.

The most predictively accurate member of the polynomial family for these noisy cubic data is the quadratic model, achieving a substantial statistical efficiency of 2.22. Efficiency declines in either direction away from the peak, but for different reasons. To the left of the peak, excessively simple models are inaccurate because they underfit real signal. To the right of the peak, excessively complex models are inaccurate because they overfit spurious noise. Optimal accuracy requires a balance between these opposing problems.

Figure 10.4 further generalizes the results for a wide range of noise levels, S/N ratios of 0.1 to 100. Beginning with familiar material from Figure 10.3, note the same results for an S/N ratio of 5 (located about seven-tenths of the way from 1 to 10 on this logarithmic abscissa), with the quadratic model most accurate with its statistical efficiency of 2.22. This figure shows that for rather accurate data having S/N ratios above 16.6, a cubic model is most predictively accurate, achieving a statistical efficiency of 1.82. But because noise increases moving to the left, progressively simpler models are best. However, the fourth-order and higher models never win, including the sixth-order model, which is the full model, equaling the actual noisy data. So which model is most predictively accurate depends on the noise level. It makes sense that as noise increases, fewer of the true model's parameters can be estimated accurately enough to be helpful, until finally only the grand mean, which is the parameter used by the constant model, can resist the onslaught of noise. On the other hand, cleaner data can support more parameters.

Often scientists encounter the entire Ockham's hill, as in Figure 10.3. But Figure 10.4 implies that it is possible to see only the left or right side of the hill. Extremely noisy data make the simplest model win, so there is a monotonic decrease in accuracy for increasingly complex models, thereby showing only the right side of Ockham's hill. Likewise, extremely accurate data make the most complex model win, so only the left side of Ockham's hill is seen.

Depending on the statistic used to express predictive accuracy, Ockham's hill may be inverted, resulting in Ockham's valley instead. Figure 10.3 shows statistical efficiency, which increases with *greater* accuracy; whereas a statistic such as the mean square prediction error increases with *worse* accuracy, so the result is Ockham's valley.

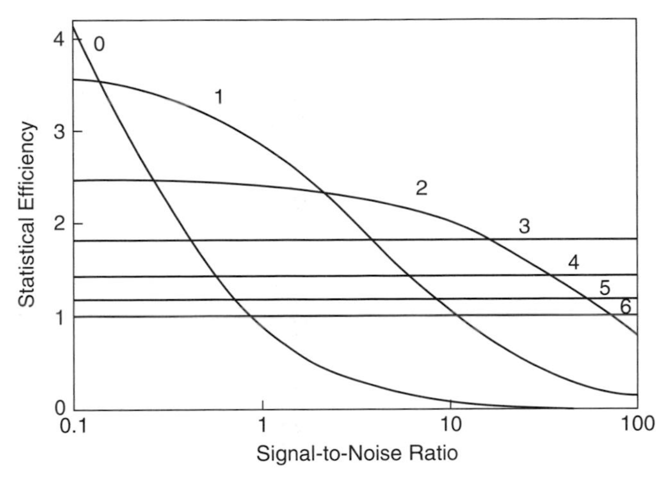

Figure 10.4 Statistical efficiency of the polynomial family over a range of S/N ratios, with the constant model (0), linear model (1), and others up to the final sixth-order polynomial (6). The constant model is most predictively accurate for extremely noisy data, with S/N below 0.15; the linear model is best for S/N from 0.15 to 2.0; the quadratic model is superior for S/N from 2.0 to 16.6; and the cubic model wins for relatively accurate data, with S/N above 16.6. With seven data points, the full model is the sixth-order equation. This full model is always most postdictively accurate but never most predictively accurate. Notice that for accurate data with an S/N ratio above 16.6, the true cubic model is most predictively accurate; but for noisier data, progressively simpler models are most accurate. Consequently, diagnosing the most predictively accurate member of a model family and determining the true model are distinguishable goals, sometimes having different answers. (Adapted from Gauch 1993 and reproduced with kind permission from *American Scientist*.)

Crop yields

The second example of parsimony at work is familiar to me from my own research from 1988 to 2012, agricultural yield trials using the AMMI model family. Plant breeders use yield trials to select superior genotypes, and agronomists use them to recommend varieties, fertilizers, and pesticides to farmers. Worldwide, several billion dollars are spent annually on yield trials. These experiments have helped plant breeders to increase crop yields, typically by about 1% to 1.5% per year for open-pollinated crops such as corn and 0.5% to 1% per year for self-pollinated crops such as soybeans. However, there is substantial and worrisome evidence that wheat and rice yield increases have slackened lately to about 0.5% per year, which is considerably less than during 1960 to 1990.

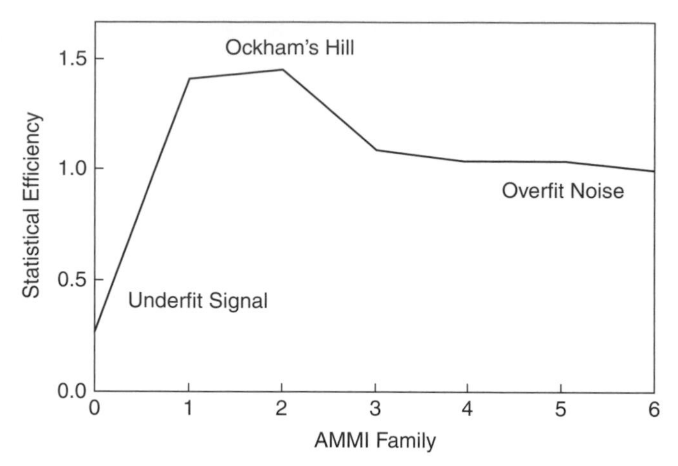

Figure 10.5 Ockham's hill for the soybean data using the AMMI family. The most predictively accurate member of the AMMI family is AMMI-2, achieving a statistical efficiency of 1.45, but AMMI-1 is almost as accurate.

The most common type of yield trial tests a number of genotypes in a number of environments that are location-year combinations, often with replication. The dataset used here is a New York State soybean trial having 7 genotypes in 10 environments with 4 replications (Gauch 1992:56). Recall that Figure 1.6 showed a photograph from this soybean yield trial.

The salient feature of this agricultural example is that neither the signal nor the noise is known exactly, quite in contrast to the easy first example with a known cubic equation and known added noise. We receive the soybean data from nature, with the signal and noise already mixed together. Indeed, these data are rather noisy, with typical errors of plus or minus 15%, so they carry only one significant digit. Because the true signal and added noise are not known separately and exactly, the method for calculating statistical efficiencies in Figure 10.5 is more complicated than for Figure 10.3, as explained in detail elsewhere (Gauch 1992:134–153). Greater accuracy for yield estimates also helps in the search for genes affecting yield (Gauch et al. 2011).

Figure 10.5 shows statistical efficiencies for the soybean data using the AMMI family. The AMMI-2 model is at the peak of Ockham's hill, achieving a statistical efficiency of 1.44, with AMMI-1 a close second. To the left of the peak, excessively simple models underfit real signal; to the right, excessively complex models overfit spurious noise.

Looking at the past, between the 1940s and the late 1970s, the principal achievement of the Green Revolution has been to increase dramatically the yields of the major grain crops – wheat, corn, and rice – in favorable environments. This achievement was crucial and it still is, accounting for a large portion of the world's food supply. But, unfortunately, there were two unintended and detrimental side effects. First, the increased productivity and profitability of major grains prompted many farmers to grow less of other grains, vegetables, and fruits, thereby restricting diets and causing deficiencies in essential vitamins and minerals, especially vitamin A, iron, and zinc. Second, the considerable neglect of marginal environments, which often require different genotypes than favorable environments, meant little benefit for the millions of persons whose food comes mainly from those poor environments.

Looking toward the future, the world's population is currently increasing about 1.2% per year. But the welcome rise out of poverty for millions, especially in China and India, implies greater demand for meat production, which uses much grain. So, crop yields need to increase somewhat more rapidly, about 1.4% per year. The good news is that plant breeders are increasingly poised to address those two deficiencies of the original Green Revolution. For numerous vegetable and fruit species, powerful new genetic tools can allow relatively small projects in several years to increase yields as much as the larger and longer projects of the Green Revolution decades ago. So, those other crops can now become more profitable and contribute to a diverse and wholesome diet, especially because breeders are also selecting for enhanced nutritional traits. Likewise, relatively small projects can more quickly address marginal environments, benefiting many of the world's poorest communities. But the bad news is that yield advances are needed despite trends in many places toward less farm land, less water, and more disease pressure – and despite challenging goals of better environmental stewardship. On balance, a sustained yield increase of 1.4% per year for the next decade or two seems attainable, until projections of world population largely level off. However, there is little margin for agricultural research to be inefficient. If a greater fraction of agricultural researchers learned how to put parsimony to work in order to get the most out of their experiments and data, numerous projects, involving many crops in many nations, would accelerate markedly.

Crime rates

The third and final example of parsimony is a sociological study of crime rates using the multiple regression model family. This study was first published by Ehrlich (1973). The present account is based on the reanalysis by Raftery (1995), using both frequentist and Bayesian methods.

Raftery explained that "Most sociological studies are observational and aim to infer causal relationships between a dependent variable and independent variables of interest." The choice of predictor variables to measure is guided by theory or background knowledge. But "theory is often weak and vague," so the usual strategy is to play it safe by including a long "laundry list" of candidate predictors in hopes of not missing anything important. Multiple regression provides a standard statistical method for deciding which candidates to include in the final model. Discarding the irrelevant candidates is important because including useless variables in the model degrades the results for parameters of genuine interest. This setup of many candidate predictors and rather weak theory is quite common, not only in sociology but also in ecology, agriculture, medicine, and many other fields.

This criminological study by Ehrlich was one of the earliest systematic efforts to address the question: Do greater punishments reduce crime rates? As Raftery recounted, there were two competing hypotheses. One hypothesis, which may be denoted H_1, is that criminal behavior is "deviant and linked to the offender's presumed exceptional psychological, social, or family circumstances." The competing hypothesis H_2 is that "the decision to engage in criminal activity is a rational choice determined by its costs and benefits relative to other (legitimate) opportunities." Ehrlich compiled extensive data on localities from 47 states in the USA. The following paragraph paraphrases Raftery's description of the second hypothesis, inserting in parentheses the numbers for the 15 candidate predictors of crime rates.

The costs of crime are related to the probability of imprisonment (X_{14}) and the average time served in prison (X_{15}), which in turn are influenced by police expenditures (X_4, X_5). The benefits of crime are related to aggregate wealth (X_{12}) and income inequality (X_{13}) in the surrounding community. The expected net payoff from the alternative of legitimate activities is related to educational level (X_3) and the availability of employment, the latter being measured by the unemployment rate (X_{10}, X_{11}) and labor force participation rate (X_6). This payoff is expected to be lower for nonwhites (X_9) and for young males (X_1). Other possible influences are southern versus northern states (X_2), the state population (X_8), and the sex ratio or number of males per female (X_7). The principal interest is in the probability and length of imprisonment, X_{14} and X_{15}, because hypothesis H_2 expects greater association with crime rates than does H_1.

Beginning with the frequentist analysis, Raftery tried three common methods for choosing statistically significant variables, including the most popular method, stepwise regression. He also tried two variants of Ehrlich's models based on sociological theory. The full model with all 15 predictors was also included as a baseline for comparison. But the results were perplexing. "There are striking differences, indeed conflicts, between the results from different models. Even the statistically chosen models, despite their superficial similarity, lead to conflicting conclusions about the main questions of interest."

Progressing to the Bayesian analysis, Raftery based model selection on the BIC approximation to the logarithm of the Bayes factor. A distinctive feature of the Bayesian approach is that it can incorporate model uncertainty by averaging over multiple models that are all supported well by the data. The 15 candidate predictors imply $2^{15} = 32{,}768$ possible models. Most of these numerous possibilities are decidedly bad, so calculations can be simplified by averaging over only the reasonably good possibilities using a criterion that Raftery terms Occam's window – although I prefer not to latinize this philosopher's name, so I call it Ockham's window. Models are excluded that (a) are 20 times less likely than the most likely model, corresponding to a BIC difference of 6; and (b) contain predictors for which there is no evidence in the sense that they have more likely sub-models nested within them that omit those predictors. For this study of crime rates, Ockham's window reduced the original 32,768 models to a very manageable 14 models. Raftery reported that this reduction is quite typical. These foremost models within Ockham's window were parsimonious, including only 5 to 8 of the 15 candidate predictors and averaging 6.4 predictors.

The Bayesian analysis showed that the probability of imprisonment (X_{14}) has a probability of 98% of having a real effect on crime rates, whereas the length of imprisonment (X_{15}) is not particularly significant (only 35%). There is strong evidence that higher crime rates are associated with both educational level (X_3) and income inequality (X_{13}) with 100% probability, as well as with young males (X_1) with 94% and nonwhites (X_9) with 83%. But there is no association with crime for aggregate wealth (X_{12}), labor force participation rate (X_6), and sex ratio (X_7). There was also evidence for a negative association between police expenditure (X_4, X_5) and crime rate, although the causal story for this was not evident.

In these findings, the importance of probability of imprisonment (X_{14}), income inequality (X_{13}), and educational level (X_3) are supportive of hypothesis H_2 that emphasizes rational deliberation. On the other hand, sizable effects for young males (X_1) and nonwhites (X_9) are supportive of H_1 that emphasizes social and family influences. These two hypotheses are just different, not mutually exclusive, so there could well be some valid aspects in both.

Regarding statistical paradigm, Raftery detailed numerous advantages of the Bayesian approach over the frequentist approach. Most important, the Bayesian analysis is superior conceptually and operationally because it alone can integrate model selection and parameter estimation. For this particular study of crime rates, a plausible argument can be made that the Bayesian analysis, which favors parsimonious models, provided clearer conclusions.

Explanation of accuracy gain

How do parsimonious models gain accuracy? First of all, it must be insisted that routinely they do. Cross-validation and related methods prove that for countless applications across science and technology. For example, to cite just one number from just the first of this chapter's examples, at an S/N ratio of 5, the parsimonious quadratic model achieves an average statistical efficiency of 2.22, and this accuracy gain is absolutely indisputable because the true signal and added noise are known exactly by construction.

Consequently, even if not one person on earth could explain how parsimonious models can be more accurate than their data, this accuracy gain would still stand as an established fact and a great opportunity. However, a fact without an explanation is unsatisfying and sometimes even unconvincing.

There are three interrelated explanations for accuracy gain by parsimonious models. They concern signal–noise selectivity, direct–indirect information, and variance–bias trade-off.

Signal–Noise Selectivity. This chapter's preview explained accuracy gain by parsimonious models in terms of signal–noise selectivity. Early model parameters capture mostly the relatively simple signal, whereas late model parameters capture mostly the relatively complex noise, as depicted in Figure 10.1. By selecting the most predictively accurate member of a model family at the peak of Ockham's hill, a signal-rich model is separated from a discarded noise-rich residual.

Direct–Indirect Information. Full and parsimonious models make use of the data in strikingly different ways. Recall the Ockham's hill for the soybean data in Figure 10.5. The data for these 7 soybean varieties tested in 10 environments with 4 replications are given in Gauch (1992:56), for a total of 280 yield measurements. For instance, the four replicates for Evans soybeans grown in Aurora, NY, in 1977 were 2,729, 2,747, 2,593, and 2,832 for an average of 2,725 kg/ha.

Suppose that upon checking these numbers against the original field notes, a typographical error is detected: the first replicate should be 2,279 rather than 2,729. What happens to the AMMI-F and AMMI-2 yield estimates upon correcting that error?

The AMMI-F yield estimates equal the actual data, namely, the averages over replicates. Hence, that one value for Evans in Aurora in 1977 changes from 2,725 to 2,613 kg/ha. Nothing else changes.

By contrast, the computation of the AMMI-2 yield estimates involves the entire data matrix. Hence, all yield estimates change, for all 7 varieties in all 10 environments, although most adjustments are rather small.

From the perspective of any given yield, such as that for Evans in Aurora in 1977, there are 4 measurements providing direct information about that yield and 276 measurements providing indirect information about other varieties or other environments or both. The full and parsimonious models are fundamentally different in what they take to be the relevant data. For each and every yield estimate, the full model AMMI-F uses its 4 replicates, whereas the parsimonious model AMMI-2 uses all 280 measurements. That is, AMMI-F uses only the direct information, whereas AMMI-2 uses both the direct and indirect information.

How much does the indirect information help the AMMI-2 yield estimates? There are 4 replicates and the statistical efficiency is 1.44. Hence, AMMI-2 using 4 replicates is as accurate as would be AMMI-F using $4 \times 1.44 = 5.76$ replicates, so the indirect information has helped as much as would adding $5.76 - 4 = 1.76$ more replications. So, the indirect 276 observations equate to 1.76 direct observations, or $276 / 1.76 \approx 157$ indirect observations are as informative as 1 direct observation. The indirect information is dilute, but it is also abundant, and therefore worth incorporating in yield estimates.

Again, how do parsimonious models gain accuracy? The most basic explanation is that they use more data – no magic, no mystery, just more data. This has been understood by statisticians ever since the seminal paper by Stein (1955). This explanation also applies to this chapter's first example, the cubic equation, because the full model uses only the 2 replicates for each of the 7 levels to estimate its values (y), whereas a parsimonious model uses all 14 observations to estimate each and every value.

Variance–Bias Trade-Off. The third and final interrelated explanation of accuracy gain by parsimonious models concerns a trade-off between variance and bias. Low variance and low bias are both desired. But increasingly complex models in a model family have more variance but less bias, so a trade-off is inevitable. Ockham's hill occurs because a modest amount of both problems is better than a huge amount of either problem. However, this is the most technical of these three explanations, so further details can be relegated to the statistical literature, such as Gauch (1992:134–153).

In review, parsimonious models gain accuracy by retaining early model parameters that selectively capture the relatively simple signal and discarding late-model parameters that selectively capture the relatively complex noise. Most fundamentally, they use available data more aggressively, extracting both direct and indirect information, and they strike an optimal trade-off between problems with variance and bias.

Philosophical reflection

This chapter's analysis of parsimony has been primarily from a scientific and technological perspective, with special interest in gaining accuracy and efficiency. But greater understanding of parsimony can emerge from adding some philosophical reflection. This section addresses two topics: parsimony and nature, and prediction and truth.

Parsimony and Nature. Recall from the historical review that parsimony has two aspects: an epistemological principle, preferring the simplest theory that fits the data, and an ontological principle, expecting nature to be simple. The epistemological aspect of parsimony has been emphasized here because it is part of scientific method. But the ontological aspect also merits attention.

So, is nature simple? For starters, understand that the reality check is itself a *simple* theory about a *simple* world. It declares that "Moving cars are hazardous to pedestrians." This is simple precisely because it applies a single dictum to all persons in all places at all times. The quintessential simplicity of this theory and its world, otherwise easily unnoticed, can be placed in bold relief by giving variants that are not so simple. For example, if nature were more complex than it actually is, more complicated variants could emerge, such as "Moving cars are hazardous to pedestrians, except for women in France on Saturday mornings and wealthy men in India and Colorado when it is raining." Although there is just one simple and sensible formulation of the reality check, obviously there are innumerable complex and ridiculous variants. Regarding cars and pedestrians, a simple world begets a simple theory. Or, to put it the other way around, a simple theory befits a simple world.

Capitalizing on this little example, meager thought and imagination suffice to see parsimony everywhere in the world – in iron atoms that are all iron, in stars that are all stars, in dogs that are all dogs, and so on. Parsimony touches our every thought. But to really understand parsimony, one must move beyond examples to principles.

Induction, uniformity, causality, intelligibility, and other scientific principles all implicate parsimony. Applications of induction to the physical world presuppose parsimony, specifically in the ontological sense expressed strongly by the law of the limited variety of nature. Likewise, the law of causality, that similar causes produce similar effects, is an aspect of simplicity. That nature is intelligible to our feeble human reason shows that some significant features of reality are moderately simple. If nature were not simple, science would lose all of its foundational principles at once.

Yet, the greatest influence of parsimony in scientific method is in the simplicity of the questions asked. Any hypothesis set that expresses a scientific question could in principle always be expanded to include more possibilities, and that action would make sense were the world more complex than it is. Were inductive logic bankrupt, were nature not uniform, were causes not followed by predictable effects, and were nature barely comprehensible, then enormously more hypotheses would merit consideration. Then, science would languish with hopelessly complicated questions that would impose impossible burdens for sufficient evidence. The beginning of science's simplicity is its simple questions.

Having argued that nature is simple, this verdict should not be interpreted simplistically! Indeed, "there is complexity to the whole idea of simplicity" (Nash 1963:182). Simon (1962) offered remarkably keen insights regarding just which aspects of nature scientists expect to be simple and just which aspects they expect to be complex. In essence, the rich complexity of life and ecosystems emerges from the frugal simplicity of basic physical and chemical laws. From general experience, scientists and engineers ordinarily have a fairly reliable general sense of how simple or complex a given system or problem is.

The verdict on parsimony is that "Ockham's Razor must indubitably be counted among the tried and useful principles of thinking about the facts of this beautiful and terrible world and their underlying causative links" (Hoffmann et al. 1996). Nevertheless, those authors also note the sensible reaction that "the very idea that Ockham's Razor is part of the scientific method seems *strange*... because... science is not about simplicity, but about complexity." The plausible resolution that those authors offer is that simple minds comprehend complex nature by means of ornate models made of simple pieces. The balance between a model's simplicity and the extent to which it approaches completeness requires a delicate and skillful wielding of Ockham's razor. The comments on Hoffmann et al. (1996) by A. Sevin concur: "Our discovery of complexity increases every day.... This good old Ockham's razor remains an indispensable tool for exploring complexity."

Prediction and Truth. Predictive success is often taken as evidence of truth. To cite one famous example, using Newton's theory of gravity, Edmond Halley (1656–1742) calculated the orbit of the impressive comet of 1682, which now bears his name, identifying it as the one that had appeared previously in 1531 and 1607, and predicting the time and place of its return in 1759. He did not live to see that return, but it did happen just as he had predicted. His striking predictive success was accepted universally as proof that his theory of comets' orbits was true, or at least very nearly true.

Generalizing from that familiar and yet representative example, predictive success is taken generally as evidence of truth, especially when numerous and diverse predictions are all correct, so that mere luck is an implausible explanation. Indeed, among theories that have attained strong and lasting acceptance among scientists, doubtless one of the most significant and consistent categories of supporting evidence is predictive success.

Nevertheless, this venerable formula, that predictive success implies truth, can be unsettled by interpreting or applying it too simplistically. Indeed, a little reflection on Figure 10.4 should be disturbing, or at least thought-provoking. By construction, a cubic equation is known to be the true model. It is sampled at seven points, with addition of random noise (at an S/N ratio of 5) to mimic measurement errors, and least-squares fits are calculated for the polynomial family. Although a cubic equation is the true model, the cubic model is less predictively accurate than another member of the polynomial family, the quadratic model. So even with the true model entered in the competition, the criterion of predictive accuracy gives the win to a false model! If such problems occur for easy cases with constructed and known models, what happens in the tough world of real scientific research? Does predictive accuracy have no reliable bearing on truth?

That a generic model, known not to be the true model, can still gain accuracy and efficiency is comforting to scientists because they often need to gain accuracy in situations where the true model is not available. But this outcome is troubling to philosophers because predictive accuracy has often been taken as an indicator of closeness to the truth.

The resolution is to make precise claims about what predictive accuracy does, or does not, optimize. More specifically, careful distinctions must be made among a model's predictions, form, and parameters. Accordingly, for this example, it is fine to claim that the quadratic model's predictions are more accurate (at an S/N ratio of 5) than those of any other member of the polynomial family, including the cubic model. Indeed, this is a fact! But it is confusing and false to interpret such a claim to mean that this model's form and parameters are accurate. Rather, a model's predictions, form, and parameters are three different things, so optimizing one is not necessarily optimizing the others.

Scientific research often has multiple steps and multiple purposes, however, which can defy simplistic accounts of parsimony's role. Indeed, a generic blackbox use of parsimony in one step without truth claims (for the model's form or parameters) can help to support the search for causal explanations in another step with truth claims and testable predictions.

Finally, this chapter's treatment of parsimony is only introductory. It has substantial limitations. A scientist's choice from among competing theories can be much more difficult than has been illustrated by any of this chapter's examples. The competing models may be of diverse forms (such as an exponential model and a trigonometric model) rather than members of a single family. More fundamentally, competing models may posit different physical entities and processes. Also, theory choice may involve criteria besides goodness of fit and parsimony, such as explanatory power, fruitfulness in generating new insights or results, and coherence with other well-accepted theories.

Despite these limitations, merely to recognize that prediction is different from postdiction, or that parsimony can increase research efficiency, may suffice to move many scientists forward. It may be suggested that gaining an elementary understanding of parsimony, combined with ordinary statistical tools, can provide 90% of the practical benefit that would emerge from exemplary understanding and impeccable statistics.

Summary

The principle of parsimony recommends that from among theories fitting the data equally well, scientists choose the simplest theory. Thus, fitting the data is not the only criterion bearing on theory choice. Building on earlier thinking by Aristotle, Grosseteste, and others, Ockham advanced our understanding of parsimony so significantly that the principle often bears his name, as Ockham's razor. In a famous paradigm shift, Copernicus chose the heliocentric theory over the geocentric theory, not on grounds of a better fit with the data but rather on grounds of greater simplicity. As that case illustrates, sometimes false theories get into trouble with parsimony before they get into trouble with more extensive or more accurate data. Accordingly, considerations of parsimony can help to place scientists on the cutting edge of their specialties.

Data are imperfect, a mixture of signal and noise. As the increasingly complex (decreasingly parsimonious) members of a model family are fitted to noisy data, the models' predictive accuracies initially rise and then subsequently decline in a response called Ockham's hill. Overly simple models underfit signal, whereas overly complex models overfit noise. Scientific experiments ordinarily have two designs, a treatment design and an experimental design, and often both provide opportunities to gain accuracy. Parsimonious models concern the treatment design, whereas replication concerns the experimental design. Best practices require exploiting both opportunities to gain accuracy.

The workings of parsimony were illustrated with three examples: a constructed cubic equation with added random noise, crop yields, and crime rates. Accuracy gain by the use of parsimonious models was explained in terms of signal and noise selectivity, direct and indirect information, and variance and bias trade-off. Basically, parsimonious models gain accuracy by utilizing more of the information in the data. Philosophers, statisticians, and scientists have contributed valuable insights into the meaning, workings, and usefulness of parsimony.

In many areas of science and technology, a better understanding of the role of parsimony in the scientific method could yield considerable gains in accuracy and efficiency. Often, a parsimonious model can increase accuracy as much as would the collection of several times as much data. Greater accuracy also means better predictions and decisions, increased repeatability, and larger returns on research investments.

Study questions

(1) What are the ontological and epistemological aspects of parsimony? In what ways are both simplicity and complexity real features of nature?

(2) What sorts of resources have enabled scientists during the past half century to put parsimony to work far better than could brilliant philosopher-scientists previously?

(3) Signal from imposed treatments is relatively simple whereas noise from uncontrolled factors is quite complex. What does this imply for recovery of signal and noise for a sequence of models of increasing complexity? Why

are models less accurate, both to the left of the peak on Ockham's hill and to the right?

(4) What are three interrelated explanations for how parsimonious models gain accuracy? Before studying this chapter, were you already aware that parsimony can increase accuracy, efficiency, and repeatability? If you are a student or professional in the sciences, are scientists in your specialty reasonably aware of opportunities to gain accuracy involving both the treatment design and the experimental design?

(5) Have you seen examples of parsimony at work in either the sciences or the humanities to gain accuracy or to clarify arguments? Do you see any promising opportunities that have not yet been exploited? Is it really true that a model can be more accurate than its data? What sort of concrete evidence can support such a claim?

11

Case studies

This chapter presents seven case studies intended to illustrate the broad applicability of the general principles of scientific method that were explained in the previous chapters. The first concerns philosophy. The middle five case studies span the physical, biological, and social sciences, including some applications to technology. The last concerns law. Hopefully, contemplating these case studies will prompt readers to discern promising new applications of the general principles of scientific method in their own disciplines, not only in the sciences but also in the humanities.

Philosophy

Recall the earlier quotation from the American Association for the Advancement of Science that "many of these fundamental values and aspects" of science "are also the province of the humanities, the fine and practical arts, and the social sciences" (AAAS 1990:xii). There is a fruitful traffic of ideas between science and philosophy and more generally between science and the "wider world of ideas" (AAAS 1990:24).

Philosophy of Science began publication in the United States in 1934, and *The British Journal for the Philosophy of Science* began in 1950. These philosophy journals are replete with superb articles and penetrating insights – although regrettably scientists rarely see such journals. Many scientists would benefit from scanning the table of contents of these journals once or twice annually to check for articles of potential interest.

Scientists base theory choice on several desirable criteria, including goodness of fit with the data, parsimony, robustness to imperfect and noisy data, predictive accuracy, generality, unification of diverse phenomena, explanatory power, testability, falsifiability, justified inferences, and coherence with knowledge of related matters (NRC 2012:48, 78–79). But the challenge with several criteria is how to combine or integrate them. An analogous situation is searching for a car with a favorable price, preferred color, and other desired features. In both situations, optimizing all criteria simultaneously is unlikely. Consequently, some account of priorities and trade-offs is needed.

Philosophers have explored this. "Philosophers have suggested many different criteria for judging scientific theories, among them that a good theory should predict, explain, and be testable. Determining the degree to which these are distinct criteria, their interrelationships, and the possible existence of other more appropriate criteria, are problems in the philosophy of science" (Grier 1975). He subsumed three criteria – prediction, explanation, and testability – in a single unified system, and found that explanation is the most comprehensive of these criteria, incorporating aspects of prediction and testability. Other philosophers have emphasized the importance of predictive success for satisfactory explanations (Douglas 2009; Cleland 2011).

A formal account of explanatory power has been developed with the great merit of providing proofs that it uniquely satisfies several compelling intuitions about what explanations mean (Schupbach and Sprenger 2011). It will be interesting to follow the future development of this philosophical project as the authors plan sequels concerning (a) the relationship between explanatory power and evidential strength, and (b) a comparison between their normative account of explanatory power and people's actual judgments of explanations.

The epistemic virtue of unification has received profound analysis by philosophers within the Bayesian paradigm for inductive inferences (Myrvold 1996, 2003; McGrew 2003; Schupbach 2005). Unification generates an evidential bonus beyond the credit earned for the explained items considered separately.

Philosophers have contributed to the literature supporting the Bayesian paradigm for statistical inferences (Leitgeb and Pettigrew 2010a, 2010b), particularly for sequential trials that inspect the data periodically (Sprenger 2009). They have also applied Bayesian analysis to evidence brought to bear on major philosophical debates (Earman 2000; Swinburne 2002, 2003). Philosophers of science also address specific topics of interest to scientists, such as climate models and global warming (Lloyd 2010; Oreskes, Stainforth, and Smith 2010).

Electronic engineering

Engineers at the National Institute of Standards and Technology developed a remarkably efficient method for testing analog-to-digital converters (ADC) and other electronic components (Souders and Stenbakken 1991). These devices have many applications in consumer electronics, especially audio systems and cell phones. Because of uncontrolled variations in the manufacturing process, ADC devices vary in their performance. Accordingly, they must be tested individually and then sorted into performance bins.

Their main example was a 13-bit ADC that measures an input voltage and expresses the result as one of 2^{13} or 8,192 different values. The conventional test used automated equipment to check each device at all 8,192 input voltages, rating the device by its maximum measurement error. The problem with that test was that 8,192 measurements were time-consuming and costly, reducing throughput and profit. Often, the testing cost was an appreciable fraction of the entire manufacturing cost. They needed a faster, cheaper test that still would maintain good reliability.

Their new strategy was to test a small number of units completely, analyze those data to discover a small subset of test voltages that would be maximally revealing of a device's performance, and then in production runs to test an enormous number of devices economically with the quick reduced test. Although these authors did not use the term "Ockham's hill," they were astutely aware of the need to strike an optimal balance between underfitting real signal and overfitting spurious noise. They developed a reduced test with only 64 measurements instead of 8,192, less than 1% of the original task. A computer applied a mathematical model to those 64 measurements to predict all 8,192 responses and, as before, the worst of those (predicted) responses served as the criterion for sorting the device into its appropriate performance bin. Figure 11.1 shows a typical result for an individual ADC device.

An interesting feature of this example is that it demonstrates benefits from parsimony for both data analysis and experimental design. Doubtless there are many potential applications in science, technology, and manufacturing for parsimonious experiments and economical tests. The long-term profitability and competitiveness of many companies will be strongly affected by whether or not their scientists learn to use clever tricks to gain efficiency.

Three other examples may be given briefly of general principles of scientific method at work in electronic engineering. Principal components analysis (PCA) was introduced in the previous chapter and has been used for fault detection and classification in the semiconductor manufacturing industry (Good, Kost, and Cherry 2010). Occasional problems emerge from "faulty measurements, misprocessed wafers, process drifts or trends, tool aging, and tool failures." Rapid fault detection is needed to avoid misprocessing hundreds of wafers and to maximize equipment uptime in order to ensure a return on large investments. Cross validation was used to select a PCA model with optimal predictive accuracy.

Image restoration seeks a good estimate of the original image from noisy, imperfect data, with applications in photographic, astronomical, ultrasound, and radar images (Seghouane 2009). A trade-off ensues between fitting the noisy data and the smoothness or parsimony of the restored image, resulting in Ockham's hill. Four statistics were compared for optimizing this trade-off, with the Bayesian Information Criterion (BIC) performing best.

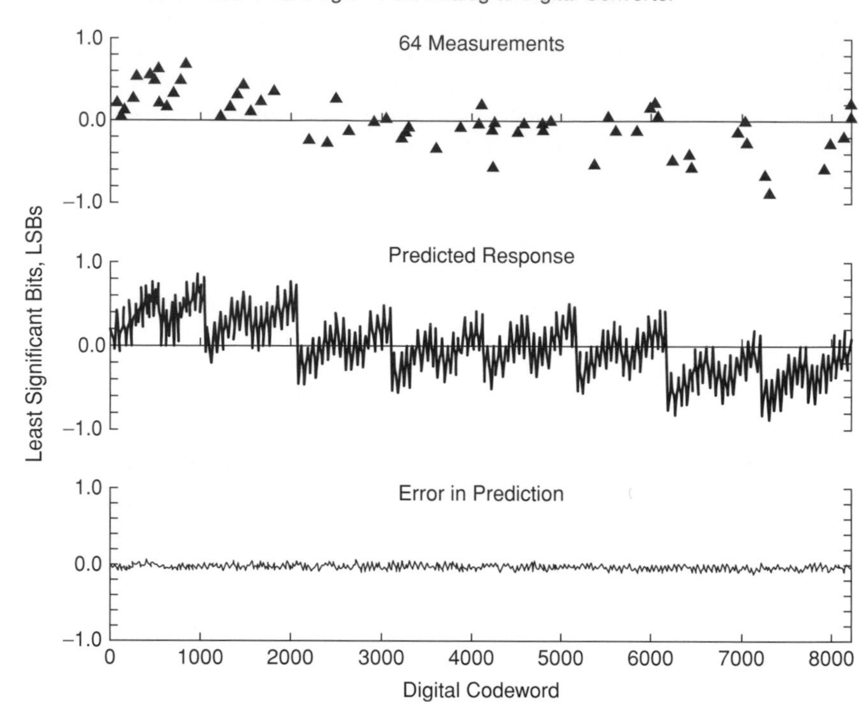

Figure 11.1 Test results from a parsimonious experiment on an analog-to-digital converter. The test requires merely 64 measurements (top) to predict the converter's errors at all 8,192 levels (middle). The differences between these predicted errors and the actual errors are minute (bottom). This quick test reduces testing and manufacturing costs. (Reproduced from Souders and Stenbakken 1991 with kind permission from the Institute of Electrical and Electronics Engineers. © 1991 IEEE.)

PCA can be applied to images for data compression, feature extraction, and time-series predictions (Lv, Yi, and Tan 2007). Too few principal components underfit signal, whereas too many incur excessive computational burdens. In a representative example, an image could be preserved with 95% precision after compression to only 4.7% as much data as the original image, or 98% precision with 12.5% as much data.

Biochemistry and pharmacology

Biochemists and pharmacologists frequently need to know a molecule's shape in order to understand how a natural molecule such as an enzyme functions or to design a new drug. Determination of the shapes of proteins or other large molecules by X-ray crystallography progresses through two stages. First, the basic shape is approximated by building the known chemical constituents to get a good fit with the experimental electron-density distribution. Second, this preliminary structure is refined by calculations that use several kinds of data. The discussion here focuses on this second stage of refining the model of a molecule's shape, particularly how the model's accuracy is assessed.

Structure refinement relies on two principal kinds of data. First and foremost are X-ray-diffraction data collected from a crystal of the macromolecule. As the working model of the molecule is revised, a predicted or expected X-ray pattern is calculated. The actual and predicted patterns are then compared by some measure of fit. Proposed adjustments of the working model are accepted or rejected depending on whether this measure indicates better or worse fit with the X-ray data.

In this comparison of data and model, there are problems, however, with both the data and the model. The data are imperfect, affected by experimental errors and limited in amount. The model is also imperfect, limited by imperfect chemical theory and finite computational power.

Accordingly, data of a second kind are also useful: a data bank of typical bond lengths and angles based on previous experiments with smaller molecules that could be measured unusually accurately. These data from other molecules are used to constrain the analysis of the X-ray data for the macromolecule of interest. For example, analysis of the X-ray data is not allowed to yield any C–C bond length that is outrageously short or long. Extensive experience shows that this combination of two kinds of data works much better than either alone. Figure 11.2 depicts the process of structure refinement, starting from an initial model and adding experimental information from known small-molecule structures and from X-ray-diffraction intensities for the molecule of interest.

Given these two kinds of data, however, the question arises about how much weight to place on each. Different weights lead to different models of the molecule. This question must be answered on a case-by-case basis because each experiment has its own unique limitations and noise level.

For decades, a measure called the R-factor has been an especially popular measure of the fit between X-ray data and a molecular model. But, more recently, crystallographers have discovered that it has serious problems. An extreme example concerns a protein structure that was presumed accurate and was published in the prestigious *Proceedings of the National Academy of Science*, but later it was found that 80% of its atoms had been placed in badly erroneous conformations (Kleywegt and Jones 1995). Also, in flagrant disregard for parsimony, the R-factor can be made arbitrarily "good" merely by adding more parameters to the model. Consequently, it has little objective meaning.

In 1992, a new concept was introduced that used the statistical method of cross-validation to generate a parsimonious, predictively accurate model

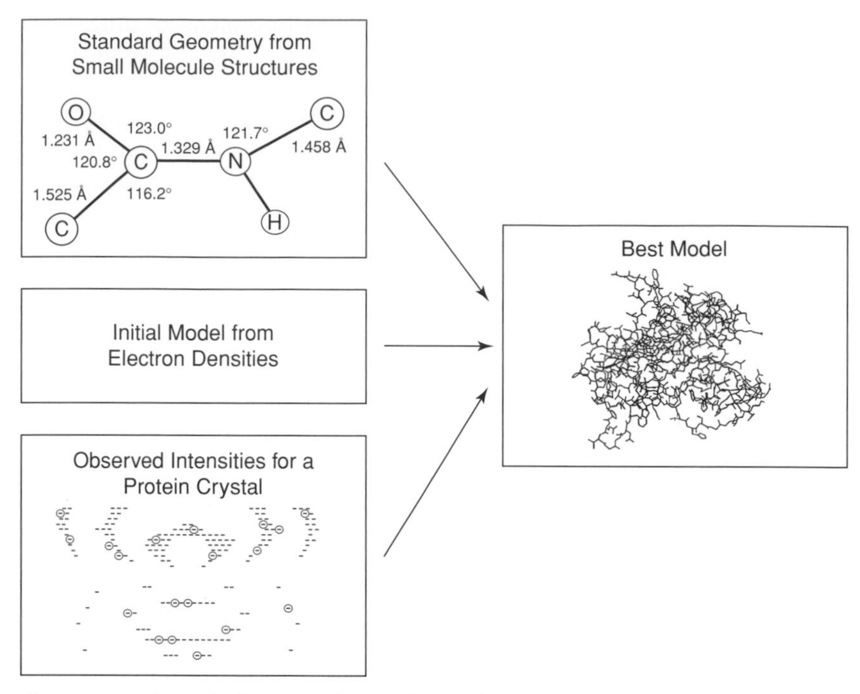

Figure 11.2 Schematic diagram indicating how indirect information about the structures of other molecules and X-ray-diffraction intensities for the molecule of interest are used during model refinement in macromolecular crystallography. The top left panel depicts angles and distances in just one small chain with several carbon, oxygen, nitrogen, and hydrogen atoms, but the actual database provides hundreds of numbers for numerous small molecules. Only average values are shown here, but the complete database also gives standard deviations to describe the ranges in actual values. No values are indicated here for the hydrogen atom because it is too small to diffract X-rays in these studies. The bottom left panel depicts an X-ray-diffraction pattern. Most of the data are allocated to the working set, and about 10% are selected at random for the test set used for cross-validation.

(Brünger 1992). The data are split into two portions: modeling data and validation data, also called the working set and test set. Typically, about 90% of the data are used to construct and refine the model, and the remaining 10%, selected more or less at random, are used to validate or check the predictive accuracy of the model. Often, this process is repeated numerous times with different randomizations and the results are averaged. (Recall that this same method of data splitting and cross-validation was cited in the previous chapter for agricultural yield-trial data.) This new measure of structure accuracy, the R_{free} value, provides reliable assessments of accuracy, as well as

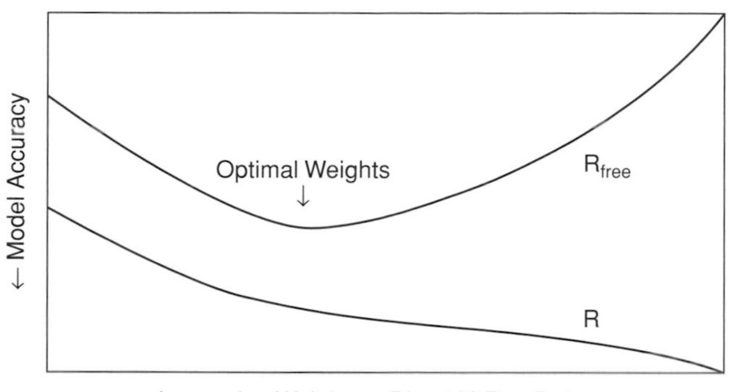

Figure 11.3 Effects of the weights placed on direct and indirect data on two measures of model accuracy. Empirical studies reveal the importance of properly weighting the relative contributions of the direct and indirect information during crystallographic refinement. The R-factor measures the postdictive agreement between the model and the diffraction data, so naturally it continues to improve as more weight is placed on the diffraction data. But R_{free} monitors the predictive accuracy of the model, using the subset of test data that were not used in refinement but rather were saved for cross-validation. The best model is that which minimizes R_{free}. Here, model accuracy increases downward, so Ockham's hill appears upside down relative to its usual depiction.

an objective criterion for choosing protocols and models during refinement. A small modification of this cross-validation procedure further improves performance (Kleywegt 2007).

Using this new R_{free} value, as the weight on the X-ray data is increased (and, correspondingly, the weight on the data bank of ideal geometries is decreased), an Ockham's hill emerges, or actually an Ockham's valley because smaller values of R_{free} indicate greater accuracy. Some intermediate weight is best, as depicted in Figure 11.3. "Too few parameters will not fit the data satisfactorily whereas too many parameters might fit noise" (Brünger 1993), just as was demonstrated in several other applications in science in Chapter 10. By contrast, the old R-factor is misleading, continuously improving as more weight is placed on the X-ray data. Hence, the predictive R_{free} value shows a nice Ockham's valley, but the postdictive R-factor keeps rewarding model complexity, just like the lines for prediction and postdiction in Figure 10.1.

Rearrangement of the usual equations used in refinement makes more obvious the deeply Bayesian nature of this combining of new and old information, that is, the X-ray data for a given molecule and the data bank of typical lengths and angles for other molecules. Even greater accuracy can be achieved with more explicit and vigorous Bayesian analyses. This makes sense because the task of structure refinement does have extensive and accurate prior information that is well worth exploiting.

Although the new paradigm introduced in 1992 is demonstrably better, it has taken time to catch on. During the following eight years from 1993 to 2000, the adoption of R_{free} in the enormous Protein Data Bank gradually rose: 1, 7, 33, 48, 71, 82, 86, and then 92% (Kleywegt and Jones 2002). At present, it is virtually 100%. Optimizing predictive accuracy with R_{free} has been crucial: "As macromolecular crystal structures are determined and refined in an increasingly automated fashion, careful assessment of the reliability and quality of the resulting models becomes increasingly important."

The larger tragedy is not that adoption has been slow since 1992 but that discovery came only in 1992. Philosophers of science had emphasized parsimony and statisticians had developed cross-validation techniques decades before 1992. Decades were wasted before cross-validation techniques were imported from statistics into this particular application in biochemistry. That unnecessary delay is curious in that surely thousands of biochemists took courses in statistics. Also, biochemists had already found cross-validation helpful in some closely related applications, such as predicting protein structure from amino-acid sequences. But the old R-factor had been received with such acceptance and complacency for decades that there was no search for a better measure of accuracy. Regrettably, the opportunity cost has been great. Had the relevance of parsimony in determining molecular shapes been appreciated a decade or two before 1992, we might now have better medicines for many diseases.

Were scientists routinely given broader training in science's method and philosophy, more frequent thought would be given to principles such as parsimony and predictive accuracy, complacency about second-rate methods would give way to energetic search for better approaches, valuable ideas would be imported faster from related fields including philosophy of science, and effective new paradigms would be accepted more quickly. There are countless potential applications of parsimony, which can be implemented by off-the-shelf statistical methods, waiting to be discovered in numerous specialties in science and technology.

As the number of protein structures solved annually continues to rise sharply, increasingly automated methods are essential. However, automated methods occasionally fail, making human intuition and creativity essential. An online protein folding game, Foldit, has been developed to enlist players worldwide to help solve difficult protein structures, even though most players are not experts in chemistry (Khatib et al. 2011). A wide range of expert attempts to solve the crystal structure of a retroviral protease had failed for more than a decade, but the online community of game players solved this structure in a few days. Human intuition complements rather than replaces computer calculations because the Foldit program uses extensive calculations based on chemical principles and data as it interacts with players. These authors concluded that "the ingenuity of game players is a formidable force that, if properly directed, can be used to solve a wide range of scientific problems."

Medicine

Medical research needs to be highly efficient, benefiting from optimal statistical designs and analyses. Efficiency can minimize the number of patients in an experimental trial who must suffer through inferior treatments, and it can maximize the number of patients who will receive superior treatments in the future. Also, given the finite resources to be divided among countless disease treatments and research projects, efficiency is crucial for containing costs. It is unethical to waste experimental subjects or to waste research funds because of second-rate statistics.

For example, consider a clinical trial that concerned a serious human disease, acute leukemia (Freireich et al. 1963). There were two treatments: 6-mercaptopurine (6-MP) and placebo. As patients were recruited into the trial, they were randomized in pairs to receive either 6-MP or placebo. Investigators recorded which patient of each pair remained in remission longer. If it was the patient receiving 6-MP, that was deemed a success; otherwise, a failure. There were 21 patient pairs.

Ethical considerations for such a serious human disease motivated researchers to adopt a fully sequential analysis, meaning that the data were analyzed anew each time another outcome became available. If 6-MP were proving to be effective, it would be unethical to continue giving the placebo to half of the experimental patients.

The fully sequential frequentist analysis used in that clinical trial had a stopping rule that would be activated when the excess of successes over failures (or the reverse) became sufficiently large to reach the 5% significance level, or else when sufficient data indicated no appreciable treatment difference. Depending on the observed data, that rule was bound to stop the experiment after results had been reported for somewhere between 9 and 67 patient pairs. It so happened that superiority of 6-MP over placebo was detected after testing 18 patient pairs. At that time, recruitment of new patients into the trial was stopped, but 3 more pairs had already begun treatment, and they were followed to observe the outcomes, giving 21 patient pairs all told.

The data were reanalyzed using a fully sequential Bayesian analysis with a noninformative uniform prior (Berry 1987). The Bayesian analysis reached the 5% level (meaning that there was only a 5% probability that the placebo was better than 6-MP) after only 10 treatment pairs – about half as many patients as in the frequentist analysis. The main reason for that sizable difference is that frequentist methods penalize final conclusions for interim looks at the data, whereas Bayesian methods do not.

Interestingly, Berry (1987) showed the results of those two analyses (plus another two analyses not discussed here) to "hundreds of statisticians and physicians" and reported that "Most wanted to stop the trial between the 8th and 12th pairs; only a few wanted to keep it going to the 18th pair," even though most of those statisticians were avowed frequentists. He then commented that "I take this as evidence that one's intuition adheres to the conditional [Bayesian] view even though one's training takes one elsewhere," and that from extensive experience, one learns "to take data at face value." This example is important because sequential clinical trials are common in medical research. Inefficient statistics exact a high price by increasing the number of patients given inferior treatments.

Generalizing beyond this example of a clinical trial regarding acute leukemia, Greenland (2008) gave several good reasons for greater use of Bayesian statistics in medical research, or epidemiology more specifically. The foremost argument for the frequentist paradigm, which was also the motivation for developing it in the first place, has been greater objectivity than the preceding Bayesian paradigm. Recall Fisher's tale of the black and brown mice and its sermonic conclusion about scientists knowing better than (Bayesian) statisticians when they lack information. To the objection that "adopting a Bayesian approach introduces arbitrariness that is not already present," Greenland responded that "In reality, the Bayesian approach makes explicit those subjective and arbitrary elements that are shared by all statistical inferences. Because these [subjective] elements are hidden by frequentist conventions, Bayesian methods are left open to criticisms that make it appear only they are using those elements."

Remarkably little clear thinking suffices to dispel such illusions. Evidence renders hypotheses more likely or less likely *than they were beforehand*, which is an intrinsic feature of evidence that cannot be changed by statistical paradigms. In other words, whatever evidence is tendered in a given study, it is always possible that other evidence could also be relevant. The Bayesian prior explicitly recognizes this other evidence. Alternatively, if an investigator prefers to address only a given collection of evidence, ignoring everything else, then results can be reported using a Bayes factor rather than posterior probabilities. Wanting a conclusion about the probabilities of hypotheses given the data, without any prior probabilities, is in the same category as wanting water without any hydrogen – it is absolutely impossible in principle. The Bayesian paradigm should be rewarded, not penalized, for making subjective elements explicit.

Consequently, a preference between frequentist and Bayesian analyses should not be based on illusions about objectivity and subjectivity. Rather, legitimate considerations include reaching conclusions efficiently and achieving computational feasibility.

An important objective in medical research is to discover the causes of disease. Rothman et al. (2008) explored causation in the context of medicine. They listed nine causal criteria called the Bradford-Hill criteria. However, they emphasized that there is no set of necessary and sufficient causal criteria: "universal and objective causal criteria, if they exist, have yet to be identified." Each criterion is subject to counterexamples, and choosing how to weight these nine criteria is highly subjective. The Bradford-Hill criteria are also used in law (Finkelstein 2009:113–115).

The difficulty in determining causes varies greatly from case to case. The causal organisms for most infectious diseases are known clearly, as well as genetic and environmental risk factors. The cause of some inherited genetic disorders is understood in complete molecular detail. However, in more complex cases, these nine criteria may be somewhat helpful. Until medical researchers have a more satisfying and settled account of causal inference, it will be strategic to import useful ideas from books on causality, such as Woodward (2003), Pearl (2009), and Losee (2011). Philosophy of science also provides important resources, such as Glynn (2011). Causality, explanation, and prediction are substantially interrelated topics.

Sociology

The general principles of scientific method have many applications in sociology. Several brief examples can illustrate this.

Social Mobility. Besides the crime rates already discussed in the previous chapter on parsimony, Raftery (1995) gave another example concerning social mobility in 16 countries. They had a 3 × 3 table of father's occupation by son's occupation for each country, with the categories being white-collar, blue-collar, and farm. The total sample size, 113,556, was quite large.

Two hypotheses suggested by sociologists were of principal interest: that mobility flows are the same in all industrialized countries and that the patterns of mobility (but not the amounts) are the same. The frequentist analysis using *p*-values selected an extremely complicated model in opposition to the parsimonious model indicated by the sociologists' hypotheses. But recall that Raftery showed that large sample sizes make *p*-values excessively likely to declare statistical significance. By contrast, model selection by BIC strongly preferred the model in agreement with sociological theory.

Although Bayesian analysis has many advantages, most notably directly assessing probabilities of the form P(hypothesis | evidence), the choice between Bayesian and frequentist analyses can be rendered somewhat less critical by consulting tables in Raftery (1995) that relate several statistical measures to a common scale of evidential strength. He listed four levels of evidence: weak, positive, strong, and very strong. They corresponded to Bayesian posterior probabilities of 50–75, 75–95, 95–99, and >99%. Other measures included the BIC difference, Bayes factor, and several common statistics. For example, strong evidence with a posterior probability of 95–99% equates to a BIC difference of

6–10, a Bayes factor of 20–150, and a p-value of 0.003 for 50 samples or 0.0001 for 10,000 samples. Note that p-values have interpretations depending on sample size but not Bayes factors and BIC differences. These tables are useful when statistical theory favors one analysis but statistical convenience favors another analysis. But they are no substitute for careful thought about whether anything besides the evidence from a particular experiment or survey is highly relevant and merits consideration.

Another study of occupational mobility concerned civilian men in the 1970s in three countries: the United States, Britain, and Japan (Goodman and Hout 2002). Occupational origin and destiny were classified using five categories: upper nonmanual, lower nonmanual, upper manual, lower manual, and farm. The total sample size was 28,887. BIC selected a relatively parsimonious model with the general tendency being the greatest mobility in the United States and the least in Japan, with Britain intermediate – although there were some exceptions.

In a commentary on that article, Xie (2002) emphasized the "tension" between parsimony and accuracy. But that commentary could have been much more insightful had there been an awareness of Ockham's hill – or even a greater understanding of BIC as a criterion that balances parsimony and accuracy, creating not a tension but a synergy in which parsimony increases (predictive) accuracy.

Obesity. As Boardman et al. (2010) explained, "The goal of health-disparities research is to identify the mechanisms that are responsible for persistent differences in health status among members of different social groups." There is great interest in racial/ethnic differences in obesity because "childhood and adolescent obesity is strongly linked to adult obesity... and because obesity is linked to increased risk of chronic health problems including type-II diabetes and hypertension." Their research combined data on body mass index (BMI), self-reported ethnicity, and genetic information on more than 5,000 single nucleotide polymorphisms for Colorado youths of ages 14 to 19, with 88 white and 23 Mexican American. PCA was used to obtain a parsimonious, manageable summary of the enormous data matrix of genetic information. The main objective was to clarify the social and genetic epidemiology of obesity.

Only the first principal component, accounting for only 3% of the variance in the genetics data, was significantly associated with BMI. After controlling for that genetic measure, ethnic differences in BMI remained large and statistically significant. Hence, "these results bolster the position of social epidemiologists who look to social context (e.g., friends, schools, workplace, residential areas) to account for important differences in the prevalence of obesity among racial and ethnic groups." They acknowledged that their sample size was modest and perhaps unrepresentative of the overall population in the United States, so genetics might be more important in some other cases. Nevertheless, their conclusion was that social factors are more important than genetic factors in the epidemiology of obesity. "Based on the results of this study, a critical examination of the built environment and cultural milieu of racial and ethnic groups will yield far more insights into body size differences across socially defined groups of people than a corresponding exploration of small (and apparently meaningless) genetic differences for this important phenotype." Specific social factors meriting attention included socioeconomic status; fast-food consumption; skipping breakfast; physical inactivity; cultural beliefs about food and ideal bodies; and government policies and market conditions that make high-calorie, low-nutrient foods cheaper than low-calorie, high-nutrient foods.

Reading and Listening Comprehension. A nice example of Ockham's valley in sociological research concerned reading and listening comprehension for a survey of 425 Taiwanese elementary students (Yang 2007). The children were surveyed regarding reading at home on a weekly basis beyond their homework assignments from school. They were also given a reading comprehension test with 10 multiple choice questions and a listening comprehension test with 10 multiple choice questions, with each of the 425 students' 20 answers categorized as correct (1) or incorrect (0). A model family deriving multiple categorical latent variables summarized the information in these data. Each model's performance was assessed by BIC to balance fit and parsimony. The model with the best (lowest) BIC score, located at the bottom of Ockham's valley, used four latent variables; whereas other models with either fewer or more latent variables performed worse, having less association with extracurricular reading at home.

Neighborhood Differences. Social scientists have known that urban neighborhoods vary greatly in crime rates, physical and mental health problems, education levels, earnings, and other social indicators. But there are questions about how social processes might cause those outcomes and how to test hypotheses about neighborhood influences. Savitz and Raudenbush (2009) sought reliable and valid measures or estimators of those social processes, using data on 343 Chicago neighborhoods.

They compared three statistical methods: the ordinary least squares estimator and Bayesian estimators with and without spatial dependence. Exploiting spatial dependence involved a weighted average of a neighborhood sample mean and the overall mean, where this "shrinkage" toward the overall mean puts parsimony to work as first discovered by Stein (1955).

Interestingly, Savitz and Raudenbush (2009) used a combination of statistical theory and empirical investigation to compare their three statistical methods. Theory supported the Bayesian analysis exploiting spatial dependence. Empirical investigation using cross validation supported the same method, revealing four advantages: (1) The Bayesian method with spatial dependence was most robust with small sample sizes; (2) its results were more consistent over time; (3) it achieved the highest correlations with those variables that sociological theory expected to be especially influential; and (4) it made the most precise and accurate predictions of future crime rates.

Economics

Economists have studied the wealth of nations as affected by market-oriented or state-interventionist policies. "Whether market allocations outperform state intervention has always been the subject of a heated debate. Not only do people around the world hold conflicting views about this issue, but these views change over time as countries learn from their own past experience, as well as the experience of other countries" (Buera, Monge-Naranjo, and Primiceri 2011).

Their analysis was Bayesian. "We formalize this connection between ideas, policies, and economic development in the context of a model where the performance of alternative policy regimes is uncertain. Policymakers start with some priors about the growth prospects of market-oriented and state-interventionist regimes, and use Bayes's law to update these priors with the arrival of new information from all countries in the world.... A country decides to pursue market-oriented policies if the perceived net impact of these policies on its gross domestic product (GDP) growth exceeds their political cost." Their data were quite extensive: prior beliefs about market-oriented and state-interventionist policies formed during 1900–1950; parameters indicating political costs for changes; data on market orientation; GDP; and several geographical, political, and civil war indicators for a panel of 133 countries over the postwar period 1950–2001.

Six models were compared for how well they fit the data, with their baseline model being the simplest. In the five more complicated models, adding variables for the presence of civil war, for membership in the Warsaw Pact under Soviet Union military control, and for recent inflation and trade deficit gave inconsequential improvement; whereas adding variables for level of education and for changes over time in political costs of market orientation did achieve incremental improvement. Poor countries were substantially less likely to have market-oriented economies than developed countries. However, comparison of all six models using BIC selected the simplest, baseline model. The results were impressive.

> We indeed find that the baseline model correctly predicts almost 97% of the policy choices observed in the data.... It accounts for 25.8%, 53.8%, 74.2%, and 77.4% of the actual policy switches within windows of ±0, 1, 2, and 3 years, respectively.

> The empirical success of the model is mainly driven by the evolution of beliefs, and not by the countries disparate and changing economic and political conditions. Indeed, if we shut down the learning mechanism – and estimate fixed beliefs, but otherwise allow the economic, political, and military controls to vary as in the data – then our model loses more than two-thirds of its predictive power. This holds despite the fact that our estimates suggest that countries are more likely to pursue market-oriented policies when they are more democratic, have higher per capita income, and are not hosting a civil war, among other things. Therefore, the inclusion of these variables is supported by the data,

but does not undermine the importance of evolving beliefs as determinants of policy changes. (Buera et al. 2011)

They also investigated the status of market-oriented and state-interventionist economies as of the end of their sample period, 2001. They found that most countries are oriented toward state-interventionist policies, with the notable exceptions being the United States, Western Europe, Hong Kong, and Singapore. The principal factor slowing adoption of market-oriented economies has been the perception by policymakers that possibly there are country-specific determinants of advantages from various policies, which blunts the perceived relevance of some successful market-oriented economies elsewhere.

They also addressed another interesting question. "'Would a global recession induce some policy reversals to interventionism?' Our answer is 'Yes.' More specifically, we estimate that approximately 5–10% of market-oriented countries would become state interventionist if the world economy experienced a negative growth shock similar to the Great Depression of the 1930s."

As this example should suggest, there are many applications in economics and business for the general principles of scientific method, such as Bayesian inference and parsimonious models. For instance, Giacomini and White (2006) noted that "Forecasting is central to economic decision-making." Using forecasts of industrial production, personal income, and consumer price index as examples, they tested various models for predictive ability. These forecasts used a large number of potential predictors, and they found that parsimony was extremely important for optimizing predictive accuracy. BIC was quite useful. Schulze (2009) has recommended that economists give greater attention to mastering scientific method.

Law

Increasingly, scientific evidence is used in law, often with statistical assessments of that evidence. Whole books and journals are devoted to scientific and statistical methods in law. "Where there are data to parse in a litigation, statisticians and other experts using statistical tools now frequently testify. And judges must understand them. In 1993, in its landmark *Daubert* decision, the Supreme Court commanded federal judges to penetrate scientific evidence and find it 'reliable' before allowing it in evidence.... The Supreme Court's new requirement made the Federal Judicial Center's *Reference Manual on Scientific Evidence*, which appeared at about the same time, a best seller. It has several important chapters on statistics" (Finkelstein 2009:vii). This section presents several brief examples of statistics in law.

Probability Fallacies. Recall that Chapter 8 distinguished the reverse conditional probabilities $P(X \mid Y)$ and $P(Y \mid X)$, relating them to each other by Bayes's theorem saying that the posterior odds equal the likelihood odds times the prior odds. In the context of law, it is important to avoid confusion between P(guilt | evidence) and P(evidence | guilt).

Legal probabilities are mostly Bayesian. . . . The more-likely-than-not standard of probability for civil cases and beyond-a-reasonable-doubt standard for criminal cases import Bayesian probabilities because they express the probabilities of events given the evidence, rather than the probabilities of the evidence, given events. Similarly, the definition of "relevant evidence" in Rule 401 of the Federal Rules of Evidence is "evidence having any tendency to make the existence of any fact that is of consequence to the determination of the action more probable or less probable than it would be without the evidence." . . . By contrast, the traditional scientific definition of relevant evidence, using classical probability, would be any "evidence that is more likely to appear if any fact of consequence to the determination of the action existed than if it didn't." (Finkelstein 2009:3)

Screening tests are used in medicine, as was illustrated in Chapter 8, and also in law. Testing for rare events increases the importance of considering the background or prior probabilities. One of Finkelstein's examples concerned drug couriers.

The failure to distinguish posterior odds and likelihood ratios has led to some confusion in the Supreme Court's jurisprudence on the constitutionality of stopping passengers to search for drugs based on a profile. In particular, debates have swirled around the question whether a profile can be a basis for "reasonable suspicion" that the person is a drug courier, which would justify a brief investigative detention, a lower standard than "probable cause" needed for arrest. Since drug couriers are rare among passengers, even profiles with impressive operating characteristics are likely to generate small posterior odds that the person is a courier. For example, imagine a profile that is quite accurate: 70% of drug couriers would fit the profile, but only 1 normal passenger in 1,000 would do so. The likelihood ratio for the profile is $0.70/0.001 = 700$, which means that the odds that a person is a courier are increased 700-fold by matching the profile. But if the rate of couriers is 1 in 10,000 passengers, the odds on the person matching the profile is a courier are only $(1/9{,}999) \times 700 = 0.07$, or a probability of 6.5%. This result is not fanciful: a study by the US Customs Service found that the hit rate for such stops was about 4%. (Finkelstein 2009:9)

Finkelstein commented that the Supreme Court and other courts in the United States and elsewhere have not yet clarified whether "reasonable suspicion" requires only a rational basis for suspicion, thereby justifying screening tests; or else a reasonable probability of crime, largely invalidating screening tests. Various courts and decisions are also divided over the propriety of instructing jurors in Bayes's theorem in order to help them distinguish between reverse conditional probabilities. There is also an unsettled discussion of whether it is preferable in litigation to report posterior probabilities that necessitate estimating prior probabilities, or else to report Bayes factors (likelihood ratios) based only on a specific body of evidence. Although the Supreme Court has ordered federal judges to assess scientific evidence accurately, there have also been somewhat worrisome empirical studies concerning how well jurors, both sophisticated and unsophisticated, can process such evidence (Boudreau and McCubbins 2009).

Predicting Votes. How will a nominee to the US Supreme Court, if confirmed, vote in the future? Cameron and Park (2009) raised this intriguing question with the following initial observation: "Commonsense and casual observation suggest a nominee's political ideology is usually a good indicator of his or her future behavior as a justice. Indeed, presidents, senators, interest groups, and the media all seem to employ this commonsense judgment, since perceived nominee ideology is so central to confirmation politics."

They used several predictors based on available information at the time of nomination. Those predictors included the political ideology of the nominating president and, depending on the nominee's career, also included votes cast as a member of Congress, votes as justices in the US Courts of Appeal, and votes as a Supreme Court justice for nominees for Chief Justice. Those predictors were combined into a single score. Another score was based on media perceptions of nominees, using content analysis of newspaper editorials. Finally, those two scores were combined by PCA, with the first component accounting for 85% of the variation. This final score for the 44 confirmed nominees from 1937 to 2006, which they named the NOMINATE-scaled perception (NSP) score, ranged from −0.603 for the liberal Justice Marshall to 0.601 for the conservative Chief Justice Rehnquist.

The main predicted variable was each justice's overall voting tendency on the political spectrum from liberal to conservative. Also predicted were voting tendencies on six specific issues: civil liberties, unions, economics, federalism, judicial power, and taxation. The NSP score was a good predictor of future voting overall and of future voting for all specific issues, except taxation.

Cameron and Park (2009) also separately analyzed the data for just 1958 to 2005, finding that the voting of confirmed justices had become much more predictable. They noted that from 1937 to 1957, only 47% of the nominees had been judges; but, from 1958 to 2005, this increased to 78% (and from 1967 to 2005, it was 87%). Not surprisingly, of their several predictors, the most informative predictor of future voting as a justice, when available, was past voting as a judge. Incidentally, they also calculated NSP scores for two justices (Roberts and Alito) whose tenure on the Supreme Court had been quite brief at the time of writing their 2009 article, leaving confirmation or disconfirmation of those predictions to the future.

Hence, statistical analysis of extensive data confirmed what common sense suspects, that indicators of ideology in the past are good predictors of voting in the future, while adding precise findings about how to combine multiple indicators in an optimal predictor. Cameron and Park concluded, "the centrality of perceived nominee ideology in Supreme Court confirmation politics is entirely understandable." Particularly because Supreme Court appointments are for life, presidents can use these nominations to extend their legacies.

Statins for Women. Lipitor (atorvastatin) has been the top-selling drug in the world, accounting for more than $12 billion in annual sales. Eisenberg and Wells (2008) performed a statistical and legal analysis of advertising by the manufacturer, Pfizer, Inc. They quoted Lipotor's advertising: "LIPITOR is clinically proven to reduce the risk of heart attack, stroke, certain kinds of heart surgeries, and chest pain in patients with several common risk factors for heart disease" where "Risk factors include family history, high blood pressure, age, low HDL ('good' cholesterol), or smoking." But, regrettably, Pfizer's advertising omitted label information relevant to women: "Due to the small number of events, results for women were inconclusive." Because their article's abstract is exceptionally clear and concise, it is worth quoting in full.

> This article presents: (1) meta-analyses of studies of cardioprotection of women and men by statins, including Lipitor (atorvastatin), and (2) a legal analysis of advertising promoting Lipitor as preventing heart attacks. The meta-analyses of primary prevention clinical trials show statistically significant benefits for men but not for women, and a statistically significant difference between men and women. The analyses do not support (1) statin use to reduce heart attacks in women based on extrapolation from men, or (2) approving or advertising statins as reducing heart attacks without qualification in a population that includes many women. The legal analysis raises the question of whether Lipitor's advertisements, which omit that Lipitor's clinical trial found slight *increased* risk for women, is consistent with Food, Drug, and Cosmetics Act and related Food and Drug Administration (FDA) regulations. The analysis suggests that FDA regulation should not preempt state law actions challenging advertising that is not supported by FDA-approved labeling. Our findings suggesting inadequate regulation of the world's best-selling drug also counsel against courts accepting the FDA's claimed preemption of state law causes of action relating to warnings and safety. Courts evaluating preemption claims should consider actual agency performance as well as theoretical institutional competence. Billions of health-care dollars may be being wasted on statin use by women but the current regulatory regime does not create incentives to prevent such behavior. (Eisenberg and Wells 2008)

As Eisenberg and Wells noted, misleading or false advertisements violate the Food, Drug, and Cosmetics Act and FDA regulations. FDA regulations state that advertisements for a prescription drug "*are* false, lacking in fair balance, or otherwise misleading" if they have one or more of 20 enumerated characteristics. The first characteristic is that the advertising contains "a representation or suggestion, not approved or permitted for use in the labeling, that a drug is... useful in a broader range of... patients... than has been demonstrated by substantial evidence or substantial clinical experience." Yet, much of Pfizer's available advertising did not mention their inconclusive results for women, although women without coronary heart disease constitute around 20% of statin recipients in many countries. The 18th and 20th characteristics of misleading advertisements were also of concern.

This statistical and legal analysis regarding Lipitor raises larger issues for drug manufacturers, medical researchers, statistical consultants, government regulators, and legal practitioners. Eisenberg and Wells concluded: "This study also has implications for reining in health-care costs. The growing multibillion dollar statins market significantly contributes to increasing health-care expenses. Our findings indicate that each year reasonably healthy women spend billions of dollars on drugs in the hope of preventing heart attacks but that scientific evidence supporting their hope does not exist."

Discussion

Students and professionals in science encounter the general principles of scientific method incessantly precisely because they *are* general. But the pervasive risk is that such encounters are processed merely as one more little technicality in getting some job done.

For example, a graduate student in engineering may apply cross validation to a dataset simply because a professor said to do this, or because of precedents in the literature, but without understanding why this is done or what benefits accrue. Were such a student asked in a thesis defense why cross validation was used, would he or she have a genuinely insightful answer or even a merely adequate answer? Would that student transfer this statistical technique to other relevant applications in future years? Or for an even simpler example that is likely to be quite sobering indeed, ask a science student why an experiment was replicated. Just what is, and is not, gained from all the extra work involved in replication? Howson and Urbach (2006) explain and compare the purposes of replication from frequentist and Bayesian perspectives.

The only effective means for learning the general principles of scientific method is to encounter them *both* as general principles *and* as specific examples. Previous chapters pursued the former component and this chapter the latter.

For instance, a student or professional who understands Ockham's hill can quickly grasp that cross validation in a particular task is doing the job of optimizing predictive accuracy, and thereby also gaining efficiency and increasing repeatability. An understanding of Ockham's hill, specifically as a general principle, stimulates new applications.

Responsibility for science education and best practices falls on professors and students alike. Professors need to communicate both the general principles of scientific method and the research techniques of a given specialty. That accords with numerous position papers on science, as reviewed in the first chapter, and with pedagogical recommendations from science educators, as reviewed in the next chapter. On the other hand, students need to be self-aware when an experiment or protocol has steps that they are mechanically performing but not genuinely understanding. They need to ask why things are done and what benefits result. Furthermore, if told to perform a frequentist analysis of the data, a student should feel free to ask whether a Bayesian analysis might give stronger and clearer results. Likewise, if not told to use cross validation or BIC or other statistical methods for optimizing predictive accuracy, a student should feel free to ask whether such methods that have proved fruitful in many other applications might well be imported into the present application.

Although this chapter included seven case studies, many more applications of science's general principles to additional fields could be cited. For instance, Clark (2005) published a compelling article on "Why environmental scientists are becoming Bayesians." He noted that "Ecologists are increasingly challenged to anticipate ecosystem change and emerging vulnerabilities." The Bayesian framework provides computational feasibility for inferences and predictions in highly complex problems, although sometimes ecologists may need help from a consulting statistician. The comprehensive accounting of variability and uncertainty in a Bayesian analysis provides guidance on what is and what is not predictable, and on which kinds of additional data would help the most.

This chapter's diverse case studies, from philosophy and law to medicine and sociology, are intended to communicate the broad relevance of science's general principles. Again, this book's message is that a disproportionately large share of future advances in science and technology will come from those researchers who have mastered their specialties like everyone else but who also have mastered the basics of science's philosophy and method. Furthermore, because of the traffic of ideas between the sciences and the humanities, understanding science's general principles is also beneficial in the humanities.

Study questions

(1) In your own discipline, whether in the sciences or the humanities, are there applications or topics having potential confusion between reverse conditional probabilities? What loss or harm might result from such confusion?

(2) Scientists in many disciplines, including sociology and medicine, often collect data on a long "laundry list" of variables that are possibly relevant. Weak theory – neither nonexistent nor strong – leads to this situation. Consequently, statistical tests are needed to determine which of these many variables are truly informative, helping predictions and explanations. Are there situations like this in your areas of interest? What are some effective statistical tests for identifying informative predictors? After key variables have been identified, how might that insight provide an opportunity for reducing the costs of subsequent extensive research?

(3) Can you think of missed opportunities to use parsimonious models of pilot data to design simplified, lower-cost tests or surveys for extensive manufacturing or application?
(4) Do you suspect that a Bayesian approach would be better than present statistical methods for some specific applications in your discipline? Give concrete examples.
(5) Why do parsimonious models and Bayesian analyses seem to catch on so slowly in many applications in agriculture, philosophy, engineering, biochemistry, medicine, sociology, law, and other disciplines? Which strategies might speed up the adoption of best practices?

12

Ethics and responsibilities

This chapter presents an exceedingly brief account of philosophical and professional ethics. It is essential and unavoidable for science and technology to be guided and constrained by an ethical vision.

The main topic, having the longest section, is science's professional ethics. All position papers on science recognize science's ethics as an essential item in the science curriculum. But that section needs to be preceded by a shorter section on philosophical ethics, which provides the broader context needed to make professional ethics meaningful.

Philosophical ethics

The ideas on philosophical ethics presented here are drawn from three exceptional books, two old and one recent: Caldin (1949), MacIntyre (1988), and Sandel (2009). Edward Caldin was a lecturer in chemistry at the University of Leeds. Alasdair MacIntyre has taught philosophy at the University of Oxford, University of Essex, Notre Dame University, Duke University, and several other universities, and is a past president of the American Philosophical Association. Michael Sandel teaches a large and popular course on ethics at Harvard University.

The first several pages of MacIntyre (1988) emphasize the challenges and controversies surrounding ethics. Most obviously, ethics raises perplexing questions, such as: "Does justice permit gross inequality of income and ownership? Does justice require compensatory action to remedy inequalities which are the result of past injustice, even if those who pay the costs of such compensation had no part in that injustice? Does justice permit or require the imposition of the death penalty and, if so, for what offences? Is it just to permit legalized abortion? When is it just to go to war?" People disagree.

But deeper than different answers are different reasons for those answers (and, indeed, people can even happen to reach the same answers for radically different reasons). People hold different worldviews.

Furthermore, besides conflicts between persons holding different worldviews, MacIntyre astutely observed that there are often inconsistencies and conflicts within a given individual. "For what many of us are educated into is, not a coherent way of thinking and judging, but one constructed out of an amalgam of social and cultural fragments inherited both from different traditions from which our culture was originally derived (Puritan, Catholic, Jewish) and from different stages in and aspects of the development of modernity (the French Enlightenment, the Scottish Enlightenment, nineteenth-century economic liberalism, twentieth-century political liberalism). So often enough in the disagreements which emerge within ourselves, as well as in those which are matters of conflict between ourselves and others, we are forced to confront the question: How ought we to decide among the claims of rival and incompatible accounts of justice competing for our moral, social, and political allegiance?" (MacIntyre 1988:2).

As if perplexing questions and external and internal conflicts are not enough, MacIntyre also noted the severe limitations of academic ethics. "Modern academic philosophy turns out by and large to provide means for a more accurate and informed definition of disagreement rather than for progress toward its resolution. Professors of philosophy who concern themselves with questions of justice and of practical rationality turn out to disagree with each other as sharply, as variously, and, so it seems, as irremediably upon how such questions are to be answered as anyone else" (MacIntyre 1988:3).

MacIntyre (1988:3)realized that some academics proffer "a genuinely neutral, impartial, and, in this way, universal point of view, freed from the partisanship and the partiality and onesidedness that otherwise affect us." But, inevitably, such an offering is itself "contentious in two related ways: its requirement of disinterestedness in fact covertly presupposes one particular partisan type of account of justice, that of liberal individualism, which it is later to be used to justify, so that its apparent neutrality is no more than an appearance, while its conception of ideal rationality as consisting in the principles which a socially disembodied being would arrive at illegitimately ignores the inescapably historically and socially context-bound character which any substantive set of principles of rationality, whether theoretical or practical, is bound to have" (MacIntyre 1988:3–4). We inhabit a culture with rival ethics and competing rationalities and worldviews, so there is no "neutral" position. That is just the way things are.

The best way to ground ethics would be to find and work within a true worldview rather than to labor with a thin least common denominator that also accommodates numerous false and incompatible worldviews. But, again, which worldview is true is not this book's question. For better or for worse, a related question is inescapable: Given the reality of lack of worldview consensus in the scientific community and the larger society, what attitude should scientists take toward discussions of ethical issues? Thankfully, this question about attitude is somewhat more manageable than the larger question about truth.

The final two pages of Sandel (2009:268–269) are helpful. He recognized the very real difficulties with discourse about ethics. "Some consider public engagement with questions of the good life to be a civic transgression, a journey beyond the bounds of liberal public reason. Politics and law should not become entangled in moral and religious disputes, we often think, for such entanglement opens the way to coercion and intolerance. This is a legitimate worry. Citizens of pluralist societies do disagree about morality and religion. Even if, as I've argued, it's not possible for government to be neutral on these disagreements, is it nonetheless possible to conduct our politics on the basis of mutual respect?" His prescription followed.

> The answer, I think, is yes. But we need a more robust and engaged civic life than the one to which we've become accustomed. In recent decades, we've come to assume that respecting our fellow citizens' moral and religious convictions means ignoring them (for political purposes, at least), leaving them undisturbed, and conducting our public life – insofar as possible – without reference to them. Often, it means suppressing moral disagreement rather than actually avoiding it. This can provoke backlash and resentment. It can also make for an impoverished public discourse, lurching from one news cycle to the next, preoccupied with the scandalous, the sensational, and the trivial.
>
> A more robust public engagement with our moral disagreements could provide a stronger, not a weaker, basis for mutual respect. Rather than avoid the moral and religious convictions that our fellow citizens bring to public life, we should attend to them more directly – sometimes by challenging and contesting them, sometimes by listening to and learning from them. There is no guarantee that public deliberation about hard moral questions will lead in any given situation to agreement – or even to appreciation for the moral and religious views of others. It's always possible that learning more about a moral or religious doctrine will lead us to like it less. But we cannot know until we try.
>
> A politics of moral engagement is not only a more inspiring ideal than a politics of avoidance. It is also a more promising basis for a just society. (Sandel 2009:268–269)

Sandel concluded that precious little ethics can be defended within the bounds of what he terms "liberal public reason." Echoing MacIntyre, conceptions of both ethics and reason are controversial, so there is no neutral, public reason. MacIntyre's resolution, which Sandel (2009:208–243) respected, is so-called communitarian ethics. As its name would suggest, communitarian ethics locates persons within a community or tradition providing resources for developing ethics. Sandel (2009:221) quoted MacIntyre, "I can only answer the question 'What am I to do?' if I can answer the prior question 'Of what story or stories do I find myself a part?'" In this book's terminology, such stories implicate one's worldview. Often a person's worldview prompts obligations, responsibilities, and even sacrifices beyond the basic requirements found in public discourse and civil law.

Communitarian ethics is agreeable to Caldin's view that science by itself cannot initiate or guarantee the good life. But presuming that life experiences have already shown the dignity of persons and the value of reason, science can stimulate and strengthen those factors contributing toward the good life:

We have argued that man has a certain definite nature, with definite potentialities; he is rational, capable of knowing truths and desiring goods, and of acting accordingly; he is also an animal, equipped with animal instincts; and he is a social being, not an isolated individual. . . . Now, in ethics as in metaphysics, we may find a key in considering the actualizing of potentialities. . . . Truthfulness and wisdom are ends befitting a being equipped with intellect; love and self-discipline become a being equipped with a rational appetite for the good; justice, honesty, faithfulness, altruism and mutual love are fitting in human beings because they are also social by nature. . . . The good life is concerned with pursuing ends consonant with our nature; with realising, fulfilling, what we are made for; this determines the moral qualities that men and society should aim at, and so gives us the guiding principles for action both at the individual and social planes. . . .

Many people have been led to think that the procedure of natural science is the royal road to truth in every field, and that what cannot be proved by science cannot be true. . . . Such a use of the great prestige of science to-day would, however, fail unless science were pursued as part of a pattern of wider scope, and the limitations of its method recognized. For science will not give us the conceptions that we need in dealing with the fundamentals of life – the great conceptions of personality, justice, love, for instance, must be drawn from elsewhere; moreover, science gives us only part of the mental training for using them. Again, although science, given its appropriate setting, plays its part in developing both intellectual and moral virtues, it does not originate those virtues; it only supports them where they exist already. It favors an intellectual climate where persons are respected; but it does not create these values, it presupposes them. . . . In relation to the great social purposes – such as peace, justice, liberty – science is instrumental; it is not normative, does not lay down what ought to be. If scientific method is applied outside its own field, if the attempt is made to use one rational method for work appropriate to another, the result is always confusion and, at worst, disaster. To make its full contribution in society, science must respect the fields of other studies; conversely, it relies upon those studies being in a healthy condition. In concrete terms, this means that a scientist who is ignorant of philosophy, history, art and literature, will not be able to speak for reason with the power that he should. . . .

Science and its technical application can both be integrated into the pattern of the good life; science forms an integral part of the good life for a scientist, and the applicability of science can be turned to good account. Potentially, then (whatever we may think of it in practice), natural science is an ally of wisdom and good living. . . . The battles against disease, destitution and ignorance are always with us, and science is a powerful ally; perhaps the most necessary ally after love and understanding and common sense" (Caldin 1949: 138, 6, 169, 159, 175, 161)

Caldin's perspective also agrees with the position of the American Association for the Advancement of Science. "Nor do scientists have the means to settle issues concerning good and evil, although they can sometimes contribute to the discussion of such issues by identifying the likely consequences of particular actions, which may be helpful in weighing alternatives" (AAAS 1989:26).

Professional ethics

The National Academy of Sciences, National Academy of Engineering, and Institute of Medicine co-published *On Being a Scientist: A Guide to Responsible Conduct in Research* (NAS 2009). This section draws principally from this little book that provides superb practical guidance. However, it provides meager philosophical grounding, perhaps because there is little consensus to appeal to, which is why the preceding section drew upon other resources. The opening comments express this book's nascent philosophical ethics and its motivation.

> The scientific enterprise is built on a foundation of trust. . . . When this trust is misplaced and professional standards of science are violated, researchers are not just personally affronted—they feel that the base of their profession has been undermined. This would impact the relationship between science and society. . . .
>
> In the past, beginning researchers learned . . . how the broad ethical values we honor in everyday life apply in the context of science. . . . This assimilation of professional standards through experience remains vitally important.
>
> However, many beginning researchers are not learning enough about the standards of science through research experiences. . . . The guide *On Being a Scientist* explores the reasons for specific actions rather than stating definite conclusions about what should or should not be done in particular situations. . . . Since all researchers need to be able to analyze complex issues of professional practice and act accordingly, every course in science and related topics and every research experience should include discussions of ethical issues. . . .
>
> Researchers have three sets of obligations that motivate their adherence to professional standards. First, *researchers have an obligation to honor the trust that their colleagues place in them*. . . . Second, *researchers have an obligation to themselves*. . . . Third, because scientific results greatly influence society, *researchers have an obligation to act in ways that serve the public*. . . . By considering all these obligations—toward other researchers, toward oneself, and toward the public—a researcher is more likely to make responsible choices. . . . Research is based on the same ethical values that apply in everyday life, including honesty, fairness, objectivity, openness, trustworthiness, and respect for others. (NAS 2009:ix–x, xv–xvi, 2–3)

Ethical violations occur with different degrees of severity. The most serious are termed "scientific misconduct." The US government recognizes three kinds of misconduct: fabrication, falsification, and plagiarism. "All research institutions that receive federal funds must have policies and procedures in place to investigate and report research misconduct, and anyone who is aware of a potential act of misconduct must follow these policies and procedures" (NAS 2009:3).

IEEE CODE OF ETHICS

WE, THE MEMBERS OF THE IEEE, in recognition of the importance of our technologies in affecting the quality of life throughout the world and in accepting a personal obligation to our profession, its members and the communities we serve, do hereby commit ourselves to the highest ethical and professional conduct and agree:

1. to accept responsibility in making decisions consistent with the safety, health and welfare of the public, and to disclose promptly factors that might endanger the public or the environment;
2. to avoid real or perceived conflicts of interest whenever possible, and to disclose them to affected parties when they do exist;
3. to be honest and realistic in stating claims or estimates based on available data;
4. to reject bribery in all its forms;
5. to improve the understanding of technology, its appropriate application, and potential consequences;
6. to maintain and improve our technical competence and to undertake technological tasks for others only if qualified by training or experience, or after full disclosure of pertinent limitations;
7. to seek, accept, and offer honest criticism of technical work, to acknowledge and correct errors, and to credit properly the contributions of others;
8. to treat fairly all persons regardless of such factors as race, religion, gender, disability, age, or national origin;
9. to avoid injuring others, their property, reputation, or employment by false or malicious action;
10. to assist colleagues and co-workers in their professional development and to support them in following this code of ethics.

Approved by the IEEE Board of Directors | February 2006

Figure 12.1 The IEEE code of ethics of the Institute of Electrical and Electronic Engineers. As the world's largest professional association for the advancement of technology, their code of professional ethics is representative of those for many scientific and technological organizations. (Reproduced with kind permission from the IEEE. © 2011 IEEE)

Lesser violations are termed "questionable research practices." Although not misconduct at the federal level, such practices are strongly discouraged and may warrant penalties at a scientist's local institution.

Most professional societies of scientists and technologists have a published code of ethics. Figure 12.1 shows a representative one. It is from the

Institute of Electrical and Electronic Engineers (IEEE), the world's largest such society.

Note the rudimentary but crucial philosophical ethics in the preamble, "in recognition of the importance of our technologies in affecting the quality of life throughout the world and in accepting a personal obligation to our profession, its members and the communities we serve . . . ". This code contains a combination of ethical values applicable to everyday life, and other values more distinctively characteristic of the scientific enterprise. Mastery of scientific method is imperative for meeting this code of ethics, especially the third item, "to be honest and realistic in stating claims or estimates based on available data" (as well as the fifth through seventh items).

Besides professional ethics, there is also a role for personal ethics. Conspicuously absent from this IEEE code of ethics for engineers is any statement about military research. Nonetheless, the personal ethics emerging from an individual's experiences and background and worldview may include or exclude weapons research. Likewise, an engineer's personal ethical values, which may augment professional ethical codes, might motivate him or her to prioritize a certain kind of research, such as research targeted to benefit resource-poor communities in the majority world – even if such work is decidedly less lucrative than many other engineering opportunities and involves increased personal risks from tropical diseases and other dangers.

The following are one-paragraph summaries of the twelve chapters of *On Being a Scientist.* That book and its list of additional resources can be consulted for greater detail (NAS 2009).

Advising and Mentoring. An advisor oversees a research project, whereas a mentor also takes a personal interest in a researcher's professional development. For instance, a mentor may suggest possible research directions and help with finding a job. Beginning researchers sometimes need several mentors to cover all of the areas of expertise they need, especially in multidisciplinary research. Good mentoring promotes the social cohesion that is so essential for the scientific community. To avoid possible confusion and disappointment, mentors and mentees should work out clear expectations together about availability, meeting times, and available equipment and funding. Although the mentee obviously benefits, the reverse also holds, that mentors benefit from exposure to new ideas and an extended network of collaborators. Mentoring is so essential to the functioning of the scientific community that it is advisable for scientific institutions to reward good mentoring and to offer training for established researchers in effective mentoring of beginning researchers.

The Treatment of Data. Because the data generated from observations of nature are the basis for scientists' conclusions, the data need to be collected and presented accurately, without drawing stronger conclusions than the data warrant. Otherwise, published errors can take on a life of their own, especially given rapid electronic communications, thereby making subsequent corrections less than entirely effective. Because scientific methods precede scientific data, methods must be described in sufficient detail to permit reviewers and readers to judge their adequacy and to replicate the research. Statistical analysis is an important tool for quantifying the accuracy of measurements and judging the significance of differences between treatments. Before publication, confidentiality may be appropriate to allow time for checking the data and conclusions. But after publication, other scientists should have access to the data so that they can verify and build on previous research. Some scientific specialties and journals have established repositories to maintain and distribute data. Technologies for digital storage are changing rapidly, so data may need to be transported to current platforms in a timely manner while this can still be done readily.

Mistakes and Negligence. Science is done by humans, who are prone to error. Even the most experienced and capable researchers can make mistakes. That is so especially at research frontiers, where available methods are being pushed to their limits, the signal is difficult to separate from the noise, and even the research questions are vague. As in life more generally, innovative scientists must take risks. But despite these understandable challenges and limitations, scientists have an obligation to their colleagues, themselves, and the public to be as careful and accurate as possible. Some errors are caused by negligence. Hasty or sloppy work, too little replication, and other faults can result in research that fails to exemplify the practices and standards of a given specialty. Often, the history of a specialty has already clarified the most frequent sorts of mistakes and the principal procedures and cross-checks that can minimize these mistakes, so researchers and reviewers should attend to this wisdom from past experience. Errors due to mistakes or negligence are often caught in the review process and thereby corrected in a timely manner. But the review process is selective and incomplete, so researchers cannot count on other persons to catch all of their mistakes and thereby keep those mistakes from further dissemination. Published mistakes can lead to other researchers wasting time and money and to the public receiving suboptimal medical treatments, defective products, and misleading ideas about nature.

Research Misconduct. Research behaviors at odds with the core principles of science are termed *research misconduct.* Such misconduct receives harsh treatment, jeopardizes the guilty party's career, and damages the overall reputation of science and thereby diminishes the potential benefits of science and technology for the public. The NAS (2009:15–18) description of research misconduct is based on a document from the US Office of Science and Technology Policy, which has been adopted by most agencies that fund scientific research. There are three kinds of misconduct. Fabrication is "making up data or results." Falsification is "manipulating research materials, equipment, or processes, or changing or omitting data or results such that the research is not accurately represented in the research record." Plagiarism is "the appropriation of another person's ideas, processes, results, or words without giving appropriate credit." In addition, the federal statement says that actions constituting misconduct must represent a "significant departure from accepted practices," must have been "committed intentionally, or knowingly, or recklessly," and must be "proven by a preponderance of evidence;" whereas it specifically excludes mere "differences of opinion." Beyond these federal guidelines, some research institutions and agencies have additional prohibitions against failure to maintain confidentiality in peer review, failure to allocate credit properly, failure to report misconduct or retaliation against those who do, and related matters. The crucial distinction between mere mistakes or negligence and serious misconduct is the intent to deceive, although a person's intentions are sometimes difficult to prove.

Responding to Suspected Violations of Professional Standards. Governments regulate some aspects of scientific research, but the scientific community is mostly self-regulating. For this to work, scientists must be willing to report suspected incidences of professional misconduct. Obviously, this is awkward, anonymity is not always possible, and reprisals from accused persons sometimes occur even though laws prohibit this. Allegations of misconduct can have serious consequences for everyone involved. Nevertheless, all scientists have an obligation to uphold the fundamental values of science, which includes reporting misconduct. All research institutions receiving federal funds must have policies and procedures to investigate potential acts of misconduct, including designating one or more officials, usually called research integrity officers. Because individual scientists and research institutions discourage inappropriate conduct, it is important to handle serious research misconduct and questionable research practices in different, appropriate manners. It is prudent to express suspicions initially in the form of questions, rather than accusations, because sometimes there is merely a misunderstanding. But such discussions do not always have a satisfactory outcome. In some cases, a preliminary and confidential conversation with a trusted friend or adviser may help, perhaps discussing only the broad outlines of the case without revealing specific details. In dealing with concerns about research misconduct, it is important to examine one's own motivations and biases, especially because inevitably others will do so. Institutional policies usually divide investigations of suspected misconduct into two stages: an initial inquiry to gather information, and a formal investigation to assess evidence and decide responses. These procedures are designed for fairness for the accused, protection for the accuser, and coordination with funding agencies.

Human Participants and Animal Subjects in Research. Research with human participants must comply with federal, state, and local regulations and codes of ethics of relevant professional societies. The intent is to minimize risks relative to expected benefits, obtain informed consent from participants or their authorized representatives, and maintain privacy and confidentiality. US federal regulations known as the Common Rule state requirements and require independent committees known as Institutional Review Boards. Research on animals is also subject to regulations and professional codes. The federal Animal Welfare Act requires humane care and treatment. Also, the US Public Health Service's *Policy on the Humane Care and Use of Laboratory Animals*, as well as the National Research Council's *Guide for the Care and Use of Laboratory Animals*, apply in many cases. Both the Animal Welfare Act and the *Policy on the Humane Care and Use of Laboratory Animals* require institutions to have Institutional Animal Care and Use Committees, which include experts in the care of animals and members of the public. The three main principles are the "three R's" of animal experiments: reduction in the number of animals used, refinement of procedures to minimize pain and distress, and replacement of conscious living higher animals with insentient materials when possible.

Laboratory Safety in Research. Government regulations and professional guidelines are intended to assure safety in research laboratories. It is estimated that in the United States, half a million workers handle hazardous biological materials every day. The short checklist presented in NAS (2009) includes appropriate usage of protective equipment and clothing, safe handling of materials in laboratories, safe operation of equipment, safe disposal of materials, safety management and accountability, hazard assessment processes, safe transportation of materials between laboratories, safe design of facilities, emergency responses, safety education of all personnel before entering the laboratory, and applicable government regulations.

Sharing of Research Results. Peer-reviewed scientific journals originated with the Royal Society of London under Henry Oldenburg's leadership. In current scientific practice, such journals provide the principal means for disseminating research results. Once published, those results can be used freely by others, but until they are so widely known and familiar as to become common knowledge, scientists who use others' results are obliged to credit their sources by means of citations. In addition, useful ideas from seminars, conference talks, and even casual conversations should be acknowledged. This allows readers to locate the original source of ideas and results and thereby to obtain additional information. Proper citation is essential, beginning with accurate author spellings, titles, years, and page numbers. Furthermore, citations should actually support the particular points claimed in a paper and identify the most relevant of the original and derivative articles. Citations are important for judging the novelty and significance of a paper. Researchers should be cautious about making results public before peer review because of the risk of preliminary findings containing errors. Also, some journals consider disclosing research on a website to constitute prior publication, which disqualifies it from subsequent publication in the journal. Scientists may be tempted to get as many little articles out of their research as possible in what are disparagingly called "least publishable units" in hopes of increasing their status or promotion prospects. But that can be counterproductive, producing a reputation for insignificant or shoddy work. The purpose of publication is principally to serve the interests of the scientific community, and beyond that the public, rather than being self-serving. To put the emphasis on quality rather than quantity, some institutions and granting agencies limit the number of papers that will be considered when evaluating a scientist for employment, promotion, or funding.

Authorship and the Allocation of Credit. The list of authors of a paper indicates who has done the work. Proper credit is important because the peer recognition generated by authorship affects scientists' careers. Establishing intended authorship early in the research process can reduce later difficulties, although some decisions may not be possible at the outset. There is a prevalent tradition that established researchers be generous in giving credit to beginning researchers. Whereas authors should have made direct and substantial contributions to the design and conduct of the research or the writing of the paper, lesser contributions can be credited by means of an acknowledgment – contributions such as providing laboratory space or useful samples, as well as helpful suggestions on drafts of the paper from colleagues and anonymous reviewers. The list of authors establishes not only credit but also accountability. Authors bear responsibility for any errors that may be found. In the case of multidisciplinary research, an author providing one kind of expertise may feel and, indeed, be incompetent to check the soundness of contributions from some other co-authors. But that can be handled by a footnote accompanying the list of authors that apportions credit and responsibility for the various components of the paper.

Intellectual Property. Scientific discoveries can have great value, for scientists in advancing knowledge, for governments in setting public policy, and for industry in developing new and better products. Intellectual property is a legal right to control the use of an idea by a patent or the expression of an idea by a copyright. These legal mechanisms attempt to balance private gains and public benefits. They give researchers, nonprofit organizations, and companies the right to profit from an idea in exchange for the property owner making the new idea public, which enables others to pursue further advances. US patent law specifies clear criteria defining who is an inventor, and it is important to include all persons who contributed substantially to an invention. Copyrights protect the expression and presentation of ideas, such as words and images in a publication; whereas they do not protect the ideas themselves, which others may use with proper attribution. Industry has the option of relying on trade secrets instead of patents. In that case, there is no obligation to make the idea public, but neither is there any protection from the idea being developed independently elsewhere. Most universities and research institutes have policies on intellectual property regarding data collection and storage, how and when results can be published, how intellectual property rights can be transferred, how patentable inventions should be disclosed, and how royalties are to be allocated among researchers and institutions. NAS (2010) provides additional information on intellectual property.

Competing Interests, Commitments, and Values. A conflict of interest is a clash among a scientist's personal, intellectual, and financial interests and his or her professional judgment. For instance, a professor who wants to commercialize some invention may be tempted to assign projects to his or her students that would expedite this commercialization, even if the students' academic interests would be served better by other projects. Researchers are usually entitled to benefit financially from their work, such as by receiving royalties or bonuses. But the prospect of financial gain should not prompt substandard experiments, biased conclusions, or exaggerated claims. Personal relationships may also cause conflicts of interest; accordingly, many funding agencies require scientists to identify others who have been their supervisors, graduate students, or postdoctoral fellows. A conflict of commitment, as distinguished from a conflict of interest, pertains to dividing time between research and other responsibilities, respecting an employers' mission and values, and representing science to the public. For instance, universities often limit the amount of time that faculty can spend on outside activities. A conflict of values can emerge from the values and beliefs that a scientist holds, including strong philosophical, religious, cultural, or political beliefs. However, as NAS (2009:46) insists, "it is clear that all values cannot – and should not – be separated from science. The desire to do good work is a human value." However, any values that compromise objectivity or introduce bias must be recognized and minimized.

The Researcher in Society. Scientists often provide expert opinion or advice to governments, universities, companies, and courts. They frequently educate the public about science policy issues and may also lobby their elected representatives or participate in political rallies or protests. When scientists become advocates on an issue, they may be perceived by their colleagues or the public as being biased. But scientists also have the right to express their convictions and to work for social change and justice, and that advocacy need not undercut a rigorous commitment to objectivity and truth. The main text of NAS (2009:48) concludes with a wonderful perspective on science's values: "The values on which science is based – including honesty, fairness, collegiality, and openness – serve as guides to action in everyday life as well as in research. These values have helped produce a scientific enterprise of unparalleled usefulness, productivity, and creativity. So long as these values are honored, science – and the society it serves – will prosper."

Finally, a prominent tool in teaching and learning professional ethics is discussion of provocative case studies, either fictitious or real. Besides the many case studies in NAS (2009), Resnik (1998:177–200) provided 50 more. Of course, in the spectrum between scientific misconduct and exemplary research, there is a nagging gray zone. To stimulate lively debate and thoughtful responses, these case studies tend to concentrate in this gray zone. But for the sake of brevity, this chapter leaves case studies to other readily available resources.

Discussion

Having listed a dozen unethical practices to avoid in the preceding section, this brief discussion turns to ethics' principal and positive objective, which is pursuing the good life. Some courses and books on science's ethics emphasize the negative, focusing largely on notorious and even criminal instances of misconduct. Such an approach to ethics fails to communicate that often the even greater challenges that we face involve encountering what is good and virtuous. This is what students need to be warned about most of all. Two little stories and one little joke may illuminate what is meant here.

The first story is about a young dog and its first bone. I was visiting a friend and his family one afternoon. Their family pet was a large dog about a year old that had never been given anything to eat except dry granulated dog food. That afternoon it was given its first bone, which had a little meat on it – the very food dogs are born to enjoy. But this bone was a greater treasure than this poor dog could handle. This dog became protective. It began snarling at my host's children if they got near. Nothing else had ever made this dog a threat to children, but encountering such a great good as this first bone did make this large dog dangerous. My friend had to intervene, alternately giving and taking the bone until finally his dog could respond better.

The second story is about Gimli and Galadriel, characters in J. R. R. Tolkien's *The Lord of the Rings*, from the chapter "Farewell to Lórien" in the first book of this trilogy. As fans of these stories will know, the dwarf Gimli has already suffered daunting perils and numerous battles by the time he meets the beautiful elf queen in Lórien, Lady Galadriel. Gimli wept as they departed Lórien and parted from Lady Galadriel, saying to his elf companion Legolas, "I have looked the last upon that which was fairest." Gimli continued: "Tell me, Legolas, why did I come on this Quest? Little did I know where the chief peril lay! Truly Elrond spoke, saying that we could not foresee what we might meet upon our road. Torment in the dark was the danger that I feared, and it did not hold me back. But I would not have come, had I known the danger of light and joy. Now I have taken my worst wound in this parting." In this story, "light and joy" are greater perils than "torment in the dark!" Is this a valid perception of human experience – of what tests us most deeply? That question is worth pondering.

Third and finally, a little joke goes like this. What is the worst party on earth? The answer is: the *great* party that you were not invited to. Note that the answer is not something like: a party with boring people, dreadful music, health-food snacks, and no beer. Implicit in this revealing joke is the wise perception that a great good missed is a worse pain than a substantial evil encountered.

Ethics is about misconduct. But first and foremost, an ethical vision defines the good life. The admirable code of ethics of the IEEE strikes this balance. It does reject dishonesty, bribery, discrimination, and such. But, more so, it emphasizes the great opportunities of scientists and technologists to promote the safety, health, and welfare of the public, to assist and support one's colleagues, and to discover and invent with skill in community.

For young students and professionals in science and technology, their greatest perils, and their greatest risks of regret in their old age, are not that they will fabricate data or steal ideas. Rather, their more likely perils are that simple laziness will diminish their contributions to society and colleagues, disabling complacency will prevent their mastery of scientific method, and reprehensible self-absorption will preclude their full appreciation of the wonder and beauty of nature.

Study questions

(1) The general public, as well as the scientific community, has a great diversity of worldview beliefs and commitments. What challenges or limitations does that diversity impose on philosophical ethics? Is there such a thing as a genuinely neutral, impartial, and universal reason for grounding ethics? Explain your answer.
(2) How does science depend on ethics and contribute to ethics?
(3) What are the three obligations of scientists identified in the professional ethics from the National Academy of Sciences? Give an example of each and explain why it is important.
(4) What are the three kinds of research misconduct? Give an example of each, fictitious or real. How does research conduct differ from questionable research practices? Give an example or two of questionable practices.
(5) Consider the IEEE code of ethics for engineers. Is there anything you would want to delete? Is there anything you would want to add? Give your reasons for each suggested change.

13

Science education

What are the benefits for scientists and others from learning the general principles of scientific method? And which scholars have best clarified and documented those benefits?

The first chapter mentioned various positive or negative assessments of these benefits by scientists and philosophers. But it noted that the most careful assessments have been by science educators, while leaving documentation of their findings to this chapter. Six benefits were listed: better comprehension, greater adaptability, greater interest, more realism, better researchers, and better teachers.

Science educators typically place various aspects of scientific method within several broader contexts, including inquiry, practice, and the nature of science (NOS). "For science educators the phrase 'nature of science,' is used to describe the intersection of issues addressed by the philosophy, history, sociology, and psychology of science as they apply to and potentially impact science teaching and learning. As such, the nature of science is a fundamental domain for guiding science educators in accurately portraying science to students" (McComas, Clough, and Almazroa, in McComas 1998:5). Educators and scientists alike perceive the importance of the NOS in the science curriculum (McComas 1998; Khine 2011).

This chapter first reviews the typical NOS concepts that have been developed primarily for K–12 (kindergarten through high school) science education. It then reviews numerous and extensive empirical investigations by science educators that document the previous six benefits. Finally, it lists 10 academic NOS concepts better suited to undergraduate and graduate students and science professionals.

Typical NOS concepts

The principal approach to teaching the NOS, particularly in the context of K–12 education, has been to develop a consensus view that reflects mainstream

Typical NOS Concepts for K–12 Education

Scientific knowledge while durable has a tentative character.

Scientific knowledge relies heavily but not entirely, on observation, experimental evidence, rational arguments, and skepticism.

There is no one way to do science (therefore, there is no universal step-by-step scientific method).

Science is an attempt to explain natural phenomena.

Laws and theories serve different roles in science; therefore students should note that theories do not become laws even with additional evidence.

People from all cultures contribute to science.

New knowledge must be reported clearly and openly.

Scientists require accurate record keeping, peer review, and replicability.

Observations are theory-laden.

Scientists are creative.

The history of science reveals both an evolutionary and revolutionary character.

Science is part of social and cultural traditions.

Science and technology impact each other.

Scientific ideas are affected by their social and historical milieu.

Figure 13.1 Fourteen typical concepts for the nature of science (NOS) for K–12 education. This list represents a consensus among science education guidelines from eight nations.

science. The National Research Council of the National Academy of Sciences has expressed its view that "Although there is no universal agreement about teaching the nature of science, there is a strong consensus about characteristics of the scientific enterprise that should be understood by the educated citizen" (NRC 2012:78). Figure 13.1 shows 14 NOS concepts derived from a comparison of 8 international science education standards, including the United States, Canada, United Kingdom, New Zealand, and Australia (McComas et al., in McComas 1998:6–7). Likewise, the science content standards for all 50 US states incorporate these NOS elements, although with varying comprehensiveness (William F. McComas, Carole K. Lee, and Sophia J. Sweeney, personal communication).

Partly because the NOS is a mandated topic and partly because it is inherently of great interest to scientists and educators, assessment instruments have been developed. Although more than 40 instruments have been developed, the Test on Understanding Science (TOUS) and the Views of Nature of Science (VNOS, in several versions for different grade and developmental levels) instruments have been especially widely used (Lederman 2007). Widely adopted instruments facilitate comparisons among the numerous empirical studies that have used them. Nevertheless, there are some legitimate questions about consensus concepts for the NOS.

An important cultural question is, "Is science universal?" (Cobern and Loving 2001). This is a fair question, given that the NOS standards and assessment instruments mentioned herein represent a rather homogeneous group of English-speaking nations. There are perplexing and sensitive issues regarding Western and indigenous perspectives on science. Some indigenous beliefs have no potential whatsoever to be of interest to mainstream science, such as the example that Cobern and Loving gave of some South Pacific islanders who attribute tides to great sea turtles leaving from or returning to their nests, rather than to the gravitational tug of the moon (and sun). On another topic such as drug discovery, both mainstream science and indigenous knowledge have contributed substantially. On yet another topic, such as identifying local crops with potential for future development into major crops by conventional and molecular plant breeding, the contributions from indigenous cultures are essential. Cobern and Loving advised that indigenous knowledge be "valued for its own merits" without confusing or diluting the methodology and knowledge of mainstream science.

A practical question regards whether NOS concepts are general, equally applicable across all sciences, or else are context specific, variously applicable in astronomy or biology or sociology or whatever. For instance, Irzik and Nola (2010) argued for a family resemblance approach to the NOS rather than a consensus approach: "Science is so rich and so dynamic and scientific disciplines are so varied that there seems to be no set of features that is common to all of them and shared only by them." They argued that this alternative is more comprehensive because it "captures the elements of NOS described by the consensus view" while it also "contains novel elements" and thereby "does justice to the differences among scientific disciplines." But recall from Chapter 1 that the American Association for the Advancement of Science expressed a nuanced position on commonality and diversity among the various scientific disciplines (AAAS 1989:25–26, 29). Accordingly, a family resemblance among the various sciences may be regarded as a component of the consensus view rather than as a competitor to it.

To address this issue about the relative importance of general and context-specific aspects of NOS concepts, Urhahne, Kremer, and Mayer (2011) conducted a large empirical study, surveying 221 secondary-school students in Germany. The students' NOS conceptions were assessed for seven concepts named: Source, Certainty, Development, Justification, Simplicity, Purpose, and Creativity. The 10 specific contexts were: Smoking, Mobile phone, Climate change, Aggression, Intelligence, Dinosaurs, Continental drift, Big Bang,

Out of Africa, and Evolution. The students' general NOS conceptions were assessed with a 40-item questionnaire and their context-specific NOS conceptions by a 10-item questionnaire for each of the 10 specific contexts. Table 7 in Urhahne et al. (2011) presented the 70 correlation coefficients between general and context-specific NOS conceptions for the 7 concepts in the 10 specific contexts.

For this concepts-by-contexts dataset, the relative importance of general and context-specific aspects of the NOS can be quantified by a standard statistical tool, the analysis of variance. It identifies three sources of variance: concept main or general effects, context main effects, and concept-by-context interaction effects. These account for 1%, 26%, and 73% of the total variance, respectively. Hence, interaction effects (73%) seem to be about three times as important as main effects (27%). However, to be more precise, only half of this interaction contains systematic patterns of interest, and the remainder is mostly noise. Hence, general and context-specific aspects of NOS were about equally important in this particular study.

Such results make sense. Although researchers in the physical, biological, and social sciences make use of all of the general principles of scientific method, the emphasis on various principles does vary somewhat among disciplines.

As evidence that a treatment of scientific method can serve diverse scientists and nonscientists alike, despite some variations in emphases, I would mention the tremendous breadth of disciplines that have publications citing this book's predecessor, *Scientific Method in Practice* (Gauch 2002). Those publications are in many sciences: anthropology, astronomy, bioinformatics, chemistry, climatology, ecology, geophysics, hydrology, morphology, natural resources, oceanography, physics, proteomics, psychology, seismology, sociology, taxonomy, toxicology, and zoology. They are also in technology: agriculture, computer science, engineering, medicine, pharmacology, public health, and space exploration. And they are in business, economics, education, law, management, music, and philosophy.

In review, there is considerable consensus on basic NOS concepts, and there are widely adopted assessment instruments. Furthermore, this NOS consensus suits multicultural settings, while respecting indigenous knowledge for its own merits, but without confusing or diluting mainstream science. There is evidence that the NOS has context-specific as well as general aspects, so different scientific disciplines such as geology and physics have somewhat different NOS emphases.

However, there are legitimate concerns about characterizations of scientific method in the typical NOS concepts developed for K–12 education. "The issue of methodology seems to be dismissed altogether by saying that there is no single method for doing science.... While it is certainly true that there is no single scientific method in the sense of a mechanical procedure that determines knowledge production step by step, there are general methodologies... that guide scientific practice in general ways. Moreover, without the idea of a scientific method or methodological rule, it is difficult to see how science can be self-corrective and provide reliable knowledge" (Irzik and Nola 2010). Indeed, the third item in Figure 13.1 is "There is no one way to do science (therefore, there is no universal step-by-step scientific method)," without any further elaboration on scientific method. Accordingly, a later section in this chapter proposes more advanced concepts for the NOS in general and scientific method in particular that are suitable for undergraduate and graduate students and science professionals. But, first, the following six sections document benefits that result from studying and mastering the general principles of scientific method.

Better comprehension

The NRC has expressed the essential goal of science education in a mere four words: "scientific knowledge with understanding" (NRC 1996:21). Hence, scientific literacy requires comprehension of both scientific content and scientific method. This literacy also relates to technological competence.

> An explicit goal of the *National Science Education Standards* is to establish high levels of scientific literacy in the United States. An essential aspect of scientific literacy is greater knowledge and understanding of science subject matter, that is, the knowledge specifically associated with the physical, life, and earth sciences. Scientific literacy also includes understanding the nature of science, the scientific enterprise, and the role of science in society and personal life. . . .
>
> The goal of science is to understand the natural world, and the goal of technology is to make modifications in the world to meet human needs. . . . Technology and science are closely related. A single problem often has both scientific and technological aspects. (NRC 1996:21–24)

Students must master scientific method and the NOS in order to make scientific knowledge their own: "As Plato insisted so long ago, education is not just the having of correct beliefs, it is the having of adequate reasons for these beliefs" (Matthews 1998). To describe the opposite situation for a student learning only science's facts, he quoted memorable words from John Locke: "Such borrowed wealth, like fairy money, though it be gold in the hand from which he received it, will be but leaves and dust when it comes to use."

One item among the "modest goals" for NOS instruction advocated by Matthews (1998) is logic: "Given that being able to reason clearly is an obvious component of being scientific, then one of the low-level nature of science objectives might be to teach some elementary formal and informal logic." Basic instruction in deductive and inductive logic can reduce common fallacies and misconceptions and can enhance reasoning, particularly the kind of reasoning needed to evaluate competing hypotheses.

Of course, in general, both reasoning ability and prior knowledge of a subject help students comprehend more advanced material, and yet surprisingly the former is often more important. For instance, what factors influence students' ability to comprehend college biology? Johnson and Lawson (1998) conducted an empirical study of 366 students enrolled in a nonmajors biology course at a community college. They pretested students to determine their reasoning ability and prior knowledge. Then some classes with about half of the students received "expository instruction" focused on facts and concepts, whereas the other classes received "inquiry instruction" giving more attention to science's logic, method, and process. The results were that "Reasoning ability but not prior knowledge or number of previous biology courses accounted for a significant amount of variance in final examination score in both instructional methods." Furthermore, the inquiry classes showed better improvement in both reasoning skills and biology comprehension.

Given that reasoning ability appears to be a significant predictor of achievement in introductory level college biology, more so than the number of previous biology courses completed or a pretest measure of domain-specific knowledge, high school biology instructors would be well advised to be more concerned with the development of their students' reasoning abilities than with making certain that they cover a wide range of specific biology concepts. The same advice would also seem to be appropriate for college level instructors who may be concerned about their students' performance in more advanced college courses. . . . The inquiry students in the present study not only showed greater improvement in reasoning ability during the semester than the expository students, they also did better on the measures of biology achievement. In other words, nothing of importance seems to be lost by switching to inquiry instruction, and much seems to be gained. (Johnson and Lawson 1998)

A similar win-win situation for scientific content and scientific method was encountered in the context of college physics. Numerous students were tested on acceleration, velocity, vectors, and forces.

It is well known that students come to the study of science with many naive prior notions about the physical world and that these notions are very difficult to change. . . . It is less well appreciated that students also come with many naive conceptions about the goals of science and about the kinds of thinking needed for science. These conceptions, imported from everyday life or from prior schooling, are even more difficult to change than students' naive notions about the physical world. Furthermore, their effects are all-pervasive, affecting greatly what students try to learn and how they go about learning it. . . .

An introductory physics course needs thus also to discuss explicitly the goals of science and the ways of thinking useful in science. These issues cannot merely be addressed by a few occasional remarks. They need to be constantly kept in students' focus, and be used as a framework within which more specific scientific knowledge and methods are embedded. (Reif 1995)

Yet another win-win combination for content and method was found by Cartier and Stewart (2000) in the context of high school genetics. "In contrast to what is popularly believed, focusing on epistemological issues in the classroom does not have to mean teaching less subject matter in exchange. In fact, helping students to appreciate the epistemological structure of knowledge in a discipline can provide them with the means to more fully understand particular knowledge claims and their interrelationships" (Cartier and Stewart 2000).

These empirical findings prompt some commentary on the *National Science Education Standards* (NRC 1996). The *Standards* include prominent curriculum requirements for the NOS or, in their terminology, for "science as inquiry." That is commendable, given the findings of science educators. However, a table on changing emphases is organized under two headings: "Less emphasis on" and "More emphasis on" (NRC 1996:113). The very first item is less emphasis on "Knowing scientific facts and information" and more emphasis on "Understanding scientific concepts and developing abilities of inquiry." This might be called the trade-off model, or the win-lose model, in which inquiry wins and information loses. Admittedly, given a fixed schedule, an hour more for the NOS does mean an hour less for science content. But that does not necessarily imply less content because NOS instruction can pave the way for more rapid coverage of more deeply comprehended material. By analogy, first sharpening a dull ax means less time spent chopping wood, but that does not necessarily mean less firewood and, indeed, the more likely outcome is more firewood. Hence, a better model, and a realistic expectation as well, would be to put both "Knowing scientific facts and information" and "Understanding scientific concepts and developing abilities of inquiry" under the column "More emphasis on."

Similarly, the more recent NRC (2012:41) position paper on science education reiterates this perception of tension or trade-off between content and method: "From its inception, one of the principal goals of science education has been to cultivate students' scientific habits of mind, develop their capability to engage in scientific inquiry, and teach them how to reason in a scientific context. There has always been a tension, however, between the emphasis that should be placed on developing knowledge of the content of science and the emphasis placed on scientific practices" or methods. But, again, extensive empirical evidence supports a positive interaction or synergy between content and method, not a negative interaction or tension.

Greater adaptability

Four terms in the educational literature have closely interrelated meanings: adaptability, versatility, creativity, and knowledge transfer. All four involve making the most of what is known by effectively connecting previous experiences to new applications.

Science and technology are experiencing rapid and pervasive changes, requiring scientists to be increasingly adaptable and flexible. Young scientists can expect to change jobs and assignments more frequently than their predecessors. They can also expect much of what they learn to have a short half-life of only a few years. This has profound implications for science education.

Accordingly, the National Science Foundation (NSF 1996) and National Academy of Sciences (NAS 1995) position papers for reshaping undergraduate and graduate science education, respectively, promote adaptability as a prominent goal. Regrettably, "We know that students rarely realize the applicability of knowledge from one context to another" (NSF 1996:3). The first of three general recommendations from the NAS (1995:76) is that "graduate programs should add emphasis on versatility; we need to make our students more adaptable to changing conditions." Versatility or adaptability has become so essential that the NAS recommends that evaluation criteria for funding education/training grants "include a proposer's plan to improve the versatility of students" (NAS 1995:80).

Lawson et al. (2000) found that hypothesis-testing skills helped even more than subject-matter knowledge in solving a new "transfer" problem. They had a large sample of 667 college undergraduates. One predictor variable was hypothesis-testing skills, assessed by a 13-item written test with each answer scored as correct or incorrect. The other predictor variable was subject-matter knowledge, assessed by a five-item multiple-choice test concerning momentum, air, helium, and balloons. The dependent variable was the score (correct or incorrect) for a new or transfer problem about a video showing the contrasting motions of a balloon filled with air and another with helium in a moving vehicle that came to an abrupt stop. Multiple regression analysis determined which predictors were significant. "The analysis revealed that hypothesis-testing skills, but not declarative [subject-matter] knowledge, accounted for a significant amount of variance" in the dependent variable, success in solving the balloon transfer problem. Also, across four levels of increasing hypothesis-testing skills, the percentages of students giving correct answers on the transfer problem rose consistently: 17%, 34%, 57%, and 65%.

Educators and psychologists have discovered ways to increase adaptability or knowledge transfer (Lobato 2006). What is crucial for transfer is not merely the number of examples shown but also the explicit and reflective attention given to general principles. In transfer problems, students tend to notice primarily the surface features held in common, whereas they must be encouraged to look beyond surface features to shared general or causal principles. That is, to solve new problems, students need to abstract, discounting the superficial details of examples and noticing the general principles. A particularly helpful strategy is comparing cases (Gentner, Loewenstein, and Thompson 2003).

The importance of knowledge transfer is especially acute in medicine because such transfer is precisely what diagnosis amounts to: applying past learning to the patient at hand. For example, Ark, Brooks, and Eva (2007) found that explicit attention on reasoning strategies increased the ability of medical students to identify key features in electrocardiograms and assign correct diagnoses. Although an extensive literature from educators and psychologists provides strategies for promoting knowledge transfer, there is evidence that much of this insight has not yet informed the pedagogy of most medical professors (Laksov, Lonka, and Josephson 2008).

Precisely because the general principles of scientific method *are* general, whereas specialized research techniques *are* specialized, these general principles are that portion of science that has the greatest potential for frequent and pervasive applications. Hence, general principles have extraordinary relevance for adaptability, versatility, creativity, and knowledge transfer. Nevertheless, this tremendous potential is not actualized automatically but rather requires an informed and disciplined implementation of effective pedagogy.

> Research clearly shows ... that transfer of NOS knowledge does not happen automatically. Students do not learn relevant NOS aspects through historical examples alone or by instruction that refers only casually to elements of the nature of science. Several researchers have pointed out that complex NOS ideas should be accompanied by explicit and reflective discussion of the underlying concepts and principles. (Urhahne et al. 2011)

NOS knowledge can become transfer knowledge, not inert knowledge, by accompanying examples with explicit and reflective discussion, by comparing rather than merely multiplying examples, and by emphasizing general principles over superficial similarities. The most obvious aspect of adaptability or transfer is application of familiar knowledge to new and similar cases. However, knowledge transfer also concerns preparation for later learning of more advanced material, as well as reconsideration and enrichment of previously learned situations in light of subsequent examples and insights.

Greater interest

As in many nations, stimulating greater student interest in science and technology is seen as a national priority for the United States.

> The understanding of, and interest in, science and engineering that its citizens bring to bear in their personal and civic decision making is critical to good decisions about the nation's future. The percentage of students who are motivated by their school and out-of-school experiences to pursue careers in these fields is currently too low for the

nation's needs. Moreover, an ever-larger number of jobs require skills in these areas, along with those in language arts and mathematics....

A rich science education has the potential to capture students' sense of wonder about the world and to spark their desire to continue learning about science throughout their lives. Research suggests that personal interest, experience, and enthusiasm—critical to children's learning of science at school or other settings—may also be linked to later educational and career choices. Thus, in order for students to develop a sustained attraction to science and for them to appreciate the many ways in which it is pertinent to their daily lives, classroom learning experiences in science need to connect with their own interests and experiences. (NRC 2012:x, 28)

Accordingly, it is important to observe that science educators have found that science in its historical and philosophical context has greater appeal for most students than does science abstracted from its context. "Incorporating the nature of science while teaching science content humanizes the sciences and conveys a great adventure rather than memorizing trivial outcomes of the process. The purpose is not to teach students philosophy of science as a pure discipline but to help them be aware of the processes in the development of scientific knowledge" (William F. McComas, in McComas 1998:13).

For example, Becker (2000) produced a sequence of "history-based" science lessons particularly to reach "those students identified as traditionally alienated from the world of science and technology." She reported that "Students who took the course found it satisfying, diverse, historical, philosophical, humanitarian, and social. They... found the historical approach to be interesting and the text enjoyable to read." The material engaged "learners of varying abilities" from a "socio-economically, ethnically, and linguistically diverse student population."

Another example is the Harvard Project Physics course, which incorporates some history and philosophy of science. More than 60 studies of the effectiveness of that course have been published, all positive: "Measures such as retention [of students] in science, participation by women, improvement on critical thinking tests and understanding of subject matter all showed improvement where the Project Physics curriculum was adopted" (Matthews 1994:6). However, although it increased interest in science, it was not effective for communicating the NOS.

In attempting to stimulate student interest in science, it helps to know that science educators have partitioned student interest into two main components termed "individual" and "situational" interest (Seker and Welsch 2006). They are quite different. Individual interest is "stable interest, enduring over time and characterized by high levels of stored knowledge and stored value" on the part of a given person. By contrast, situational interest "is generated primarily by certain conditions and/or concrete objects... in the environment" and it "tends to be short-lived, lasting only as long as the situation provides interest." It can be stimulated by puzzles, games, stories about the personal lives of scientists, role-playing, drama writing, movies, and other forms of entertainment. Obviously, long-term individual interest is more important than ephemeral situational interest. However, if situational interest is stimulated for extended periods of time, it may eventually promote individual interest.

Students with high individual interest and those with low individual interest had considerably different reactions to different instructional approaches. Students with high interest reacted positively to material on the history and nature of science. But others with low interest "expected to memorize what the science teacher told them and what the textbook emphasized" and disliked historical review showing that ideas develop over time and even are sometimes corrected or discarded. For instance, one student complained, "If someone is going to come along in the next few years and come up with something new, then what's the point? I'm not learning this stuff if it will change."

A particularly important part of the NOS for interesting a diversity of students in science is its simplest, common-sense elements. Warren et al. (2001) contrasted perspectives on the relationship between science and common sense as being either "fundamentally discontinuous" or "fundamentally continuous" and found the latter much more conducive to learning science: "Our work... shows that children, regardless of their national language or dialect, use their everyday language routinely and creatively to negotiate the complex dilemmas of their lives and the larger world. Likewise, in the science classroom children's questions and their familiar ways of discussing them do not lack complexity... or precision; rather, they constitute invaluable intellectual resources which can support children as they think about and learn to explain the world around them scientifically." Culturally sensitive "everyday sensemaking" is a powerful bridge to scientific interests and patterns of thought.

Although the skills involved in understanding the scientific method and the NOS promote understanding science, at a deeper level, one's interest in cultivating those skills must be promoted by a desire to seek the truth. An important study by Ben-Chaim, Ron, and Zoller (2000) explored numerous skills and dispositions related to scientific thinking. They reported that "truth-seeking is courageous intellectual honesty, a major attribute of CT [critical thinking], which, in turn, is an important component of HOCS [higher-order cognitive skills]."

More realism

Realism about science requires understanding the accuracy, confidence, and scope of any given scientific result. But this topic inherently and integrally involves scientific method because methods precede and enable results. The level of certainty accorded to any given scientific result ranges from highly tentative to confident fact, and the scope ranges from particular situations to great generality.

> Science is replete with ideas that once seemed promising but have not withstood the test of time, such as the concept of the "ether" or the *vis vitalis* (the "vital force" of life). Thus any new idea is initially tentative, but over time, as it survives repeated testing, it can acquire the status of a fact—a piece of knowledge that is unquestioned and uncontested, such as the existence of atoms. . . .
>
> Scientific knowledge is a particular kind of knowledge with its own sources, justifications, ways of dealing with uncertainties, and agreed-on levels of certainty. When students understand how scientific knowledge is developed over systematic observations across multiple investigations, how it is justified and critiqued on the basis of evidence, and how it is validated by the larger scientific community, the students then recognize that science entails the search for core explanatory constructs and the connections between them. They come to appreciate that alternative interpretations of scientific evidence can occur, that such interpretations must be carefully scrutinized, and that the plausibility of the supporting evidence must be considered. . . .
>
> Decisions must also be made about what measurements should be taken, the level of accuracy required, and the kinds of instrumentation best suited to making such measurements. As in other forms of inquiry, the key issue is one of precision—the goal is to measure the variable as accurately as possible and reduce sources of error. The investigator must therefore decide what constitutes a sufficient level of precision and what techniques can be used to reduce both random and systematic error. . . .
>
> Under everyday circumstances, . . . Newton's second law accurately predicts changes in the motion of a single macroscopic object of a given mass due to the total force on it. But the second law is not applicable without modification at speeds close to the speed of light. Nor does it apply to objects at the molecular, atomic, and subatomic scale or to an object whose mass is changing at the same time as its speed. (NRC 2012: 79, 251, 59–60, 114)

Realism about science requires a discerning grasp of science's powers and limits. As already documented in Chapter 6, awareness of science's powers and limits has been widely recognized as a key component of scientific literacy (AAAS 1989:26, 1990:20–21; NRC 1996:21). For science and technology undergraduates, the NRC (1999:34) posed six specific questions that they should be able to answer: "How are the approaches that scientists employ to view and understand the universe similar to, and different from, the approaches taken by scholars in other disciplines outside of the natural sciences? What kinds of questions can be answered by the scientific and engineering methods, and what kinds of questions lie outside of these realms of knowledge? How does one distinguish between science and pseudoscience? Why are scientists often unable to provide definitive answers to questions they investigate? What are risk and probability, and what roles do they play when one is trying to provide scientific answers to questions? What is the difference between correlation and causation?"

What science educators have found is that an understanding of science's method powerfully promotes the educational goal of developing realistic ideas about science. This makes sense given that powers and limits are largely consequences of methods: "Understanding how science operates is imperative for evaluating the strengths and limitations of science" (William F. McComas, in McComas 1998:12). Also, "The ability to distinguish good science from parodies and pseudoscience depends on a grasp of the nature of science" (Matthews 2000:326).

Empirical studies by science educators have shown that good understanding of scientific method promotes realistic beliefs not only for ordinary scientific inquiries but also for scientific matters having substantial worldview import. For example, Lawson and Weser (1990) conducted a large empirical study of 954 college students in a nonmajors biology course. Scientific reasoning was assessed by students' answers to 16 questions involving conservation of mass; volume displacement; control of variables; and reasoning about proportions, probabilities, combinations, and correlations. Students who performed poorly were referred to as "intuitive" thinkers, moderately well as "transitional" thinkers, and quite well as "reflective" thinkers. Scientific beliefs were examined with seven questions regarding special creation, orthogenesis, the soul, constitutive nonreductionism, vitalism, teleology, and nonemergentism. Students were pretested for both scientific reasoning and scientific beliefs, and then post-tested after the biology course by readministering the same questionnaires. These data were subjected to PCA to characterize the main patterns, as well as statistical tests to detect specific differences between intuitive and reflective students.

Their main finding was: "As predicted, the results showed that the less skilled reasoners were more likely to initially hold the nonscientific beliefs and were less likely to change those beliefs during instruction. It was also discovered that less skilled reasoners were less likely to be strongly committed to the scientific beliefs." Their suggested explanation was that reflective students hold beliefs more strongly than intuitive students "because they have acquired the reasoning skills that enable them to consider the alternative beliefs and the evidence in a hypothetico-deductive manner to arrive at firm conclusions." There were numerous positive interactions between reasoning level and strength of confidence in scientific beliefs. Their two main conclusions were: "(1) teachers should be aware that students may hold beliefs that are inconsistent, even contradictory to scientific beliefs and that these beliefs may be difficult to alter; (2) some students lack the reasoning skills that appear to be necessary to comprehend the arguments and evidence in favor of scientific beliefs; thus instruction should focus on ways of improving student reasoning skills as well as teaching scientific conceptions."

Better researchers

Research competence for undergraduate and graduate science majors is a major educational objective. Again, my thesis is that the winning combination for scientists is strength in both the research technicalities of a given specialty and the general principles of scientific method. Hence, it is virtually axiomatic that a deep understanding of scientific method develops better researchers. Furthermore, what has already been documented in this chapter regarding better comprehension, adaptability, interest, and realism makes for better researchers.

Ryder and Leach (1999) conducted an empirical investigation of research competence among science undergraduates in the UK as a function of their understanding of the general principles of scientific method. They found that such understanding "can have a major impact on students' activities during investigative [research] projects," specifically on "students' decisions about whether to repeat an experiment, whether to question an earlier interpretation of their data as a result of new evidence, where to look for other scientific work related to their project and what counts as a conclusion to their investigation." On the other hand, inadequate or naïve conceptions of scientific method "constrain student learning" and made it impossible for students to achieve some of their research objectives. More exactly, they found that mere diligence sufficed for students to function as competent technicians, collecting accurate data by means of a prescribed protocol; but real understanding of scientific method was needed for students to function as creative scientists, understanding how data and theory interact to support knowledge claims.

Interestingly, they investigated not only the students but also the students' research supervisors. For one student with a particularly naïve and limiting view of scientific method, they found no evidence that the student's supervisor recognized the problem or implemented any strategy to remedy the deficiency. Professors and supervisors must realize that teaching more and more about research technicalities is not the required remedy when the real problem is lack of understanding of some basic principle of scientific inquiry. Clearly, problems of the latter sort are quite common.

Hunter, Laursen, and Seymour (2007) conducted an empirical study of summer undergraduate research at four liberal arts colleges with a sample of 138 students, 71 faculty, and 9 administrators. They found that typical undergraduate research prompts quite limited growth in NOS understanding. "Many faculty and student observations reported gains in applying their knowledge and skills to research work, although fewer mentioned increases in higher order thinking skills, particularly the development of a complex epistemological understanding of science or the ability to define a research question and develop experimental design." The process of developing a real understanding of the NOS by means of undergraduate research experience "is neither easy nor guaranteed." This result echoes Ryder and Leach's conclusion that it is easier to produce competent technicians than creative scientists. It also reinforces the unanimous finding in countless studies that progress in understanding the NOS, even in the context of authentic research experience, will not occur unless this topic receives intentional, explicit, reflective attention.

Sadler et al. (2010) provided an ambitious review of 53 empirical studies of research apprenticeships for secondary and college students as well as K–12 teachers. Research apprenticeships promoted positive outcomes for interest and career aspirations in science, knowledge of specific science content, technical and statistical skills, confidence in one's ability to do scientific research, and dispositions for effective communication and collaboration. By contrast, on the matter of principal concern here, again extensive empirical evidence showed that research experience prompted rather little growth in NOS understanding.

> Apprenticeship programs are often touted as effective vehicles for learning about the NOS. The basic argument is that opportunities to work with scientists offer new perspectives on how science is done and the nature of scientific knowledge. This argument is based on an implicit model of NOS learning. That is, learners will gain sophisticated understandings of NOS by virtue of participating in authentic science activities without instruction dedicated to the realization of this goal. Many science education researchers have challenged the notion that implicit approaches for teaching about NOS results in meaningful learning. These critics have argued that explicit approaches that support learner development of sophisticated NOS ideas through targeted instruction and learner reflection are more likely to result in NOS learning. It is possible to partner apprenticeship experiences with explicit NOS instruction. (Sadler et al. 2010)

Of their 53 studies, 20 reported on NOS learning, with 19 relying on implicit models and only 1 implementing explicit instruction. Schwartz, Lederman, and Crawford (2004) interviewed 13 preservice teachers involved in a course that combined a research setting with explicit NOS instruction. The VNOS instrument was used to assess NOS understanding. In stark contrast with the miserable performance of implicit models, this explicit approach resulted in 11 of its 13 participants gaining in NOS understanding. Explicit instruction, rather than research apprenticeship alone, was the key influence. More recently, Bautista and Schussler (2010) implemented explicit NOS instruction successfully within an introductory college biology laboratory.

In review, extensive empirical studies by science educators support two principal findings about the relationship between NOS understanding and research competence. First, greater understanding of the NOS in general and scientific method in particular *does* develop better researchers, especially by clarifying how data and theory interact to support knowledge claims. Second, regarding the reverse direction, research experience *does not* easily or automatically increase NOS understanding but rather progress with the NOS requires explicit and reflective instruction, whether or not that instruction occurs within the context of scientific research. Science educators should bear in mind these two findings as they seek to give scientists their winning combination of strength in both the research technicalities of a given specialty and the general principles of scientific method.

Better teachers

Regarding science and technology faculty, the NSF (1996:ii) advises that "It is important to assist them to learn [and teach] not only science facts, but, just as important, the methods and processes of research." Likewise, the NRC (1996:21) requires that science teachers understand the NOS, including the "modes of scientific inquiry, rules of evidence, ways of formulating questions, and ways of proposing explanations," and the NRC (1999:41, 2012:253–260, 319–323) reaffirms those requirements. In addition to items such as "physical science" and "life science," one of the eight categories of content standards for science is the "history and nature of science" (NRC 2001:148).

Matthews (1994:199–213, 2000:321–351) documented that knowledge of the history, philosophy, and method of science can strengthen teachers. Such knowledge "can improve teacher education by assisting teachers to develop a richer and more authentic understanding of science, . . . can assist teachers [to] appreciate the learning difficulties of students, because it alerts them to the historic difficulties of scientific development and conceptual change, . . . [and] can contribute to the clearer appraisal of many contemporary educational debates that engage science teachers and curriculum planners" (Matthews 1994:7). The NRC concurs.

> Teaching science . . . requires that teachers have a strong understanding of the scientific ideas and practices they are expected to teach, including an appreciation of how scientists collaborate to develop new theories, models, and explanations of natural phenomena. Rarely are college-level science courses designed to offer would-be science teachers, even those who major in science, the opportunity to develop these understandings. Courses designed with this goal are needed. (NRC 2012:256)

But unfortunately, science teachers often lack a sophisticated and consistent understanding of scientific method and inquiry (Abd-El-Khalick and Lederman 2000; Lederman 2007). Naturally, teacher weaknesses in the NOS have direct implications for student weaknesses. "What is . . . clear from recent research is that rarely do students receive instruction in the K–12 grades that contribute to better understandings of NOS unless their teachers have had some kind of professional development for teaching NOS" (Akerson et al. 2011).

But the good news is that professional development can work. "Sustained professional development improves teachers' abilities to teach science, and recent research shows that professional development can improve teachers' conceptualizations of nature of science (NOS) to those in line with recommendations of national reforms as well as their abilities to explicitly teach NOS to their own elementary students" (Cullen, Akerson, and Hanson 2010). Indeed, as one might well expect, there is extensive empirical evidence that high-quality teaching increases student achievement (NRC 2001:44–65).

Nevertheless, extensive empirical evidence also shows that teacher NOS knowledge does not automatically translate into student NOS knowledge. "Research on the translation of teachers' conceptions into classroom practice, however, indicates that even though teachers' conceptions of NOS can be thought of as a *necessary* condition, these conceptions, nevertheless, should not be considered *sufficient*" (Lederman 2007). Pedagogy and curriculum are additional factors. To learn and then to teach the NOS, explicit and reflective attention on this topic is needed.

Increasingly, many nations and states are mandating that science educators take courses in the NOS (Matthews 2000:321–351). Also there is a trend toward mandating that assessments of scientific literacy include questions about the NOS (Norman G. Lederman et al., in McComas 1998:331–350). But such mandates must be implemented carefully: "I have become more convinced . . . that improving evaluation of science instruction is probably the most important part of the equation for improving science education. But, if evaluative processes and materials are not built upon the best knowledge about the nature of science and applied appropriately for the developmental level of students, through school and university, they will fall short of what is needed" (Robinson 1998).

Finally, the best motivation for science teachers to learn and teach the NOS is awareness of demonstrated benefits. As already mentioned, national standards for science education, mandated assessments of scientific literacy, and other factors are sure to increase the attention given to the NOS. But for science educators, true motivation, as contrasted with perfunctory compliance, arises from known benefits. Indeed, "reports indicating positive changes in students' views and actions regarding the nature of science are needed to bolster teachers' confidence that attention to these issues will reap the desired effects" (William F. McComas et al., in McComas 1998:29). Accordingly, this chapter's documentation of six benefits from explicit and reflective study of the NOS and scientific method has great relevance for motivating teacher interest in the NOS, and thereby student interest also.

Academic NOS concepts

The importance of science literacy is widely appreciated, especially in the K–12 curriculum. But Moore (1998), then-president of the scientific research society

Sigma Xi, asked the incisive question, "But what kind of scientific literacy is important?" His reply was that "The basic concepts measured by survey questions may have little to do with the kinds of knowledge that are needed for competitiveness in the technological world." Accordingly, it would be regrettable if "Basic literacy and numeracy take precedence over more sophisticated scientific understanding in the priorities of education." The very real value of basic scientific literacy for "simple good citizenship in a technological world" for all K–12 students should not be allowed to supersede or dwarf the equally real value of advanced scientific literacy for college students and science professionals. Neither elementary nor advanced education should be neglected.

Accordingly, in contrast to the typical NOS concepts for K–12 education that Figure 13.1 showed, Figure 13.2 shows academic NOS concepts suitable for college students and science professionals. There is almost no overlap between these two lists of concepts.

These ten academic NOS concepts have two main sources. One is the literature in science, statistics, and philosophy, which informs topics such as the PEL model, deductive and inductive reasoning, and parsimony and accuracy. The other is position papers on science education, which motivate topics such as the relationship between the sciences and the humanities, science's presuppositions, and science's worldview import. These 10 items are representative, not exhaustive. Additional topics in this book on scientific method include explication and defense of science's rationality, science's ethics and responsibilities, and effective pedagogy for science education.

Between Figures 13.1 and 13.2, there is one intentional and direct conflict: The first item in the typical NOS concepts is "Scientific knowledge while durable has a tentative character," whereas the corresponding last item in the academic NOS concepts is "Much scientific knowledge is certain, some is probable, and some is speculative; and scientists ordinarily make careful, justified, and realistic claims for their findings." In support for this shift toward more confident language, there has been a change from the excessive emphasis on tentativeness in the AAAS (1989, 1990) and NRC (1996) position papers to explicit recognition of "fact" and "knowledge that is unquestioned and uncontested" in the more recent NRC (2012:79) position paper, as already quoted at greater length in this chapter's section on more realism. Indeed, "a focus on practices (in the plural) avoids the mistaken impression that . . . uncertainty is a universal attribute of science. In reality, practicing scientists employ a broad spectrum of methods, and although science involves many areas of uncertainty as knowledge is developed, there are now many aspects of scientific knowledge that are so well established as to be unquestioned foundations of the culture and its technologies. It is only through engagement in the practices that students can recognize how such knowledge comes about and why some parts of scientific theory are more firmly established than others" (NRC 2012:44).

Academic NOS Concepts

Science is one of the liberal arts. There is a rich traffic of ideas between the sciences and the humanities. Philosophy of science is especially relevant for scientists.

Scientific inquiry involves the general principles of scientific method pervading all sciences and the research techniques of a given specialty. General principles need be mastered only once, whereas specialized techniques change from project to project. The winning combination for scientists is mastering both.

The PEL model asserts that when fully disclosed, any argument supporting a scientific conclusion necessarily has premises of three kinds: presuppositions, evidence, and logic.

Science's presuppositions are installed by philosophical reflection on any exemplar of rudimentary common sense, such as "Moving cars are hazardous to pedestrians."

Science's presuppositions and logic are rooted in common sense, not worldview distinctives, so they are worldview independent. But empirical and public evidence from the sciences and the humanities can have worldview import, and personal experience is also potentially relevant although not public evidence. Worldview conclusions based on worldview-independent presuppositions and logic and on worldview-informative empirical and public evidence are the prerogative of individual scientists; but they are not the prerogative of scientific institutions because the scientific community lacks consensus.

Deduction reasons from a given model to expected observations. Predicate logic requires only several axioms, and arithmetic, probability, and other deductive systems such as geometry and calculus can be added with several more axioms.

Induction reasons from actual observations to inferred model. The probability of evidence given hypothesis, $P(E|H)$, is related to its reverse conditional probability, $P(H|E)$, by Bayes's theorem, which also involves the prior probability $P(H)$. Confusing these reverse conditional probabilities is a fallacy, as is ignoring the influence of prior probabilities.

Evidence is evaluated by multiple criteria, including accurate fit with the data, parsimony, robustness to noise, predictive accuracy, generality, unification of diverse phenomena, explanatory power, testability, and coherence with other knowledge.

For noisy data being fitted by a model family, which represents an extremely common situation in science and technology, the optimal tradeoff is at the peak of Ockham's hill, with less complex models underfitting signal and more complex models overfitting noise. A parsimonious model can be more accurate than its data, thereby increasing repeatability and accelerating progress.

Much scientific knowledge is certain, some is probable, and some is speculative; and scientists ordinarily make careful, justified, and realistic claims for their findings.

Figure 13.2 Ten basic concepts for the nature of science (NOS) at an academic level. University undergraduate and graduate students and science professionals need to understand the NOS at a higher level than the typical NOS concepts developed for K–12 education.

Reviews of the NOS literature make it abundantly clear that contemporary science education is dominated by the opinion that all scientific knowledge is tentative (Lederman 1992, 2007; Abd-El-Khalick and Lederman 2000). Although atypical of most other science educators, Ernst von Glasersfeld (in Tobin 1993:26) followed his perceptions of science's woes to their logical conclusion of radical skepticism: "To conclude that, because we have a perceptual experience which we call 'chair,' there must be a chair in the 'real' world is to commit the realist fallacy. We have no way of knowing what is or could be beyond our experiential interface."

What explains such sentiments that *all* science (and perhaps all common sense too) is tentative and revisable? Recall from Chapter 4 that there has been a historical shift in philosophy of science from the excessively confident philosophy begun by logical empiricists in the 1920s to the excessively diffident philosophy emerging in the 1960s. That shift prompted the insightful remark by Callebaut (1993:xv) that "Philosophy of science as currently practiced is a reaction against a reaction" – a pendulum swinging from one extreme to the opposite extreme. Correspondingly, there has been a shift in science education, decade by decade from 1960 to 2000, from confident to tentative views of scientific knowledge (Abd-El-Khalick and Lederman 2000). Accordingly, with apologies to Callebaut, one might judge that science education as currently practiced is a reaction against a reaction – whereas it ought to be a response to nature and thereby a service to students and a foundation for technology. Fortunately, the recent NRC (2012) position paper on science education insists that uncertainty is not a universal attribute of science and, more pointedly, that science possesses unquestioned and uncontested knowledge of unobservables such as atoms.

Persistence in teaching the disingenuous opinion that all science is tentative and revisable seems likely to have two deleterious side effects. First, just as too much exposure to violence or entertainment tends to produce insensitivity, a pervasive and incessant tentativeness causes unresponsiveness. By contrast, the focused and occasional detection of specific blunders in actual scientific inquiry does prompt response and correction. Second, telling students that an official precept of the NOS is that all science is tentative and revisable, when many of these same students perceive from their science courses that numerous scientific facts are certain and unrevisable, constitutes an invitation to confusion and insincerity. It sends the message that the NOS is in the realm of silly philosophical games rather than actual scientific practices.

A promising direction for further research in science education is empirical study of the benefits for college students and science professionals from studying more advanced academic NOS concepts. Interesting questions include: When scientists with and without having received academic NOS instruction are compared, what differences emerge regarding research, teaching, and administrative abilities? When a researcher's awareness of the general principles of scientific method (such as Ockham's hill) enables an important innovation within his or her own scientific specialty or technological application, what factors affect how quickly it catches on? What makes for effective pedagogy and stimulating classroom dynamics when teaching academic NOS to undergraduate and graduate students?

By their frequent use in numerous empirical studies, the TOUS and VNOS instruments for assessing understanding of typical NOS concepts have helped to standardize expectations and to facilitate comparisons among studies. Accordingly, development of a corresponding instrument for academic NOS concepts could be quite helpful. Items such as those in Figure 13.2 could be considered for inclusion.

Finally, this chapter on science education concludes with an expression of science education at its very best. By "liberal," Matthews simply means encompassing the humanities.

> At its most general level the liberal tradition in education embraces Aristotle's delineation of truth, goodness, and beauty as the ideals that people ought to cultivate in their appropriate spheres of endeavor. That is, in intellectual matters truth should be sought, in moral matters goodness, and in artistic and creative matters beauty. Education is to contribute to these ends: it is to assist the development of a person's knowledge, moral outlook and behavior, and aesthetic sensibilities and capacities. For a liberal, education is more than the preparation for work. . . . The liberal tradition seeks to overcome intellectual fragmentation. Contributors to the liberal tradition believe that science taught from such a perspective, and informed by the history and philosophy of the subject, can engender understanding of nature, the appreciation of beauty in both nature and science, and the awareness of ethical issues unveiled by scientific knowledge and created by scientific practice. . . . The liberal tradition maintains that science education should not just be an education or training *in* science, although of course it must be this, but also an education *about* science. Students educated in science should have an appreciation of scientific methods, their diversity and their limitations. They should have a feeling for methodological issues, such as how scientific theories are evaluated and how competing theories are appraised, and a sense of the interrelated role of experiment, mathematics and religious and philosophical commitment in the development of science. . . . The liberal approach requires a great deal from teachers; this needs to be recognized and provided for by those who educate and employ teachers. (Matthews 1994:1–3, 6)

Summary

Science educators typically place various aspects of scientific method within several broader contexts, including inquiry, practice, and the nature of science (NOS). Comparisons among standards for K–12 science education, both national and international, have produced a consensus view of the NOS. The

TOUS and VNOS instruments for assessing student understanding of the NOS use that consensus, and their wide adoption has facilitated comparisons among studies.

From hundreds of empirical studies, science educators have found six benefits from explicit and reflective study of the NOS. (1) Understanding of scientific method promotes learning of scientific content, so rather than being competitive goals, there is a win-win situation for content and method. (2) Understanding the general principles of scientific method promotes greater adaptability, transferring knowledge to solve new problems. (3) A humanities-rich version of science that includes historical and philosophical elements from the NOS stimulates greater student interest in science. (4) Understanding scientific method and the NOS promotes more realism about the accuracy and scope of specific findings as well as the powers and limits of science more generally. (5) Mastering scientific method is necessary for developing better researchers who are not merely competent technicians but rather are actually creative scientists. (6) Grasping the NOS results in better teachers of both the NOS and science content.

College students and science professionals need to master academic NOS concepts rather than the typical NOS concepts intended for K–12 students. Ten basic items merit inclusion at the academic level. (1) Science is among the liberal arts and there is a rich traffic of ideas between the sciences and the humanities. (2) The winning combination for scientists is mastering both the general principles of scientific method pervading all sciences and the research techniques of a given specialty. (3) The PEL model asserts that when fully disclosed, any argument supporting a scientific conclusion necessarily has premises of three kinds: presuppositions, evidence, and logic. (4) Science's presuppositions are installed by philosophical reflection on any exemplar of rudimentary common sense, such as "Moving cars are hazardous to pedestrians." (5) Although science's presuppositions and logic are rooted in common sense and thereby are worldview independent, empirical and public evidence from the sciences and the humanities can have worldview import, and personal experience is also potentially relevant although not public evidence. (6) Deduction reasons from a given model to expected observations. (7) Induction reasons from actual observations to inferred model. (8) Evidence is evaluated by multiple criteria, including accurate fit with the data, parsimony, robustness to noise, predictive accuracy, generality, unification of diverse phenomena, explanatory power, testability, and coherence with other knowledge. (9) A parsimonious model, at the peak of Ockham's hill, can be more accurate than its data, thereby increasing repeatability and accelerating progress. (10) Much scientific knowledge is certain, some is probable, and some is speculative; and scientists ordinarily make careful, justified, and realistic claims for their findings.

Study questions

(1) Regarding the first benefit of NOS instruction, better comprehension, what do you think of the win-win situation depicted for NOS and science? Is the empirical evidence convincing? Do you find this win-win situation surprising or expected?

(2) What are some empirical findings by science educators showing that NOS understanding promotes adaptability, interest, and realism?

(3) What are some empirical findings by science educators showing that NOS understanding makes for better researchers and better teachers?

(4) What are three or four academic NOS concepts relevant for enhancing perspective on science? Explain each briefly.

(5) What are three or four academic NOS concepts relevant for increasing scientific productivity? Explain each briefly.

14

Conclusions

This concluding chapter focuses on only two topics. The first section returns to the topic of science's contested rationality to resolve the issues pondered in Chapter 4, using the physics of motion as a concrete example. The second section provides a concise overview of scientific method.

Other chapters ended with five study questions, but this chapter ends with ten exit questions covering material from this entire book. These questions correspond to the 10 items in Figure 13.2 showing an academic curriculum on the nature of science in general and scientific method in particular. Ten questions cannot possibly cover all of the material in this book. Nevertheless, success in answering these representative questions well is indicative of a philosopher-scientist possessing a sophisticated, academic understanding of scientific method. In turn, understanding science helps in developing technology and appreciating nature.

Motion and rationality

Recall from Chapter 4 that Sir Karl Popper, Thomas Kuhn, and other prominent philosophers have challenged the very foundations of science, citing four deadly woes: (1) Science cannot prove any theory either true or false. (2) Observations are theory-laden, and theory is underdetermined by data. (3) Successive paradigms are incommensurable. (4) What makes a belief scientific is merely that it is what scientists say. The upshot of these woes is that science is alleged to be irrational or arational, unable to claim any truths about the physical world, principally because nature does not significantly constrain theory choice. These four woes were introduced in Chapter 4, but resolution has had to wait until this final chapter after first having developed all of the essential components of scientific method in the intervening chapters.

Is science in big trouble or not? Most fundamentally, does nature constrain theory? This section approaches these questions in a concrete way by exploring

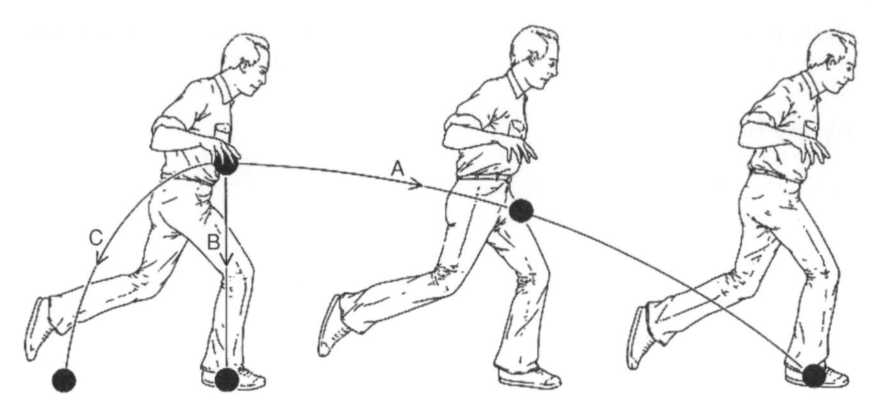

Figure 14.1 A ball is dropped by a running person. Students were asked if the ball would fall forward with the runner (path A), drop straight down (B), or fall backward (C). Most college students surveyed gave incorrect answers. (This drawing by Michael Goodman is reproduced with his kind permission.)

an area of physics accessible to every reader through the performance of some simple experiments involving the motions of familiar objects at ordinary speeds, far below that of light, such as a dropped or thrown ball. Accordingly, this topic will afford readers a suitable arena for testing the thesis that nature constrains theory and the antithesis that it does not.

McCloskey (1983) surveyed high school and college students regarding their intuitive beliefs about motion. He also recounted both the medieval theory of impetus and the modern theory of inertia. Most important, he reported experiments on the actual motions of objects. Briefly, impetus theory says that when a mover sets an object in motion, that imparts an internal force or impetus that will act continuously to keep the object moving until the impetus has dissipated. By contrast, inertia theory says that an object will continue in its state of rest or its state of uniform motion in a straight line unless acted on by an external force. These incompatible theories predict different motions or paths in various settings. From informal experiences with moving objects, people tend to form ideas that are much like the medieval impetus theory.

As a specific instance of motion, imagine a running person who drops a ball from waist height. What path will the ball follow as it drops? Figure 14.1 shows the three possibilities: the ball will fall forward with the runner (path A), drop straight down (B), or fall backward (C). Which path do you expect? Be sure to write or remember your response for comparison with the answer given momentarily.

This section's analysis of the ball's path will progress through five stages. First, the subjective dimension of persons' beliefs and actions will be surveyed. Second, the objective physics of a ball's actual path will be explained by

Newtonian mechanics. Third, the historical dimension of questions and discoveries about motion will be reviewed, from Aristotle to Newton. Fourth, educational challenges and strategies will be discussed. Fifth and finally, there will be some philosophical reflection.

First, regarding persons' intuitive beliefs about the motion of a dropped ball, McCloskey surveyed numerous college students. The percentages espousing the three hypotheses in Figure 14.1 were as follows:

H_A: The ball will travel forward, landing ahead of the point of release (45%).

H_B: The ball will fall straight down, landing under the point of release (49%).

H_C: The ball will travel backward, landing behind the point of release (6%).

Note that hypothesis H_B was the most popular answer, the expectation that the ball would fall straight down. McCloskey was also interested in the relationship between belief and action. Accordingly, besides surveying subjects' verbal responses to physics questions, he monitored subjects' actual actions with physical objects. College students were given the task of dropping a ball while moving forward so that it would hit a target marked on the floor. Performance of that simple task should reveal each subject's belief about the path that the ball would take. There were three possible actions, corresponding to the three hypotheses. The percentages of students performing those three actions were as follows:

A_A: Drop the ball before reaching the target (13%).

A_B: Drop the ball directly over the target (80%).

A_C: Drop the ball after reaching the target (7%).

Note that these survey results, based on actions, generally were similar to the verbal survey. Again, A_B was most popular, and A_C least popular. Which action would you perform to make a dropped ball hit the target marked on the floor? Again, be sure to write or remember your response.

Second, let us now turn from the subjective dimension of persons' beliefs about motion to the objective dimension of a ball's actual path. The correct answer is hypothesis H_A, corresponding to action A_A: The ball travels forward with the runner as it falls, so it must be dropped before reaching the target marked on the floor. Was your intuitive belief true? Was your recommended action correct?

Using Newton's laws of motion, McCloskey explained the ball's path. When the ball is dropped by the runner, its inertia causes it to continue moving forward at the same speed as the runner, because (ignoring air resistance) no force is acting to change its horizontal velocity. As the ball moves forward, it also moves down at a steadily increasing velocity because of acceleration due to gravity. The forward and downward motions combine in a path that closely approximates a parabola, path A in Figure 14.1. This parabolic path was discovered by Galileo in 1604 and published in 1638.

Regrettably, the correct answer was a minority position. Only 45% of the students surveyed held the correct belief, and worse yet, only 13% performed the correct action. Hence, for even such a simple instance of motion, McCloskey found that subjective beliefs and objective facts are not in agreement as commonly as one might suppose: "One might expect that as a result of everyday experience people would have reasonably accurate ideas about the motion of objects in familiar situations.... It seems that such is not the case."

Third, the historical dimension of beliefs about motion is also surprising. Aristotle had two theories of motion, one for celestial objects and another for terrestrial objects, but both were incorrect on many counts. Medieval impetus theory was better but still rather problematic. "What path do moving objects take?" was an archetypical science question for millennia, along with "What are things made of?" Both questions have proved amazingly difficult. The erosion of the medieval impetus theory by the work of William of Ockham and the advances in experimental and mathematical methods wrought by Robert Grosseteste positioned Galileo to construct the theory of inertia and to discover the parabolic trajectory of a falling object with an initial horizontal motion. Subsequently, Newton subsumed those findings in his grand, unified mechanics, including his first law of motion regarding inertia. So, from the time that Aristotle posed simple questions about motion until a satisfactory solution emerged from Galileo and Newton, two millennia had passed. Contributions from many of the world's great intellects, advances by many successive generations, and two millennia were required to answer the seemingly simple question posed in Figure 14.1. Even now, centuries after Newton, most people still have erroneous beliefs about motion that reflect the medieval impetus theory more than Newtonian mechanics. Of course, no one is now taught that fallacious medieval theory as a fact in school, yet from informal experiences with moving objects, apparently most people formulate an intuitive theory that generally agrees with impetus theory.

Fourth, this contrast between subjective beliefs and objective truths presents interesting educational challenges. Besides the motion of a dropped ball, McCloskey studied intuitive beliefs about motions of balls and pucks in various other settings. In all cases, misconceptions were common.

Given the commonness of errors about motion, science educators must develop effective methods for correcting these errors. How can misconceptions be dispelled? How can students be convinced of the correct laws of motion? The obvious answer, as McCloskey noted, is instruction in Newtonian mechanics. However, intuitive beliefs are surprisingly difficult to modify. For example, before taking a physics course, only 13% of college students took the correct action of dropping the ball before reaching the target; but after the physics course, this figure increased to 73%, which still left erroneous preconceptions intact in 27% of the students. In another survey, 93% of high school students believed in impetus, and a physics course reduced those erroneous assumptions only slightly, to 80%. Neither standard lectures nor laboratory demonstrations seem to be adequately effective for overcoming false preconceptions.

The problem is that new experiences are routinely interpreted within an existing paradigm: "It appears that students often rely on the intuitive impetus theory as a framework for interpreting new course material. As a result the material can be misinterpreted and distorted to fit the intuitive preconceptions" (McCloskey 1983). Consequently, educators must seek better instruction methods.

Better progress seems to result from having students articulate their beliefs, after which the differences between their initial beliefs and Newtonian mechanics can be sorted out. Implicit presuppositions may resist instruction and experiments, but explicit beliefs frequently respond. Often, it is not enough just to be told the truth; rather, one must also recognize one's current beliefs and how they differ from new beliefs being offered by the instructor. In the confrontation between hypotheses and data, the hypotheses must be recognized explicitly. Explicit hypotheses plus experimental evidence can equal realistic learning.

Although most people have many misconceptions about motion, and although the historical path to correct knowledge was long and tortuous, the facts are now readily available. Indeed, the laws of motion are empirical and testable matters, requiring little time and resources to perform simple experiments, such as the one depicted in Figure 14.1. The actual motions are thus matters of public knowledge because the required experiments are trivial and because Newton's laws of motion have been well known for centuries. Repetition of the experiments by various persons in various places dropping various objects can give results as certain as desired. True knowledge of motion is empirical, testable, public, objective, and certain.

Furthermore, knowledge about motion is important. Moving objects impact human welfare in a thousand arenas, from sports to warfare. Correct ideas work, whereas erroneous preconceptions fail. Ideas have consequences. Truth matters. For example, strategies other than releasing the ball before reaching the target invariably result in failure to hit the target. That is just the way the world is.

Fifth and finally, this example of a dropped ball's motion prompts some philosophical reflection. Astonishingly, in his interview in *Scientific American*, the noted philosopher Thomas Kuhn fondly recounted the great "Eureka!" moment in his career, when he concluded that "Aristotle's physics 'wasn't just bad Newton'" but rather "it was just different" (Horgan 1991). Well, what does it mean to say that Aristotle and Newton are "just different" when a projectile launched by Aristotle's theory would miss its target, whereas a projectile launched by Newton's theory would hit its target?

Aristotle's impetus theory of motion is bad science. It was bad science in antiquity because it failed to instantiate Aristotle's own standard of truth as correspondence between belief and reality. It remains bad science now because any science teacher who taught impetus theory as a fact would be grossly incompetent.

Recall from Chapter 4 the four deadly woes that allegedly challenge science. What is their meaning or credibility for this specific case of moving objects? Supposedly, these four woes pervade all of science, which necessarily includes the current case. So, do these woes unsettle our presumed scientific knowledge of motion, or does our valid scientific knowledge of motion unsettle these woes?

(1) The first challenge is that science cannot prove any theory either true or false. A substantial contribution to this verdict is the objection, endorsed by Karl Popper, that inductive logic is bankrupt. But consider the three hypotheses about a dropped ball's motion (see Figure 14.1). It is within the reader's grasp, given a ball and a few minutes, to prove one of these hypotheses true and two of them false. It is simply ludicrous to claim that science never finds any objective, final truth. More generally, millions of experiments have proved inertia theory true and impetus theory false.
(2) The second challenge is that observations are theory-laden and that theory is underdetermined by data. The intent of that challenge is to say that theory and observation are so profoundly intertwined that observation cannot constrain or guide theory choice. But, manifestly, the three hypotheses in Figure 14.1 and observations of a ball's actual path are two different matters, so observations can constrain theory.
(3) The third allegation is that successive paradigms are incommensurable, so competing theories are just different. A person who began with the most popular hypothesis that a ball dropped by a runner would fall straight down (path B in Figure 14.1), but then considered the two alternative possibilities and performed experiments revealing that a dropped ball would travel forward with a runner (path A), is a person who has undergone a paradigm shift. But, after shifting to H_A, there is no inability to understand one's former belief in H_B or other persons' current beliefs in H_B. Indeed, anyone should be capable of understanding all three competing hypotheses, H_A, H_B, and H_C. The difference between them is not that adherents of any given hypothesis cannot understand what adherents of others are saying. Rather, the difference between them is that H_A is supported by empirical evidence and physics theory, and $\mathrm{H_A}$ enables persons who act in accordance with it to accomplish certain tasks; whereas H_B and H_C are false hypotheses, commending misguided actions that will lead to failure, inevitably. Of course, different hypotheses or theories or paradigms can be commensurable and comparable!
(4) The fourth and final charge is that what makes a belief scientific is merely that it is what scientists say. Is the only difference between H_A and the other hypotheses that H_A is what physicists say, whereas laypersons often

espouse other hypotheses? Of course not. The greater difference is that H_A describes the actual path that a ball will travel, whereas the other hypotheses do not. At issue here are not only subjective beliefs but also objective facts; not only persons' beliefs but also the paths of falling objects. What makes H_A scientific is that it presents knowledge gained by the scientific method that accurately describes the true state of nature regarding a ball's motion.

The thrust of this example is that science's truths emerge when hypotheses are confronted by data. Even Galileo endorsed the impetus theory in his early writings, around 1590, before his experiments with physical objects forced a revision of his beliefs. The ultimate instruction about moving objects is actual experimentation with moving objects, particularly when the student articulates explicitly his or her own preconceptions and understands clearly the various possible hypotheses.

Furthermore, scientists have learned millions of other things about the natural world that feature just as much objectivity and certainty as does this section's example of a ball's motion. For example, the equivalent conductivity of an aqueous solution of hydrochloric acid at a concentration of 5.0 mol l^{-1} and temperature of 30°C is 180.2 Ω^{-1} cm^2 $equiv^{-1}$ (Lide 1995:5–87). That ordinary scientific finding is true and certain. Further research may supply a more accurate measurement of this particular value, a deeper understanding of electric currents in liquids, and many more relevant facts and insights, thereby proliferating and enhancing various applications in technology and industry. But this value of 180.2 has already been measured competently and accurately, and it will not change appreciably.

It might be objected that this section's reply to the four challenges underlying anti-realist interpretations of science misses the mark because the real action concerns the status of unobservables, such as atoms, rather than observables, such as balls. However, that objection cannot be sustained.

It is simply false that this realist/anti-realist debate concerns only unobservables. Inexorably, the logic involved in advocating these four woes can reach any claim about an external physical reality, regardless of how ordinary or exotic the object might be. Indeed, philosopher Karl Popper was quite forthcoming in extending his skepticism to ordinary objects, such as a glass of water, as mentioned in Chapter 5. Also, bear in mind that Thomas Kuhn's own example of incommensurable paradigms was precisely this section's example, the motion of ordinary objects according to Aristotle and Newton.

Regarding unobservables such as atoms, recall the section on water in Chapter 3 that rehearsed the familiar fact that a water molecule is composed of two hydrogen atoms and one oxygen atom, H_2O. Readers who have experienced no difficulties with that elementary chemistry knowledge would seem to be quite poor candidates for thinking that unobservables pose any fundamental or insurmountable problems for science. Also recall from the previous chapter that mainstream science deems "the existence of atoms" to be a "fact... that is unquestioned and uncontested" (NRC 2012:79).

Consequently, the realist/anti-realist debate can be discussed just as productively and just as fairly in the context of balls as in that of atoms. The real issues are far more fundamental than the trifling distinction between observables and unobservables. Nor is that distinction even very meaningful, for everything is observed with some degree of indirectness. Also, a given object can change status because of perspectival differences that are without ontological interest, such as a runner approaching from a distance who is first unobservable, then a moving speck, then a human being, and finally a recognized friend.

On balance, there are individual cases in science for which the four woes do apply legitimately, or at least one of them does. For instance, some theories are incommensurable for various reasons, including the theories being too vague to support decisive tests, or the required observations being too expensive, or the required calculations being too difficult even with the fastest computers. Such limitations are in the realm of ordinary scientific investigations, not scary philosophical stories. What is denied here is that incommensurable paradigms and the other three woes are pervasive features of natural science that negate all claims of rationality, truth, objectivity, and realism. Properly discerned, the four woes are quite limited problems applying occasionally to special circumstances, rather than pervasive and catastrophic problems.

Understand that this section is not arguing that no resources exist that can overturn and undo science. This section clarifies the content of those resources rather than denying their existence. Indeed, radical skepticism provides resources that can undo science.

What resources can unsettle "A dropped ball travels forward with the runner," or "Water is H_2O," or "Mendel's short peas resulted from a mutation in the gene for gibberellin 3β-hydroxylase"? Nothing less than radical skepticism's resources that also unsettle "A moving car is hazardous to pedestrians" or "Here is a glass of water" would suffice. But radical skepticism is outside common sense, outside mainstream science, and outside mainstream philosophy. Whether such peculiar resources interest or grip a given person is a matter of individual taste and choice. Likewise, given reason's double office of regulating belief and guiding action, whether radical skepticism can be held with sincerity is a matter of individual conscience and conviction.

But, again, this book's project is to presuppose common sense and then build scientific method, not to refute the skeptic and thereby establish common sense. As explained in Chapter 5, mainstream science displays philosophical sophistication by disclosing the existence and comprehensibility of the physical world as a presupposition of science rather than claiming it as a conclusion of science. If this section's example of a ball's motion and the NRC position on atoms' existence make sense, then the four woes in Chapter 4 fail and science's four bold claims in Chapter 2 stand. Science achieves rationality, truth, objectivity, and realism.

At an academic level, the contest over science's rationality concerns matters such as these four woes. This book has now said all that it will say about those academic woes.

But, at a popular level, the contest over science's rationality concerns other matters. There are two principal objections, one philosophical and the other historical. These popular objections also merit a response because they have huge cultural impact, perhaps even more so than the academic objections. What enables these objections to become so popular, rather than the academic objections, is their simplicity. They are easily learned in a few minutes and easily repeated in a few sentences – even by a lackluster and lazy high school student.

The popular philosophical objection is that only naïve people believe that science has any certain truths, whereas the philosophers know better. Anyone in the know realizes that all scientific findings are tentative and revisable. Besides, many national science education standards express this realistic assessment of science's limits, and many science teachers present it as the official right answer. We are only human, after all.

It would be interesting, however, to ask such an objector whether he or she has ever read even one page of the primary literature in philosophy of science, which appears in journals such as *Philosophy of Science* and *The British Journal for the Philosophy of Science.* If not, one might suggest taking just an hour to scan some of the abstracts and pages in such journals. I submit that this modest exercise would reveal that although science's pervasive and perpetual tentativeness is a fringe opinion among some professional philosophers, overwhelmingly mainstream philosophy aligns with mainstream science in affirming much truth in scientific knowledge. Those persons who have heard of Socrates and Aristotle, but not Pyrrho of Elis and Sextus Empricus, might easily grasp that there has been a difference between mainstream philosophy and fringe philosophy all along over the centuries.

The other popular objection is historical. It claims that the history of science reveals countless confident beliefs of scientists that subsequently have been overturned and revised. Given that checkered history, why should we think that science now possesses any final truths immune to revision? A favorite story concerns the earth-centered cosmology that scientists held for millennia until Copernicus and Galileo.

But this seeming simple story has substantial content about intellectual history in ancient and medieval times, contemporary cultural consensus on a sun-centered solar system, and linguistic and logical skills needed to perceive the direct contradiction between sun-centered and earth-centered solar systems. This observation prompts a question. For anyone who readily can manage these historical, cultural, and logical skills, why would some basic astronomy be utterly unmanageable? That incongruity demands something like a systematic and coherent explanation.

For starters, this historical knowledge has already invoked the same presupposition about the existence and comprehensibility of the physical world that is necessary and sufficient for science, so presuppositions cannot be the problem. Furthermore, the objector's historical and logical prowess might be put to better use. One might ask whether there is a relevant historical difference in our having, but ancient persons lacking, the technology to send probes to all of the planets, and even send people to the moon and return them safely along with moon rocks. Might that difference help to explain why the ancient earth-centered solar system was revisable, whereas the contemporary sun-centered solar system is not revisable?

Having addressed academic and popular objections to science's rationality and truth, it must be mentioned that some persons exhibit an acute and comprehensive emotional aversion to truth claims. Emotional aversion has quite different sources and manifestations than intellectual aversion, and it may have complicated and obscure origins in a given individual. But common features seem to be the feeling that truth imposes constraints on one's freedom and provokes intolerance toward others, rather than that truth increases success in dealing with reality and augments opportunities to serve others effectively. Of course, some persons may have both intellectual and emotional objections to prospects of actually finding any settled truth. Perhaps the most common state is a conflicted heart, both loving and hating truth, depending on how convenient or inconvenient a given truth might seem.

But this brief book on scientific method cannot energetically engage this topic of truth's desirability. Fortunately, two admirably concise books have already addressed attitudes toward truth and reality with insight and poignancy: the classic *Ideas Have Consequences* by Weaver (1948) and the recent *Fear of Knowledge* by Boghossian (2006). Is truth the enemy or ally of wisdom, humility, liberty, dignity, love, gratitude, and life? This is a pivotal issue that affects heart attitudes toward truth and its attainability. The best answers draw resources from the sciences and the humanities.

Finally, for university students, my experience as a teacher of scientific method is that exposure to a wide range of perspectives on science's rationality and truth is a fruitful means – and perhaps the very best means – for stimulating real thought and earned conclusions. This favorable educational outcome is doubly likely when ideal classroom dynamics result from including students representing a wide diversity of majors across the sciences and the humanities. On the other hand, for K–12 students, my settled conviction is that neither extremist perspective – that all science is tentative and revisable or that science alone delivers knowledge that is reliable and true – represents best practices. Rather, mainstream science and mainstream philosophy, as well as common sense, support a strong case that best practices in science education at all levels include understanding and teaching this simple verity: much science is true and certain, some is probable, and some is speculative. This balanced perspective on science respects human powers and limits, fosters scientific interests, understands scientific method and practice, recognizes technological opportunities and risks, enhances citizenship and governance in our technological world, welcomes complementary insights on nature from the humanities, and promotes an appreciation of nature.

Summary of scientific method

In two words, the business of scientific method is *theory choice* – the choice of what to believe about the physical world. The presuppositions, evidence, and logic that comprise the scientific method have been explored in earlier chapters. Here, a brief overview will draw much from the insightful and concise account of scientific method by Box, Hunter, and Hunter (1978:1–15).

Figure 14.2 shows the basic elements of data collection and analysis in scientific research. Starting at the top of this figure, the object under study is some part of the natural world. The scientist's objective is to find the truth about this physical thing, as emphasized in Chapters 2 and 3. The currently favored hypothesis H_i regarding this object influences the design of a relevant experiment to generate empirical observations. When designing an experiment, a scientist must consider the prospective value of its data for discriminating between the competing hypotheses, together with the experiment's costs and risks. Informative experiments concentrate attention on situations for which the competing hypotheses predict different outcomes.

Naturally, different scientists may choose different research strategies: "Notice that, on this view of scientific investigation, we are not dealing with a *unique* route to problem solution. Two equally competent investigators presented with the same problem would typically begin from different starting points, proceed by different routes, and yet could reach the same answer. What is sought is not uniformity but convergence" (Box et al. 1978:5). To illustrate that point, they considered the game of "twenty questions," in which no more than twenty questions, each having only a yes or no answer, are allowed for determining what object someone is thinking of. Supposing that the object to be guessed was Abraham Lincoln's stovepipe hat, they then listed the questions and answers as the game was played by two different teams. One team asked 11 questions, and the other asked 8 mostly different questions, but both teams reached the same, right answer. A unique path to discovery is not required, so there is room for creativity, but some subject-matter knowledge and strategy are needed. In science, efficient experimental designs and aggressive data analyses are the means for asking and answering questions posed to nature.

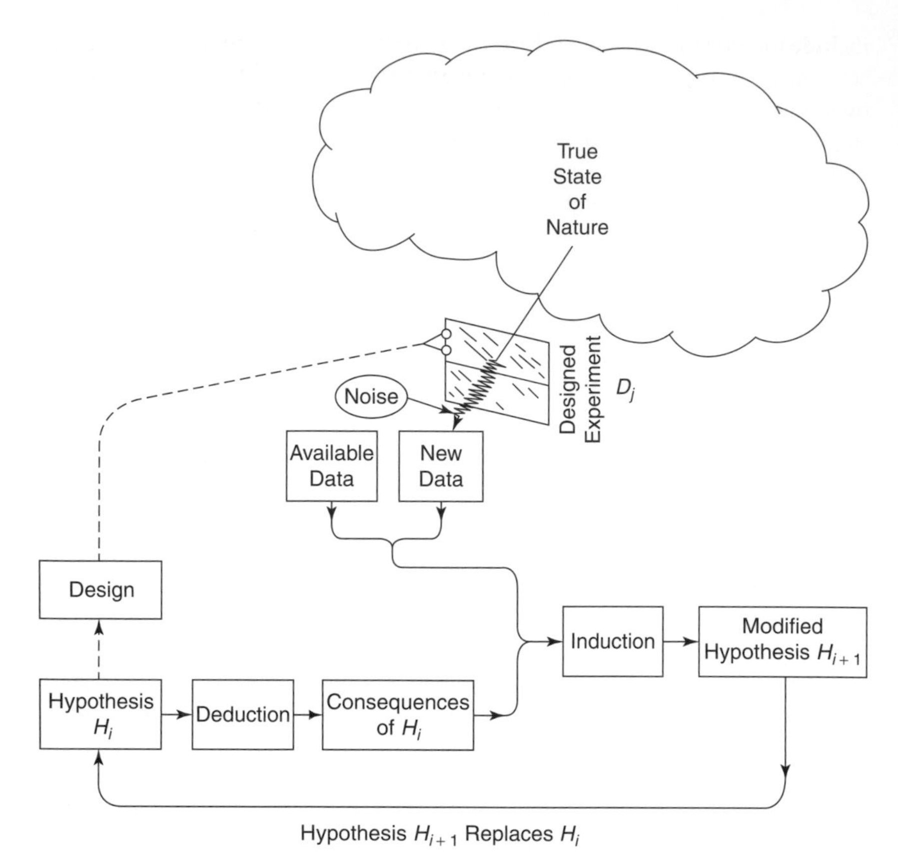

Figure 14.2 An overview of the scientific method. Deduction and induction are used to bring data to bear on theory choice. The goal is for theory choice to converge on the true state of nature. (Reproduced from Box et al. 1978:4, with kind permission from John Wiley & Sons.)

The scientist's window on the world is a designed experiment. The experiment is not perfect, however, for its data contain some noise. Good designs minimize the deleterious effects of experimental errors. Besides the new data being generated in the current experiment, usually there are additional data available from previous research. A particular strength of Bayesian analysis is its explicit formalism for combining prior and new information in inductive inferences, as elaborated in Chapter 9. Particularly because the data are noisy, parsimonious models are favored, as explained in Chapter 10. Good inductive methods will extract as much usable information from the data as possible, minimize distortions due to noise, and expose any problems that would indicate that new hypotheses and paradigms need to be considered.

Moving to the bottom left of Figure 14.2, there is a competition between two hypotheses, the currently favored H_i and the newly modified H_{i+1}. Deductive logic can be used to derive the consequences or predictions of the competing hypotheses, as explored in Chapter 7. Then, inductive logic can be used to compare the various hypotheses' predictions against the data to determine which hypothesis is true or most likely to be true. In the example shown here, the evidence favors the new H_{i+1}, which then replaces H_i as the currently favored hypothesis.

Regarding the PEL model, this figure does mention experimental evidence and deductive and inductive logic, but it does not mention presuppositions. Nevertheless, presuppositions are in play and are indispensable. Presuppositions enable scientists to focus on a small roster of common-sense hypotheses, such as H_i and H_{i+1}, without hindrance from an endless roster of other wild hypotheses, as explained in Chapter 5. Presuppositions are needed to render any hypothesis sensible, but they do not render any hypothesis either more or less credible than others, as does evidence, so scientists routinely present evidence but ignore presuppositions. But to give a scientific conclusion with full disclosure, the presuppositions must also be exhibited.

Figure 14.2 depicts a common situation with just two hypotheses, the currently favored hypothesis and a newer one. But sometimes there are multiple working hypotheses, such as H_A, H_B, and H_C in Figure 14.1. Also, the hypotheses may be competing on more or less equal footing, with none decidedly favored by the scientific community. In any case, whether there are two or many hypotheses, and whether or not one is favored, the overall scheme remains basically the same. Deduction generates the predicted outcomes for each hypothesis, experiments determine actual outcomes, and induction weighs hypotheses in light of the data. But the size of the roster of hypotheses does affect how much work will be needed to rule out all of the false alternatives, and a scientist's hunch about which hypothesis is true does affect which experiments will be tried first and how quickly the answer will be reached.

Sometimes scientists win secrets from nature gradually, so over the years or centuries, H_i gives way to H_{i+1}, then H_{i+2}, then H_{i+3}, and so on indefinitely. In other cases, definitive and accurate truth is found and with success, the research can stop, leaving a settled theory. Interest then can move on to other topics that are still open. Sometimes researchers specify a predetermined stopping rule to intervene when an experiment has generated enough data to support a definitive or adequate result. Knowing when to stop is important when the data are expensive, research conclusions are needed to guide pressing applications, or a scientific specialty has numerous other interesting questions that have not yet been researched.

In this figure's overview of scientific method, the key feature is that the hypotheses are confronted by data, leading to convergence on the truth about physical reality. Although the conjectured state of nature – the hypothesis or theory – may be inexact or even false, "the data themselves are generated by the true state of nature" (Box et al. 1978:5). Thereby nature constrains theory. That is why science works. That is how science finds truth.

Exit questions

(1) Is science one of the liberal arts? How are the sciences and the humanities related in general and in understanding nature more specifically? As delineated by position papers on science from major scientific organizations, what attitude does mainstream science take toward scientism, the idea that only science produces reliable and respectable knowledge?
(2) This book's main thesis is that the winning combination for scientists is mastering both the general principles of scientific method pervading all sciences and the research techniques of a given specialty. Would you agree? What differences might there be in learning these two aspects of scientific research? Do you feel that both have been treated adequately in your own educational experience?
(3) What does the PEL model assert about the kinds of premises needed to support any scientific conclusion? Explain your answer with a concrete example.
(4) Science needs presuppositions about the existence and comprehensibility of the physical world. How can these presuppositions be legitimated? Why is the existence and comprehensibility of the physical world a presupposition of science rather than a conclusion?
(5) What potential does science have for worldview import? In what manner are science's presuppositions, logic, and evidence involved in such worldview import? Are the humanities and personal experience also relevant in reaching worldview convictions? How do the prerogatives of individual scientists and of scientific institutions differ as regards arguing for specific worldview conclusions based on science?
(6) Deduction reasons from what to what? What are two or three of the axioms for predicate logic and also for probability?
(7) Induction reasons from what to what? Let H denote a hypothesis and E some evidence. How does Bayes's theorem relate the reverse conditional probabilities $P(H \mid E)$ and $P(E \mid H)$? What is an example of the fallacy of confusing reverse conditional probabilities? Also, what is an example of the fallacy of ignoring the influence of prior probabilities?
(8) What are six or seven criteria used in evaluating scientific theories? Explain each briefly. To what extent does nature constrain theory choice?
(9) How are accuracy and parsimony related by Ockham's hill? Why are models less accurate on each side of the peak of Ockham's hill? How can a model be more accurate than its data?

(10) Science's four bold claims are rationality, truth, objectivity, and realism. Are these claims justified for science considered as an overall enterprise? Typically, how careful and justified are knowledge claims for individual, specific findings? In your own opinion, how successful is contemporary science in describing and understanding nature?

References

AAAS (American Association for the Advancement of Science). 1989. *Science for All Americans: A Project 2061 Report on Literacy Goals in Science, Mathematics, and Technology*. Washington, DC: AAAS.

1990. *The Liberal Art of Science: Agenda for Action*. Washington, DC: AAAS.

1993. *Benchmarks for Science Literacy*. Oxford, UK: Oxford University Press.

2000. *Designs for Science Literacy*. Oxford, UK: Oxford University Press.

Abd-El-Khalick, F., and Lederman, N. G. 2000. Improving science teachers' conceptions of nature of science: A critical review of the literature. *International Journal of Science Education* 22:665–701.

Adler, M. J. 1978. *Aristotle for Everybody: Difficult Thought Made Easy*. New York: Macmillan.

Akerson, V. L., Buck, G. A., Donnelly, L. A., Nargund-Joshi, V., and Weiland, I. S. 2011. The importance of teaching and learning nature of science in the early childhood years. *Journal of Science Education and Technology* 20: 537–549.

Annas, J., and Barnes, J. 1985. *The Modes of Scepticism: Ancient Texts and Modern Interpretations*. Cambridge, UK: Cambridge University Press.

Anscombe, G. E. M., and von Wright, G. H. (eds.). 1969. *On Certainty* (by Ludwig Wittgenstein). New York: Harper & Row.

Ark, T. K., Brooks, L. R., and Eva, K. W. 2007. The benefits of flexibility: The pedagogical value of instructions to adopt multifaceted diagnostic reasoning strategies. *Medical Education* 41:281–287.

Audi, R. (ed.). 1999. *The Cambridge Dictionary of Philosophy*, 2nd edn. Cambridge, UK: Cambridge University Press.

Banner, M. C. 1990. *The Justification of Science and the Rationality of Religious Belief*. Oxford, UK: Oxford University Press.

Barnard, G. A. 1958. Studies in the history of probability and statistics. IX. Thomas Bayes's essay towards solving a problem in the doctrine of chances. *Biometrika* 45:293–315.

Bautista, N. U., and Schussler, E. E. 2010. Implementation of an explicit and reflective pedagogy in introductory biology laboratories. *Journal of College Science Teaching* 40:56–61.

Bayes, T. 1763. An essay towards solving a problem in the doctrine of chances. *Philosophical Transactions of the Royal Society* 53:370–418.
Beauchamp, T. L. (ed.). 1999. *David Hume: An Enquiry Concerning Human Understanding*. Oxford, UK: Oxford University Press.
Becker, B. J. 2000. MindWorks: Making scientific concepts come alive. *Science & Education* 9:269–278.
Ben-Chaim, D., Ron, S., and Zoller, U. 2000. The disposition of eleventh-grade science students toward critical thinking. *Journal of Science Education and Technology* 9:149–159.
Benjamin, A. S., and Hackstaff, L. H. 1964. *Saint Augustine: On Free Choice of the Will*. Indianapolis, IN: Bobbs-Merrill.
Berger, J. O. 1985. *Statistical Decision Theory and Bayesian Analysis*, 2nd edn. New York: Springer-Verlag.
Berger, J. O., and Berry, D. A. 1988. Statistical analysis and the illusion of objectivity. *American Scientist* 76:159–165.
Bering, J. M. 2006. The cognitive psychology of belief in the supernatural. *American Scientist* 94:142–149.
Berry, D. A. 1987. Interim analysis in clinical trials: The role of the likelihood principle. *American Statistician* 41:117–122.
Blackburn, S. 1994. *The Oxford Dictionary of Philosophy*. Oxford, UK: Oxford University Press.
Boardman, J. D., Blalock, C. L., Corley, R. P., et al. 2010. Ethnicity, body mass, and genome-wide data. *Biodemography and Social Biology* 56:123–136.
Boehner, P. 1957. *Ockham Philosophical Writings*. New York: Nelson.
Boghossian, P. 2006. *Fear of Knowledge: Against Relativism and Constructivism*. Oxford, UK: Oxford University Press.
Boudreau, C., and McCubbins, M. D. 2009. Competition in the courtroom: When does expert testimony improve jurors' decisions? *Journal of Empirical Legal Studies* 6:793–817.
Box, G. E. P., Hunter, W. G., and Hunter, J. S. 1978. *Statistics for Experimenters: An Introduction to Design, Data Analysis, and Model Building*. New York: John Wiley & Sons, Ltd.
Boyd, R., Gasper, P., and Trout, J. D. (eds.). 1991. *The Philosophy of Science*. Cambridge, MA: MIT Press.
Broad, C. D. 1952. *Ethics and the History of Philosophy*. London: Routledge & Kegan Paul.
Broad, W. J. 1979. Paul Feyerabend: Science and the anarchist. *Science* 206:534–537.
Brown, H. I. 1987. *Observation and Objectivity*. Oxford, UK: Oxford University Press.
Brown, S. C. (ed.). 1984. *Objectivity and Cultural Divergence*. Cambridge, UK: Cambridge University Press.
Brünger, A. T. 1992. Free R value: A novel statistical quantity for assessing the accuracy of crystal structures. *Nature* 355:472–475.
1993. Assessment of phase accuracy by cross validation: The free R value. *Acta Crystallographica, Section D* 49:24–36.
Buera, F. J., Monge-Naranjo, A., and Primiceri, G. E. 2011. Learning the wealth of nations. *Econometrica* 79:1–45.

Burks, A. W. 1977. *Chance, Cause, Reason: An Inquiry into the Nature of Scientific Evidence.* Chicago, IL: University of Chicago Press.

Cajori, F. 1947. *Sir Isaac Newton's Mathematical Principles of Natural Philosophy and His System of the World.* Berkeley, CA: University of California Press.

Caldin, E. F. 1949. *The Power and Limits of Science: A Philosophical Study.* London: Chapman & Hall.

Callebaut, W. 1993. *Taking the Naturalistic Turn, or How Real Philosophy of Science Is Done.* Chicago, IL: University of Chicago Press.

Callus, D. A. (ed.). 1955. *Robert Grosseteste, Scholar and Bishop: Essays in Commemoration of the Seventh Centenary of His Death.* Oxford, UK: Oxford University Press.

Cameron, C. M., and Park, J.-K. 2009. How will they vote? Predicting the future behavior of Supreme Court nominees, 1937–2006. *Journal of Empirical Legal Studies* 6:485–511.

Carey, S. S. 2012. *A Beginner's Guide to Scientific Method,* 4th edn. Belmont, CA: Wadsworth.

Cartier, J. L., and Stewart, J. 2000. Teaching the nature of inquiry: Further developments in a high school genetics curriculum. *Science & Education* 9:247–267.

Carus, P. 1902. *Immanuel Kant: Prolegomena to Any Future Metaphysics That Can Qualify as a Science.* La Salle, IL: Open Court.

Chatalian, G. 1991. *Epistemology and Skepticism: An Enquiry into the Nature of Epistemology.* Carbondale, IL: Southern Illinois University Press.

Clark, J. S. 2005. Why environmental scientists are becoming Bayesians. *Ecology Letters* 8:2–14.

Cleland, C. E. 2011. Prediction and explanation in historical natural science. *The British Journal for the Philosophy of Science* 62:551–582.

Cobern, W. W., and Loving, C. C. 2001. Defining "science" in a multicultural world: Implications for science education. *Science Education* 85:50–67.

Collins, F. S. 2006. *The Language of God: A Scientist Presents Evidence for Belief.* New York: Free Press.

Collins, H. 2009. We cannot live by scepticism alone. *Nature* 458:30–31, correspondence 458:702–703.

Collins, H., and Pinch, T. 1993. *The Golem: What Everyone Should Know About Science.* Cambridge, UK: Cambridge University Press.

Couvalis, G. 1997. *The Philosophy of Science: Science and Objectivity.* London: Sage.

Cox, D. R., and Hinkley, D. V. 1980. *Problems and Solutions in Theoretical Statistics.* New York: Chapman & Hall.

Craig, W. L., and Moreland, J. P. (eds.). 2009. *The Blackwell Companion to Natural Theology.* Chichester, UK: Wiley-Blackwell.

Crombie, A. C. 1962. *Robert Grosseteste and the Origins of Experimental Science.* Oxford, UK: Oxford University Press.

Cullen, T. A., Akerson, V. L., and Hanson, D. L. 2010. Using action research to engage K–6 teachers in nature of science inquiry as professional development. *Journal of Science Teacher Education* 21:971–992.

Cuneo, T., and van Woudenberg, R. (eds.). 2004. *The Cambridge Companion to Thomas Reid.* Cambridge, UK: Cambridge University Press.

Dales, R. C. 1973. *The Scientific Achievement of the Middle Ages.* Philadelphia: University of Pennsylvania Press.

Dalgarno, M., and Matthews, E. (eds.). 1989. *The Philosophy of Thomas Reid.* Dordrecht: Kluwer.

Dampier, W. C. 1961. *A History of Science and Its Relations with Philosophy and Religion.* Cambridge, UK: Cambridge University Press.

Davies, B., and Leftow, B. (eds.). 2006. *Thomas Aquinas: Summa Theologiae, Questions on God.* Cambridge, UK: Cambridge University Press.

Dawkins, R. 1996. *The Blind Watchmaker: Why the Evidence of Evolution Reveals a Universe Without Design.* New York: Norton.

2006. *The God Delusion.* Boston, MA: Houghton Mifflin.

Derry, G. N. 1999. *What Science is and How it Works.* Princeton, NJ: Princeton University Press.

Dixon, T., Cantor, G., and Pumfrey, S. (eds.). 2010. *Science and Religion: New Historical Perspectives.* Cambridge, UK: Cambridge University Press.

Douglas, H. E. 2009. Reintroducing prediction to explanation. *Philosophy of Science* 76:444–463.

Earman, J. 2000. *Hume's Abject Failure: The Argument against Miracles.* Oxford, UK: Oxford University Press.

Easterbrook, G. 1997. Science and God: A warming trend? *Science* 277:890–893.

Ecklund, E. H. 2010. *Science vs. Religion: What Scientists Really Think.* Oxford, UK: Oxford University Press.

Edwards, P. (ed.). 1967. *The Encyclopedia of Philosophy,* 8 vols. New York: Macmillan.

Eggen, P.-O., Kvittingen, L., Lykknes, A., and Wittje, R. 2012. Reconstructing iconic experiments in electrochemistry: Experiences from a history of science course. *Science & Education* 21:179–189.

Ehrlich, I. 1973. Participation in illegitimate activities: A theoretical and empirical investigation. *Journal of Political Economy* 81:521–565.

Einstein, A. 1954. *Ideas and Opinions.* New York: Crown.

Eisenberg, T., and Wells, M. T. 2008. Statins and adverse cardiovascular events in moderate-risk females: A statistical and legal analysis with implications for FDA preemption claims. *Journal of Empirical Legal Studies* 5:507–550.

Finkelstein, M. O. 2009. *Basic Concepts of Probability and Statistics in the Law.* New York: Springer.

Fisher, R. A. 1973. *Statistical Methods and Scientific Inference,* 3rd edn. New York: Macmillan. (Originally published 1956.)

Frank, P. G. (ed.). 1957. *Philosophy of Science: The Link Between Science and Philosophy.* Englewood Cliffs, NJ: Prentice-Hall.

Frege, G. 1893. *Grundgesetze der Arithmetik.* Jena: Hermann Pohle.

Freireich, E. J., Gehan, E., Frei, E., et al. 1963. The effect of 6-mercaptopurine on the duration of steroid-induced remissions in acute leukemia: A model for evaluation of other potentially useful therapy. *Blood* 21:699–716.

Friedman, K. S. 1990. *Predictive Simplicity: Induction Exhum'd.* Elmsford Park, NY: Pergamon.

Gauch, H. G. 1988. Model selection and validation for yield trials with interaction. *Biometrics* 44:705–715.

1992. *Statistical Analysis of Regional Yield Trials: AMMI Analysis of Factorial Designs.* New York: Elsevier.

1993. Prediction, parsimony, and noise. *American Scientist* 81:468–478, correspondence 507–508.

2002. *Scientific Method in Practice.* Cambridge, UK: Cambridge University Press; Chinese edition, 2004, Beijing: Tsinghua University Press.

2006. Winning the accuracy game. *American Scientist* 94:133–141, correspondence 94:196, addendum 94:382.

2009a. Science, worldviews, and education. *Science & Education* 18:667–695; also published in Matthews, M. R. (ed.). 2009. *Science, Worldviews and Education*, pp. 27–48. Heidelberg: Springer Verlag.

2009b. Responses and clarifications regarding science and worldviews. *Science & Education* 18:905–927; also published in Matthews, M. R. (ed.). 2009. *Science, Worldviews and Education*, pp. 303–325. Heidelberg: Springer Verlag.

Gauch, H. G., Rodrigues, P. C., Munkvold, J. D., Heffner, E. L., and Sorrells, M. 2011. Two new strategies for detecting and understanding QTL × environment interactions. *Crop Science* 51:96–113.

Gelman, A., Carlin, J. B., Stern, H. S., and Rubin, D. B. 2004. *Bayesian Data Analysis*, 2nd edn. New York: Chapman & Hall.

Gentner, D., Loewenstein, J., and Thompson, L. 2003. Learning and transfer: A general role for analogical encoding. *Journal of Educational Psychology* 95:393–408.

Giacomini, R., and White, H. 2006. Tests of conditional predictive ability. *Econometrica* 74:1545–1578.

Gimbel, S. (ed.). 2011. *Exploring the Scientific Method: Cases and Questions.* Chicago, IL: University of Chicago Press.

Glynn, L. 2011. A probabilistic analysis of causation. *British Journal for the Philosophy of Science* 62:343–392.

Godfrey-Smith, P. 2003. *Theory and Reality: An Introduction to the Philosophy of Science.* Chicago, IL: University of Chicago Press.

Good, R. P., Kost, D., and Cherry, G. A. 2010. Introducing a unified PCA algorithm for model size reduction. *IEEE Transactions on Semiconductor Manufacturing* 23:201–209.

Goodman, L. A., and Hout, M. 2002. Statistical methods and graphical displays for analyzing how the association between two qualitative variables differs among countries, among groups, or over time: A modified regression-type approach. *Statistical Methodology* 28:175–230.

Goodman, N. 1955. *Fact, Fiction, and Forecast.* Cambridge, MA: Harvard University Press.

Gottfried, K., and Wilson, K. G. 1997. Science as a cultural construct. *Nature* 386:545–547.

Gower, B. 1997. *Scientific Method: An Historical and Philosophical Introduction.* London: Routledge.

Grayling, A. C. 2011. How we form beliefs. *Nature* 474:446–447.

Greenland, S. 2008. Introduction to Bayesian statistics. In: Rothman, K. J., Greenland, S., and Lash, T. L. (eds). *Modern Epidemiology*, 3rd ed., pp. 328–344. New York: Wolters Kluwer.

Grier, B. 1975. Prediction, explanation, and testability as criteria for judging statistical theories. *Philosophy of Science* 42:373–383.

Gross, P. R., Levitt, N., and Lewis, M. W. (eds.). 1996. *The Flight from Science and Reason.* Baltimore, MD: Johns Hopkins University Press.

Gustason, W. 1994. *Reasoning from Evidence: Inductive Logic.* New York: Macmillan.

Guyer, P. (ed.). 1992. *The Cambridge Companion to Kant.* Cambridge, UK: Cambridge University Press.

Hamilton, A. G. 1978. *Logic for Mathematicians.* Cambridge, UK: Cambridge University Press.

Hamilton, W. (ed.). 1872. *The Works of Thomas Reid, D.D.*, 7th edn. Edinburgh: MacLachlan & Stewart.

Himsworth, H. 1986. *Scientific Knowledge and Philosophic Thought.* Baltimore, MD: Johns Hopkins University Press.

Hodson, D. 2009. *Teaching and Learning about Science: Language, Theories, Methods, History, Traditions and Values.* Rotterdam: Sense Publishers.

Hoff, P. D. 2009. *A First Course in Bayesian Statistical Methods.* New York: Springer.

Hoffmann, R., Minkin, V. I., and Carpenter, B. K. 1996. Ockham's razor and chemistry. *Bulletin de la Société chimique de France* 133(2):117–130.

Horgan, J. 1991. Profile: Thomas S. Kuhn, reluctant revolutionary. *Scientific American* 264(5):40, 49.

1992. Profile: Karl R. Popper, the intellectual warrior. *Scientific American* 267(5): 38–44.

1993. Profile: Paul Karl Feyerabend, the worst enemy of science. *Scientific American* 268(5):36–37.

2005. Clash in Cambridge: Science and religion seem as antagonistic as ever. *Scientific American* 293(3):24B, 26–28.

Howson, C. 2000. *Hume's Problem: Induction and the Justification of Belief.* Oxford, UK: Oxford University Press.

Howson, C., and Urbach, P. 2006. *Scientific Reasoning: The Bayesian Approach*, 3rd edn. La Salle, IL: Open Court.

Hunter, A.-B., Laursen, S. L., and Seymour, E. 2007. Becoming a scientist: The role of undergraduate research in students' cognitive, personal, and professional development. *Science Education* 91:36–74.

Irwin, T. H. 1988. *Aristotle's First Principles.* Oxford, UK: Oxford University Press.

Irzik, G., and Nola, R. 2010. A family resemblance approach to the nature of science for science education. *Science & Education* 20:591–607.

Jefferys, W. H., and Berger, J. O. 1992. Ockham's razor and Bayesian analysis. *American Scientist* 80:64–72.

Jeffrey, R. C. 1983. *The Logic of Decision*, 2nd edn. Chicago, IL: University of Chicago Press.

Jeffreys, H. 1973. *Scientific Inference*, 3rd edn. Cambridge, UK: Cambridge University Press.

1983. *Theory of Probability*, 3rd edn. Oxford, UK: Oxford University Press.

Johnson, M. A., and Lawson, A. E. 1998. What are the relative effects of reasoning ability and prior knowledge on biology achievement in expository and inquiry classes? *Journal of Research in Science Teaching* 35:89–103.

Joo, Y., Wells, M. T., and Casella, G. 2010. Model selection error rates in nonparametric and parametric model comparisons. In Berger, J. O., Cai, T. T., and Johnstone, I. M.

(eds.). *Borrowing Strength: Theory Powering Applications – A Festschrift for Lawrence D. Brown*, pp. 166–183. Beachwood, OH: Institute of Mathematical Statistics.

Kass, R. E., and Raftery, A. E. 1995. Bayes factors. *Journal of the American Statistical Association* 90:773–795.

Kemeny, J. G. 1959. *A Philosopher Looks at Science.* New York: Van Nostrand.

Khatib, F., DiMaio, F., Foldit Contenders Group, Foldit Void Crushers Group, et al. 2011. Crystal structure of a monomeric retroviral protease solved by protein folding game players. *Nature Structural & Molecular Biology* 18:1175–1177.

Khine, M. S. (ed.). 2011. *Advances in the Nature of Science Research: Concepts and Methodologies.* New York: Springer.

King, P. 1995. *Augustine: Against the Academicians; The Teacher.* Indianapolis, IN: Hackett.

Kleywegt, G. J. 2007. Separating model optimization and model validation in statistical cross-validation as applied to crystallography. *Acta Crystallographica Section D* 63:939–940.

Kleywegt, G. J., and Jones, T. A. 1995. Where freedom is given, liberties are taken. *Structure* 3:535–540.

2002. Homo Crystallographicus – Quo Vadis? *Structure* 10:465–472.

Koertge, N. (ed.). 1998. *A House Built on Sand: Exposing Postmodernist Myths about Science.* Oxford, UK: Oxford University Press.

Kolmogorov, A. N. 1933. *Grundbegriffe der Wahrscheinlichkeitsrechnung.* Berlin: Julius Springer.

Kuhn, T. S. 1970. *The Structure of Scientific Revolutions*, 2nd edn. Chicago, IL: University of Chicago Press. (Originally published 1962.)

Lakatos, I., and Musgrave, A. (eds.). 1968. *Problems in the Philosophy of Science.* Amsterdam: North-Holland.

(eds.). 1970. *Criticism and the Growth of Knowledge.* Cambridge, UK: Cambridge University Press.

Laksov, K. B., Lonka, K., and Josephson, A. 2008. How do medical teachers address the problem of transfer? *Advances in Health Sciences Education* 13: 345–360.

Larson, E. J., and Witham, L. 1999. Scientists and religion in America. *Scientific American* 281(3):88–93.

Lawson, A. E., Clark, B., Cramer-Meldrum, E., Falconer, K. A., Sequist, J. M., and Kwon, Y.-J. 2000. Development of scientific reasoning in college biology: Do two levels of general hypothesis-testing skills exist? *Journal of Research in Science Teaching* 37:81–101.

Lawson, A. E., and Weser, J. 1990. The rejection of nonscientific beliefs about life: Effects of instruction and reasoning skills. *Journal of Research in Science Teaching* 27:589–606.

Lederman, N. G. 1992. Students' and teachers' conceptions of the nature of science: A review of the research. *Journal of Research in Science Teaching* 29:331–359.

2007. Nature of Science: Past, Present, and Future. In: Abell, S. K., and Lederman, N. G. (eds.). *Handbook of Research on Science Education*, pp. 831–879. Mahwah, NJ: Lawrence Erlbaum Associates.

Lehrer, K. 1989. *Thomas Reid.* London: Routledge.

Leitgeb, H., and Pettigrew, R. 2010a. An objective justification of Bayesianism. I: Measuring inaccuracy. *Philosophy of Science* 77:201–235.

2010b. An objective justification of Bayesianism. II: The consequences of minimizing inaccuracy. *Philosophy of Science* 77:236–272.

Leplin, J. 1997. *A Novel Defense of Scientific Realism.* Oxford, UK: Oxford University Press.

Lide, D. R. (ed.). 1995. *CRC Handbook of Chemistry and Physics*, 76th edn. Boca Raton, FL: CRC Press.

Lindberg, D. C. 2007. *The Beginnings of Western Science: The European Scientific Tradition in Philosophical, Religious, and Institutional Context, Prehistory to A.D. 1450*, 2nd edn. Chicago, IL: University of Chicago Press.

Lindberg, D. C., and Numbers, R. L. 2003. *When Science and Christianity Meet.* Chicago, IL: University of Chicago Press.

Lloyd, E. A. 2010. Confirmation and robustness of climate models. *Philosophy of Science* 77:971–984.

Lobato, J. 2006. Alternative perspectives on the transfer of learning: History, issues, and challenges for future research. *The Journal of the Learning Sciences* 15: 431–449.

Losee, J. 2001. *A Historical Introduction to the Philosophy of Science*, 4th edn. Oxford, UK: Oxford University Press.

2011. *Theories of Causality: From Antiquity to the Present.* New Brunswick, NJ: Transaction Publishers.

Lv, J. C., Yi, Z., and Tan, K. K. 2007. Determination of the number of principal directions in a biologically plausible PCA model. *IEEE Transactions on Neural Networks* 18:910–916.

MacDonald, S. 1998. Natural theology. In: Craig, E. (ed.). *Routledge Encyclopedia of Philosophy*, vol. 6, pp. 707–713. London: Routledge.

MacIntyre, A. 1988. *Whose Justice? Which Rationality?* Notre Dame, IN: University of Notre Dame Press.

MacKay, D. J. C. 1992. Bayesian interpolation. *Neural Computation* 4:415–447.

Marrone, S. P. 1983. *William of Auvergne and Robert Grosseteste: New Ideas of Truth in the Early Thirteenth Century*. Princeton, NJ: Princeton University Press.

Matthews, M. R. 1994. *Science Teaching: The Role of History and Philosophy of Science.* London: Routledge.

1998. In defense of modest goals when teaching about the nature of science. *Journal of Research in Science Teaching* 35:161–174.

2000. *Time for Science Education: How Teaching the History and Philosophy of Pendulum Motion Can Contribute to Science Literacy.* Dordrecht: Kluwer.

McClellan, J. E., and Dorn, H. 2006. *Science and Technology in World History: An Introduction*, 2nd edn. Baltimore, MD: Johns Hopkins University Press.

McCloskey, M. 1983. Intuitive physics. *Scientific American* 248(4):122–130.

McComas, W. F. (ed.). 1998. *The Nature of Science in Science Education: Rationales and Strategies.* Dordrecht: Kluwer.

McGrayne, S. B. 2011. *The Theory That Would Not Die: How Bayes' Rule Cracked the Enigma Code, Hunted Down Russian Submarines & Emerged Triumphant from Two Centuries of Controversy.* New Haven, CT: Yale University Press.

McGrew, T. 2003. Confirmation, heuristics, and explanatory reasoning. *The British Journal for the Philosophy of Science* 54:553–567.

McKeon, R. 1941. *The Basic Works of Aristotle.* New York: Random House.

McQuarrie, A. D. R., and Tsai, C.-L. 1998. *Regression and Time Series Model Selection.* River Edge, NJ: World Scientific.

Medawar, P. B. 1969. *Induction and Intuition in Scientific Thought.* Philadelphia: American Philosophical Society.

Merton, R. K. 1973. *The Sociology of Science: Theoretical and Empirical Investigations.* Chicago, IL: University of Chicago Press.

Merton, R. K., and Sztompka, P. 1996. *On Social Structure and Science.* Chicago, IL: University of Chicago Press.

Meyling, H. 1997. How to change students' conceptions of the epistemology of science. *Science & Education* 6:397–416.

Moore, J. H. 1998. Public understanding of science – and other fields. *American Scientist* 86:498.

Mukerjee, M. 1998. Undressing the emperor. *Scientific American* 278(3):30, 32.

Myrvold, W. C. 1996. Bayesianism and diverse evidence. *Philosophy of Science* 63:661–665.

2003. A Bayesian account of the virtue of unification. *Philosophy of Science* 70: 399–423.

NAS (National Academy of Sciences). 1995. *Reshaping the Graduate Education of Scientists and Engineers.* Washington, DC: National Academies Press.

NAS (and National Academy of Engineering and Institute of Medicine). 2009. *On Being a Scientist: A Guide to Responsible Conduct in Research*, 3rd edn. Washington, DC: National Academies Press.

NAS. 2010. *Managing University Intellectual Property in the Public Interest.* Washington, DC: National Academies Press.

Nash, L. K. 1963. *The Nature of the Natural Sciences.* Boston, MA: Little, Brown.

NCEE (National Commission on Excellence in Education). 1983. *A Nation at Risk: The Imperative for Educational Reform.* Washington, DC: US Department of Education.

Niaz, M. 2011. *Innovating Science Teacher Education: A History and Philosophy of Science Perspective.* London: Routledge.

Nola, R., and Sankey, H. 2007. *Theories of Scientific Method.* Montreal: McGill-Queen's University Press.

Norton, D. F. (ed.). 1993. *The Cambridge Companion to Hume.* Cambridge, UK: Cambridge University Press.

NRC (National Research Council). 1996. *National Science Education Standards.* Washington, DC: National Academies Press.

1997. *Science Teaching Reconsidered: A Handbook.* Washington, DC: National Academies Press.

1999. *Transforming Undergraduate Education in Science, Mathematics, Engineering, and Technology.* Washington, DC: National Academies Press.

2001. *Educating Teachers of Science, Mathematics, and Technology: New Practices for the New Millennium.* Washington, DC: National Academies Press.

2012. *A Framework for K–12 Science Education: Practices, Crosscutting Concepts, and Core Ideas.* Washington, DC: National Academies Press.

NSF (National Science Foundation). 1996. *Shaping the Future: New Expectations for Undergraduate Education in Science, Mathematics, Engineering, and Technology.* Arlington, VA: National Science Foundation.

NSTA (National Science Teachers Association). 1995. *A High School Framework for National Science Education Standards.* Arlington, VA: National Science Teachers Association.

O'Hear, A. 1989. *An Introduction to the Philosophy of Science.* Oxford, UK: Oxford University Press.

Oreskes, N., Stainforth, D. A., and Smith, L. A. 2010. Adaptation to global warming: Do climate models tell us what we need to know? *Philosophy of Science* 77: 1012–1028.

Palmer, H. 1985. *Presupposition and Transcendental Inference.* New York: St. Martin's Press.

Pearl, J. 2009. *Causality: Models, Reasoning, and Inference*, 2nd edn. Cambridge, UK: Cambridge University Press.

Piepho, H.-P., and Gauch, H. G. 2001. Marker pair selection for mapping quantitative trait loci. *Genetics* 157:433–444.

Pirie, M. 2006. *How to Win Every Argument: The Use and Abuse of Logic.* New York: Continuum.

Polanyi, M. 1962. *Personal Knowledge: Towards a Post-Critical Philosophy.* Chicago, IL: University of Chicago Press.

Popper, K. R. 1945. *The Open Society and Its Enemies.* London: Routledge & Kegan Paul.

1968. *The Logic of Scientific Discovery.* New York: Harper & Row. (Originally published 1959.)

1974. *Conjectures and Refutations: The Growth of Scientific Knowledge*, 5th edn. New York: Harper & Row.

1979. *Objective Knowledge: An Evolutionary Approach.* Oxford, UK: Oxford University Press.

Potter, M. 2000. *Reason's Nearest Kin: Philosophies of Arithmetic from Kant to Carnap.* Oxford, UK: Oxford University Press.

Raftery, A. E. 1995. Bayesian model selection in social research. *Sociological Methodology* 25:111–163.

Reif, F. 1995. Understanding and teaching important scientific thought processes. *Journal of Science Education and Technology* 4:261–282.

Resnik, D. B. 1998. *The Ethics of Science: An Introduction.* London: Routledge.

Robert, C. P. 2007. *The Bayesian Choice: From Decision-Theoretic Foundations to Computational Implementation*, 2nd edn. New York: Springer.

Robinson, J. T. 1998. Science teaching and the nature of science. *Science & Education* 7:617–634. (Reprint of 1965 article.)

Rosenberg, A. 2012. *Philosophy of Science: A Contemporary Introduction*, 3rd edn. London: Routledge.

Rosenthal-Schneider, I. 1980. *Reality and Scientific Truth: Discussions with Einstein, von Laue, and Planck.* Detroit, MI: Wayne State University Press.

Rothman, K. J., Greenland, S., Poole, C., and Lash, T. L. 2008. Causation and causal inference. In: Rothman, K. J., Greenland, S., and Lash, T. L. (eds.). *Modern Epidemiology*, 3rd edn., pp. 5–31. New York: Wolters Kluwer.

Ryder, J., and Leach, J. 1999. University science students' experiences of investigative project work and their images of science. *International Journal of Science Education* 21:945–956.

Sadler, T. D., Burgin, S., McKinney, L., and Ponjuan, L. 2010. Learning science through research apprenticeships: A critical review of the literature. *Journal of Research in Science Teaching* 47:235–256.

Salmon, W. C. 1967. *The Foundations of Scientific Inference.* Pittsburgh, PA: University of Pittsburgh Press.

1984. *Logic*, 3rd edn. Englewood Cliffs, NJ: Prentice-Hall.

Sandel, M. J. 2009. *Justice: What's the Right Thing to Do?* New York: Farrar, Straus and Giroux.

Savitz, N. V., and Raudenbush, S. W. 2009. Exploiting spatial dependence to improve measurement of neighborhood social processes. *Sociological Methodology* 39:151–183.

Schilpp, P. A. (ed.). 1951. *Albert Einstein: Philosopher-Scientist.* New York: Tudor Publishing.

(ed.). 1974. *The Philosophy of Karl Popper*, 2 vols. Evanston, IL: Library of Living Philosophers.

Schulze, W. 2009. Field experiments in food and resource economics research: Discussion. *American Journal of Agricultural Economics* 91:1279–1280.

Schupbach, J. N. 2005. On a Bayesian analysis of the virtue of unification. *Philosophy of Science* 72:594–607.

Schupbach, J. N., and Sprenger, J. 2011. The logic of explanatory power. *Philosophy of Science* 78:105–127.

Schwartz, R. S., Lederman, N. G., and Crawford, B. A. 2004. Developing views of nature of science in an authentic context: An explicit approach to bridging the gap between nature of science and scientific inquiry. *Science Education* 88:610–645.

Schwarz, G. 1978. Estimating the dimension of a model. *The Annals of Statistics* 6:461–464.

Seghouane, A.-K. 2009. Model selection criteria for image restoration. *IEEE Transactions on Neural Networks* 20:1357–1363.

Seker, H., and Welsh, L. C. 2006. The use of history of mechanics in teaching motion and force units. *Science & Education* 15:55–89.

Shermer, M. 2004. God's number is up. *Scientific American* 291(1):46.

Simon, H. A. 1962. The architecture of complexity. *Proceedings of the American Philosophical Society* 106:467–482.

Snow, C. P. 1993. *The Two Cultures and the Scientific Revolution.* Cambridge, UK: Cambridge University Press. (Originally published 1959.)

Sokal, A. 1996. Transgressing the boundaries: Toward a transformative hermeneutics of quantum gravity. *Social Text* 46/47:217–252.

2008. *Beyond the Hoax: Science, Philosophy, and Culture.* Oxford, UK: Oxford University Press.

Souders, T. M., and Stenbakken, G. N. 1991. Cutting the high cost of testing. *IEEE Spectrum* 28(3):48–51.

Sprenger, J. 2009. Evidence and experimental design in sequential trials. *Philosophy of Science* 76:637–649.

Stalker, D. (ed.). 1994. *Grue! The New Riddle of Induction.* La Salle, IL: Open Court.

Stehr, N., and Meja, V. (eds.) 2005. *Society & Knowledge: Contemporary Perspectives in the Sociology of Knowledge & Science,* 2nd edn. New Brunswick, NJ: Transaction Publishers.

Stein, C. 1955. Inadmissibility of the usual estimator for the mean of a multivariate normal distribution. *Proceedings of the Third Berkeley Symposium on Mathematical Statistics and Probability* 1:197–206.

Stewart, I. 1996. The interrogator's fallacy. *Scientific American* 275(3):172–175.

Stirzaker, D. 1994. *Elementary Probability.* Cambridge, UK: Cambridge University Press.

Stove, D. C. 1982. *Popper and After: Four Modern Irrationalists.* Elmsford Park, NY: Pergamon.

Swinburne, R. 1997. *Simplicity as Evidence of Truth.* Milwaukee, WI: Marquette University Press.

(ed). 2002. *Bayes's Theorem.* Oxford, UK: Oxford University Press.

2003. *The Resurrection of God Incarnate.* Oxford, UK: Oxford University Press.

Taper, M. L., and Lele, S. R. (eds.). 2004. *The Nature of Scientific Evidence: Statistical, Philosophical, and Empirical Considerations.* Chicago, IL: University of Chicago Press.

Theocharis, T., and Psimopoulos, M. 1987. Where science has gone wrong. *Nature* 329:595–598.

Thompson, W. C., and Schumann, E. L. 1987. Interpretation of statistical evidence in criminal trials: The prosecutor's fallacy and the defense attorney's fallacy. *Law and Human Behavior* 11:167–187.

Tobin, K. (ed.). 1993. *The Practice of Constructivism in Science Education.* Hillsdale, NJ: Lawrence Erlbaum.

Trigg, R. 1980. *Reality at Risk: A Defence of Realism in Philosophy and the Sciences.* Totowa, NJ: Barnes & Noble.

1993. *Rationality and Science: Can Science Explain Everything?* Oxford, UK: Blackwell.

Turner, M. 2010. No miracle in the universe. *Nature* 467:657–658.

Urbach, P. 1987. *Francis Bacon's Philosophy of Science.* La Salle, IL: Open Court.

Urhahne, D., Kremer, K., and Mayer, J. 2011. Conceptions of the nature of science – are they general or context specific? *International Journal of Science and Mathematics Education* 9:707–730.

Wade, N. 1977. Thomas S. Kuhn: Revolutionary theorist of science. *Science* 197:143–145.

Waldrop, M. M. 2011. Faith in science. *Nature* 470:323–325.

Warren, B., Ballenger, C., Ognowski, M., Rosebery, A. S., and Hudicourt-Barnes, J. 2001. Rethinking diversity in learning science: The logic of everyday sense-making. *Journal of Research in Science Teaching* 38:529–552.

Weaver, R. M. 1948. *Ideas Have Consequences.* Chicago, IL: University of Chicago Press.

Weisheipl, J. A. (ed.). 1980. *Albertus Magnus and the Sciences.* Toronto: Pontifical Institute of Mediaeval Studies.

Whitehead, A. N. 1925. *Science and the Modern World.* Cambridge, UK: Cambridge University Press.

Whitehead, A. N., and Russell, B. 1910–13. *Principia Mathematica,* 3 vols. Cambridge, UK: Cambridge University Press.

Williams, L. P., and Steffens, H. J. 1978. *The History of Science in Western Civilization. Vol. II of The Scientific Revolution.* Lanham, MD: University Press of America.

Wolpert, L. 1993. *The Unnatural Nature of Science.* Cambridge, MA: Harvard University Press.

Woodward, J. 2003. *Making Things Happen: A Theory of Causal Explanation.* Oxford, UK: Oxford University Press.

Woolhouse, R. S. 1988. *The Empiricists.* Oxford, UK: Oxford University Press.

Xie, Y. 2002. Comment: The essential tension between parsimony and accuracy. *Sociological Methodology* 28:231–236.

Yang, C.-C. 2007. Confirmatory and structural categorical latent variables models. *Quality & Quantity* 41:831–849.

Index